Spatial Information Science for Natural Resource Management

Suraj Kumar Singh
Suresh Gyan Vihar University, Jaipur, India

Shruti Kanga
Suresh Gyan Vihar University, Jaipur, India

Varun Narayan Mishra
Suresh Gyan Vihar University, Jaipur, India

A volume in the Advances in Environmental Engineering and Green Technologies (AEEGT) Book Series

IGI Global
PUBLISHER of TIMELY KNOWLEDGE

Published in the United States of America by
IGI Global
Engineering Science Reference (an imprint of IGI Global)
701 E. Chocolate Avenue
Hershey PA, USA 17033
Tel: 717-533-8845
Fax: 717-533-8661
E-mail: cust@igi-global.com
Web site: http://www.igi-global.com

Library of Congress Cataloging-in-Publication Data

Names: Singh, Suraj Kumar, 1983- editor. | Kanga, Shruti, 1984- editor. | Mishra, Varun Narayan, 1986- editor.
Title: Spatial information science for natural resource management / Suraj Kumar Singh, Shruti Kanga, and Varun Narayan Mishra, editors.
Description: Hershey, PA : Engineering Science Reference, an imprint of IGI Global, [2020] | Includes bibliographical references and index. | Summary: "This book explores coordinated approaches to sustainable development and management of natural resources to keep a balance of the environment, ecology, and human livelihood. Featuring coverage on a wide range of topics including crop yield estimation, ecosystem services, and land information systems, this book covers interdisciplinary techniques in monitoring and managing natural resources"-- Provided by publisher.
Identifiers: LCCN 2019059566 (print) | LCCN 2019059567 (ebook) | ISBN 9781799850274 (hardcover) | ISBN 9781799851547 (paperback) | ISBN 9781799850281 (ebook)
Subjects: LCSH: Natural resources--Management. | Geospatial data--Computer processing. | Sustainable development.
Classification: LCC HC85 .S63 2020 (print) | LCC HC85 (ebook) | DDC 333.70285--dc23
LC record available at https://lccn.loc.gov/2019059566
LC ebook record available at https://lccn.loc.gov/2019059567

This book is published in the IGI Global book series Advances in Environmental Engineering and Green Technologies (AEEGT) (ISSN: 2326-9162; eISSN: 2326-9170)

British Cataloguing in Publication Data
A Cataloguing in Publication record for this book is available from the British Library.

All work contributed to this book is new, previously-unpublished material.
The views expressed in this book are those of the authors, but not necessarily of the publisher.

For electronic access to this publication, please contact: eresources@igi-global.com.

Advances in Environmental Engineering and Green Technologies (AEEGT) Book Series

ISSN:2326-9162
EISSN:2326-9170

Editor-in-Chief: Sang-Bing Tsai, Zhongshan Institute, University of Electronic Science and Technology of China, China & Wuyi University, China; Ming-Lang Tseng, Lunghwa University of Science and Technology, Taiwan; Yuchi Wang, University of Electronic Science and Technology of China Zhongshan Institute, China

MISSION

Growing awareness and an increased focus on environmental issues such as climate change, energy use, and loss of non-renewable resources have brought about a greater need for research that provides potential solutions to these problems. Research in environmental science and engineering continues to play a vital role in uncovering new opportunities for a "green" future.

The **Advances in Environmental Engineering and Green Technologies (AEEGT)** book series is a mouthpiece for research in all aspects of environmental science, earth science, and green initiatives. This series supports the ongoing research in this field through publishing books that discuss topics within environmental engineering or that deal with the interdisciplinary field of green technologies.

COVERAGE

- Sustainable Communities
- Policies Involving Green Technologies and Environmental Engineering
- Alternative Power Sources
- Renewable Energy
- Pollution Management
- Radioactive Waste Treatment
- Waste Management
- Biofilters and Biofiltration
- Air Quality
- Green Technology

IGI Global is currently accepting manuscripts for publication within this series. To submit a proposal for a volume in this series, please contact our Acquisition Editors at Acquisitions@igi-global.com or visit: http://www.igi-global.com/publish/.

The Advances in Environmental Engineering and Green Technologies (AEEGT) Book Series (ISSN 2326-9162) is published by IGI Global, 701 E. Chocolate Avenue, Hershey, PA 17033-1240, USA, www.igi-global.com. This series is composed of titles available for purchase individually; each title is edited to be contextually exclusive from any other title within the series. For pricing and ordering information please visit http://www.igi-global.com/book-series/advances-environmental-engineering-green-technologies/73679. Postmaster: Send all address changes to above address.

Titles in this Series

For a list of additional titles in this series, please visit:
https://www.igi-global.com/book-series/advances-environmental-engineering-green-technologies/73679

Energetics Perspective on the Environmental and Human Impact of Buildings
Teodora Melania Şoimoşan (Technical University of Cluj-Napoca, Romania) and Ligia Mihaela Moga (Technical University of Cluj-Napoca, Romania)
Engineering Science Reference • © 2020 • 300pp • H/C (ISBN: 9781799835721) • US $195.00

Molecular Plant Breeding and Genome Editing Tools for Crop Improvement
Pradip Chandra Deka (Sir Padampat Singhania University, India)
Engineering Science Reference • © 2020 • 321pp • H/C (ISBN: 9781799843122) • US $195.00

Population Dynamics and Conservation of Red Deer Emerging Research and Opportunities
Junaid Ahmad Malik (Govt. Degree College, Bijbehara, India)
Engineering Science Reference • © 2020 • 200pp • H/C (ISBN: 9781799828112) • US $175.00

Smart Design and Economic Perspectives of Renewable Energy Systems Emerging Research and Opportunities
Ali Mohamed Eltamaly (Mansoura University, Egypt) and Mohamed Abdelaziz Mohamed (Minia University, Egypt)
Engineering Science Reference • © 2020 • 165pp • H/C (ISBN: 9781799841746) • US $165.00

Analyzing the Physics of Radio Telescopes and Radio Astronomy
Kim Ho Yeap (Universiti Tunku Abdul Rahman, Malaysia) and Kazuhiro Hirasawa (University of Tsukuba, Japan)
Engineering Science Reference • © 2020 • 324pp • H/C (ISBN: 9781799823810) • US $225.00

IGI Global
PUBLISHER of TIMELY KNOWLEDGE

701 East Chocolate Avenue, Hershey, PA 17033, USA
Tel: 717-533-8845 x100 • Fax: 717-533-8661
E-Mail: cust@igi-global.com • www.igi-global.com

Editorial Advisory Board

Table of Contents

Preface xviii

Acknowledgment xxii

Chapter 1
Crisp and Fuzzy AHP in GIS-MCDA for Wildlife Habitat Suitability Analysis 1
Suman Sinha, Amity University, Kolkata, India

Chapter 2
GIS-based Multi-Criteria Analysis for Delineation of Groundwater Potential Zones: A Case Study from Jodhpur District, Rajasthan, India 24
Raghib Raza, Quantum Asia Pvt. Ltd., India
Gajendra Kumar Chawla, Watershed Development and Soil Conservation Department, Jodhpur, India
Chandra Shekhar Dwivedi, Central University of Jharkhand, India

Chapter 3
Evaluation of Multi-Temporal Sentinel-1 Dual Polarization SAR Data for Crop Type Classification 44
Thota Sivasankar, NIIT University, India
Pavan Kumar Sharma, Amnex Infotechnologies Pvt. Ltd., India
M. N. S. Ramya, Independent Researcher, USA
Pithani Venkatesh, Skymet Weather Services Pvt. Ltd., India
G. D. Bairagi, M.P. Council of Science and Technology, India

Chapter 4
Geo-Spatial Technology for Land Resources Management in Nigeria 62
Ugonna Chimnonyerem Nkwunonwo, University of Nigeria, Enugu, Nigeria

Chapter 5
Investigation of Urban Environment Using Tasseled Cap Transformation: Vegetation and Urban Indices in Indian Sub-Continental City (Kharagpur).......88
Asraful Alam, University of Calcutta, India
Arijit Ghosh, Sidho-Kanho-Birsha University, Purulia, India
Lakshminarayan Satpati, University of Calcutta, India

Chapter 6
Efficacy of Advanced Remote Sensing (Hyperspectral and LIDAR) in Enhancing Forest Resources Management 97
Laxmikant Shrama, Central University of Rajasthan, India
Rajit Gupta, Central University of Rajasthan, India
Rajani Kant Verma, Central University of Rajasthan, India

Chapter 7
Synergy of Satellite-Derived Drought Indices for Agricultural Drought Quantification and Yield Prediction..122
Dipti Ladli, Central University of Jharkhand, India
Kanhaiya Lal, Central University of Jharkhand, India
Kiran Jalem, National Institute of Rural Development and Panchayati Raj, Hyderabad, India
Avinash Kumar Ranjan, National Institute of Technology, Rourkela, India

Chapter 8
Comparative Flood Inundation Mapping Utilizing Multi-Temporal Optical and SAR Satellite Data Over North Bihar Region: A Case Study of 2019 Flooding Event Over North Bihar...149
Gaurav Tripathi, Department of Geoinformatics, Central University of Jharkhand, Brambe, India
Arvind Chandra Pandey, Department of Geoinformatics, Central University of Jharkhand, Brambe, India
Bikash Ranjan Parida, Department of Geoinformatics, Central University of Jharkhand, Brambe, India
Achala Shakya, Department of Computer Engineering, National Institute of Technology, Kurukshetra, India

Chapter 9
Inferring Relationship of Landslides, Tectonics, and Climate: Tons Valley, NW Himalaya 169
Imlirenla Jamir, Department of Geology, Nagaland University, India
Pranaya Diwate, Centre for Climate Change and Water Research, Suresh Gyan Vihar University, India
Vipin Kumar, Department of Geology, University of Liege, Belgium
Gambhir Singh Chauhan, Department of Geology, H.N.B. Garhwal University, India

Chapter 10
Sentinel SAR Data and In-Situ-Based High-Resolution Above-Ground Carbon Stocks Estimation Within the Open Forests of Ramgarh District 180
Akash Anand, Department of Geoinformatics, Central University of Jharkhand, India

Chapter 11
Status and Opportunities for Forest Resources Management Using Geospatial Technologies in Northeast India 206
Kasturi Chakraborty, North Eastern Space Applications Centre, India
Thota Sivasankar, NIIT University, India
Junaid Mushtaq Lone, Geoinformatics Programme, Department of Earth Sciences, University of Kashmir, India
K. K. Sarma, North Eastern Space Applications Centre, India
P. L. N. Raju, North Eastern Space Applications Centre, India

Chapter 12
Statistical Analysis of Major Flood Events During 1980-2015 in Middle Ganga Plain, Ganga River Basin, India 225
Aman Arora, Department of Geography, Faculty of Natural Sciences, Jamia Millia Islamia, New Delhi, India
Masood Ahsan Siddiqui, Department of Geography, Faculty of Natural Sciences, Jamia Millia Islamia, New Delhi, India
Manish Pandey, University Center for Research and Development (UCRD), Chandigarh University, Mohali, India

Chapter 13
Inland Saline Wetlandscapes: The Missing Links for 4th Ramsar Strategic Plan (2016-2024) in India 242
Laxmikant Sharma, Central University of Rajasthan, India
Rajashree Naik, Central University of Rajasthan, India
Alok Raj, Central University of Rajasthan, India

Chapter 14
Geomorphic Changes Related to Anthropogenic Interference Along the Ganga River From Rishikesh to Haridwar, Uttarakhand, India: Geomorphic Response of the Ganga to Tehri Dam 269
Manish Pandey, University Center for Research and Development (UCRD), Chandigarh University, Mohali, India
Aman Arora, Department of Geography, Faculty of Natural Sciences, Jamia Millia Islamia, New Delhi, India
Rajesh Kumar, Center for the Study of Regional Development, Jawaharlal Nehru University, India
Vijendra Kumar Pandey, Department of Geography, Kirori Mal College, University of Delhi, New Delhi, India
Akshay Kumar, Department of Remote Sensing, Birla Institute of Technology, Mesra, India

Compilation of References 288

About the Contributors 348

Index 354

Detailed Table of Contents

Preface xviii

Acknowledgment xxii

Chapter 1
Crisp and Fuzzy AHP in GIS-MCDA for Wildlife Habitat Suitability Analysis 1
Suman Sinha, Amity University, Kolkata, India

Geographic information system-based multi-criteria decision analysis (GIS-MCDA) is a process of decision making where geographical data and value judgments are integrated. Analytic hierarchy process (AHP) is a useful technique in MCDA for determining weights. This study focuses on the evaluation of GIS-MCDA using different uncertainty levels in AHP. Best suitable sites for tiger habitats are located and analyzed in Sariska Wildlife Reserve, India using crisp and fuzzy AHP in GIS-MCDA, and thereafter, an optimal habitat suitability model is proposed. The percentage deviation over the uncertainty levels ranges slightly over 5%. The relative difference between CAHP and FAHP is nearly 2.7%. Chi-square test reveals relationship between the degree of uncertainty and the difference between the maps. For real-world situations with increased variability, fuzzification is preferred and shows the best results. The worldwide declining status of the tigers is a serious threat to the overall biodiversity, and the methods adopted in this study thus target their conservation and management.

Chapter 2
GIS-based Multi-Criteria Analysis for Delineation of Groundwater Potential Zones: A Case Study from Jodhpur District, Rajasthan, India 24
Raghib Raza, Quantum Asia Pvt. Ltd., India
Gajendra Kumar Chawla, Watershed Development and Soil Conservation Department, Jodhpur, India
Chandra Shekhar Dwivedi, Central University of Jharkhand, India

In the present study, delineation of ground water potential in Jodhpur district, Rajasthan is framed by using geospatial technique (i.e., remote sensing and GIS) and with multi-criteria leadership (MCDM) procedure. The analytical network process (ANP) is one of the suitable strategies that makes the research workable for delineation methodically, and corporate the analytic hierarchy process (AHP) as an uncommon case. For proper value to recognize the ground water potential zone in Jodhpur, Rajasthan, the AHP and ANP methods are utilized to decide the values of different parameters and their classes. The AHP values are then connected in a direct mix into raster calculator to get five distinctive groundwater potential zones in the investigation region, to be specific as 'very poor' (2052.0 km2), 'poor' (4225.9 km2), 'moderate' (6355.1 km2), 'good' (6451.2 km2), and 'very good' (3301.7 km2). It has been presumed that about 9752.9 km2 area of Jodhpur district has very good to good groundwater potential, which is about 43.56% of the complete study area.

Chapter 3
Evaluation of Multi-Temporal Sentinel-1 Dual Polarization SAR Data for Crop Type Classification 44
Thota Sivasankar, NIIT University, India
Pavan Kumar Sharma, Amnex Infotechnologies Pvt. Ltd., India
M. N. S. Ramya, Independent Researcher, USA
Pithani Venkatesh, Skymet Weather Services Pvt. Ltd., India
G. D. Bairagi, M.P. Council of Science and Technology, India

India is one of the highly populated countries, and its economy mainly depends on agriculture. The crop type classification is an essential requirement for ensuring food security, crop monitoring, and to understand the environmental consequences of cultivated ecosystems. This study exploits freely available multi-temporal SAR data for discriminating crop types, such as wheat, gram, and mustard, over Ashok Nagar district, Madhya Pradesh, India. Nine Sentinel-1 dual-polarized data acquired from January 2018 to April 2018 in interferometric wide swath mode are used. Class separability analysis using Bhattacharyya Distance (BD) has been performed for multi-temporal VV and VH backscatter, log-ratio, and Radar Vegetation Index (RVI) to quantify the ability to distinguish temporal profiles of crops. RVI has shown the significant result in class separability analysis in comparison with other parameters. Crop type classification map has been generated using a support vector machine classifier with overall accuracy and Kappa coefficient of 96.32% and 0.95, respectively.

Chapter 4
Geo-Spatial Technology for Land Resources Management in Nigeria 62
Ugonna Chimnonyerem Nkwunonwo, University of Nigeria, Enugu, Nigeria

There is little discussion regarding natural resources management issues in Nigeria, unlike many other places around the world, where such issues have made significant contributions to research growth. This is a troubling situation that complicates Nigeria's present need to address her increasing sequence of aggravations related to land cover modifications. Sustainable use of land, water, solid minerals, and forest is difficult and overwhelms local efforts. Demographic pressures and the corresponding need for developments and societal livability have a proclivity to overuse land resources and impact negatively on their present quality and future regeneration. Traditional and indigenous approaches are still the bases of natural resources management. The cumulative challenges of these issues with other prevalent anomalies are increasingly compromising Nigeria's land resources base. However, geo-spatial technology with its potential for policy and decision-support is being set forth to address these challenges and to fill the current gaps in knowledge of natural resources management.

Chapter 5
Investigation of Urban Environment Using Tasseled Cap Transformation: Vegetation and Urban Indices in Indian Sub-Continental City (Kharagpur).......88
Asraful Alam, University of Calcutta, India
Arijit Ghosh, Sidho-Kanho-Birsha University, Purulia, India
Lakshminarayan Satpati, University of Calcutta, India

Urban settlements have more complex environments, in unremitting fruition, where most of the world population lives. Most of the cities in developing countries have been developed without a rationale, and the life conditions are repeatedly insufferable. For this research work, NDVI is particularly used to assess the status of vegetation cover. Tasseled cap is another index that creates three band images for this study. Brightness, greenness, wetness are the three bands that represent the area under consideration. The present study aims particularly at comparing high NDVI area and greenness values given by tasseled cap and low NDVI values and high brightness values and status of urban environment. Based on the overlapping of tasseled cap image, an NDVI image is observed in which most of the area of healthy vegetation is located in the north west and south east part of Kharagpur city, which extended from south west to north east and north to south respectively.

Chapter 6
Efficacy of Advanced Remote Sensing (Hyperspectral and LIDAR) in Enhancing Forest Resources Management 97
Laxmikant Shrama, Central University of Rajasthan, India
Rajit Gupta, Central University of Rajasthan, India
Rajani Kant Verma, Central University of Rajasthan, India

Sustainable management of natural forest resources is a vital requirement in the changing climatic conditions on Earth. Two advances techniques, hyperspectral remote sensing (HRS) and LIDAR (light detection and ranging) remote sensing (LRS), provide more enhanced and accurate measurements than that of conventional optical remote sensing (ORS). Hyperspectral sensor like AVIRIS, which has hundreds of narrow bands, have advantages over a broadband multispectral sensor. In addition, the fusion of HRS and LRS can play an essential role in assessing biophysical and biochemical variables of forest species. In this chapter, the authors reviewed the extant literature and tried to understand the position of HRS, LRS, and their integration with the machine and deep learning algorithms for the effective monitoring and management of natural forest resources. Further, scopes and challenges are also discussed to enhance the effectiveness of these techniques in natural forest resources management.

Chapter 7
Synergy of Satellite-Derived Drought Indices for Agricultural Drought Quantification and Yield Prediction...122
Dipti Ladli, Central University of Jharkhand, India
Kanhaiya Lal, Central University of Jharkhand, India
Kiran Jalem, National Institute of Rural Development and Panchayati Raj, Hyderabad, India
Avinash Kumar Ranjan, National Institute of Technology, Rourkela, India

The present study was conducted over Jharkhand state (India) for assessing the drought condition and corresponding yield of paddy (district-level) during Kharif 2018. Vegetation drought indices, namely Vegetation Condition Index (VCI), Temperature Condition Index (TCI), Vegetation Health Index (VHI), and vegetation indices (VI) anomaly, were derived from different VI (i.e., NDVI, EVI) to assess the paddy health condition during drought year (2018) and non-drought year (2017). Later, the correlation between the DES-based yield data and derived drought indices (for the year 2017) were made to develop the district-level paddy yield model for the drought year 2018. The key results of the study shown that VCI derived from EVI data was found to be more reasonable to depict the drought condition, wherein ~21% area was under severe drought condition, 43% area under moderate drought condition, and 36% area under no drought condition. In addition, the yield prediction model derived from VCI (EVI-based) was found to be promising for predicting the paddy yield for Kharif 2018 with fair R2 of 0.53.

Chapter 8
Comparative Flood Inundation Mapping Utilizing Multi-Temporal Optical and SAR Satellite Data Over North Bihar Region: A Case Study of 2019 Flooding Event Over North Bihar 149
Gaurav Tripathi, Department of Geoinformatics, Central University of Jharkhand, Brambe, India
Arvind Chandra Pandey, Department of Geoinformatics, Central University of Jharkhand, Brambe, India
Bikash Ranjan Parida, Department of Geoinformatics, Central University of Jharkhand, Brambe, India
Achala Shakya, Department of Computer Engineering, National Institute of Technology, Kurukshetra, India

Floods are investigated to be the utmost frequent and destructive phenomena among all other types of natural calamities worldwide. Thus, flood events need to be mapped to understand their impact on the affected region. The present case study is intended to examine and analyze the flood events occurred in July-August 2019 over the Northern Bihar region situated in Kosi and Gandak river basins. Furthermore, a comparative study was carried out to map the satellite based near real time flood inundation using multi-temporal Sentinel–1A (SAR) and MODIS NRT Flood data (optical and 3-day composite). Optical (MODIS) and Sentinel-1 SAR data were acquired to compare their flood inundation extent and the result shows overestimation in MODIS flood data due to varying spatial resolutions.

Chapter 9
Inferring Relationship of Landslides, Tectonics, and Climate: Tons Valley, NW Himalaya 169
Imlirenla Jamir, Department of Geology, Nagaland University, India
Pranaya Diwate, Centre for Climate Change and Water Research, Suresh Gyan Vihar University, India
Vipin Kumar, Department of Geology, University of Liege, Belgium
Gambhir Singh Chauhan, Department of Geology, H.N.B. Garhwal University, India

Landslides, despite being the surficial impression of climate-tectonic-erosion linkage, are rarely explored in this context in Himalaya. The need for such study becomes more crucial in the evaluation of the regional hillslope denudation budget. We are of the understanding that the distributional pattern of landslides can reveal the relative significance of tectonic and climate. To test this hypothesis, ~ 55 landslides of the Tons River valley, Himalaya along with the tectonic and climate proxies are used in the present study. Steepness index and valley floor width to valley height ratio are used to infer the tectonic regime whereas; Tropical Rainfall Measurement Mission based daily rainfall data and swath profile of Normalized Difference Vegetation Index are used to deduce spatial variability in climate. The study revealed the possible existence of a positive feedback system in the Higher Himalaya Crystalline and the simultaneous role of tectonic-climate in the Lesser Himalaya Crystalline. The LHS is found to possess a zone of landslide cluster, possibly due to local fault.

Chapter 10
Sentinel SAR Data and In-Situ-Based High-Resolution Above-Ground Carbon Stocks Estimation Within the Open Forests of Ramgarh District.........180
Akash Anand, Department of Geoinformatics, Central University of Jharkhand, India

The present study deals with an approach to estimate the above ground biomass (AGB) to assess the total carbon stock of forest cover present in Ramgarh district using remote sensing and GIS techniques. Due to the fact that biomass estimation is one of the most influential biophysical parameters in traditional carbon sequestration techniques, satellite remote sensing plays an important role in AGB and carbon stock estimation. Presently, AGB is estimated using Sentinel1A SAR data in conjunction with in-situ field data, which is conducted in 20 different sites within the forest area. Biomass is calculated for each plot, and a correlation analysis is performed with the backscatter value obtained from SAR data to generate an allometric equation that is used to calculate the AGB and carbon stock for the entire forest cover. Both Polarization VV and VH are correlated with field data in which cross-polarized backscatter value shown a stronger correlation of 0.75 (R2 Value). C-band is proved to be the best band for the estimation of biomass and carbon stock in tropical mixed forests.

Chapter 11
Status and Opportunities for Forest Resources Management Using Geospatial Technologies in Northeast India 206
Kasturi Chakraborty, North Eastern Space Applications Centre, India
Thota Sivasankar, NIIT University, India
Junaid Mushtaq Lone, Geoinformatics Programme, Department of Earth Sciences, University of Kashmir, India
K. K. Sarma, North Eastern Space Applications Centre, India
P. L. N. Raju, North Eastern Space Applications Centre, India

The forest resource of North East Region (NER) of India is a store house of several unique, endangered, endemic, medicinal plant, bamboo, etc. species in diverse forest type and high forest density. Several authors and organizations have contributed to the study of the richness and diversity distributed in different forest types and forest density. This chapter attempts to highlight the uniqueness of the forest of NER and the role of geospatial technology and presents various interesting studies pertaining to the region as an input to forest resource assessment. Remote sensing and GIS have an important role in NER forest resource assessment, management, and conservation. Various studies carried out with the help of remote sensing and GIS technology have highlighted the ongoing forest degradation and deforestation taking place in this region due to developmental activity and economic benefits. There is continuous improvement in the forest estimates from coarse resolution satellite data to unmanned aerial vehicles (UAV) in the recent times.

Chapter 12
Statistical Analysis of Major Flood Events During 1980-2015 in Middle Ganga Plain, Ganga River Basin, India 225
Aman Arora, Department of Geography, Faculty of Natural Sciences, Jamia Millia Islamia, New Delhi, India
Masood Ahsan Siddiqui, Department of Geography, Faculty of Natural Sciences, Jamia Millia Islamia, New Delhi, India
Manish Pandey, University Center for Research and Development (UCRD), Chandigarh University, Mohali, India

To understand the vicious nature of extreme flood events for the most flood prone region of Ganga River Basin, this study uses 36 years (1980-2015) of flood records from Dartmouth Flood Observatory (DFO) and the Centre for Research on the Epidemiology of Disasters (CRED) Emergency Events Database (EM-DAT). Further, the Water Level (WL) data collected from Central Water Commission (CWC) for same period are utilized to compare with the data of DFO and EM-DAT to identify the major flood events recorded in the Middle Ganga Plain (MGP). The final dataset comprises of 15 attributes (parameters) and is prepared of identified 99 flood instances for statistical analysis. The descriptive statistical analysis is performed for the following parameters: severity class, flood duration in days, affected flood area, flood magnitude, total number of deaths, and total count of displaced people. The graphical representation of all selected parameters provides an insight of common flood events, which lie between ±95% confidence level and exclude the major events as outliers.

Chapter 13
Inland Saline Wetlandscapes: The Missing Links for 4th Ramsar Strategic Plan (2016-2024) in India...242
Laxmikant Sharma, Central University of Rajasthan, India
Rajashree Naik, Central University of Rajasthan, India
Alok Raj, Central University of Rajasthan, India

Wetland ecosystems are one of the highly productive ecosystems in the world. These ecosystems have been deteriorating at a faster rate. Ramsar Convention is putting enormous effort to protect, maintain, and restore these ecosystems. Currently, the fourth phase of Strategic Plans of Ramsar Convention is going on, in which saline wetlandscapes can play vital role to attain 19 targets of this plan. In India there are 27 Ramsar sites in all the biogeographic zones; however, research work has been carried out in the past five years in only eight Ramsar sites. Currently, four years are available for the strategic plans to encourage more wetland researches. The chapter presents a case study of Sambhar Salt Lake, a Ramsar site of India that is on the verge of extinction. Normalized Difference Water Index has been calculated for three decades in 1992, 2009, and 2019, revealing the declining phases of the lake.

Chapter 14
Geomorphic Changes Related to Anthropogenic Interference Along the Ganga River From Rishikesh to Haridwar, Uttarakhand, India: Geomorphic Response of the Ganga to Tehri Dam 269
Manish Pandey, University Center for Research and Development (UCRD), Chandigarh University, Mohali, India
Aman Arora, Department of Geography, Faculty of Natural Sciences, Jamia Millia Islamia, New Delhi, India
Rajesh Kumar, Center for the Study of Regional Development, Jawaharlal Nehru University, India
Vijendra Kumar Pandey, Department of Geography, Kirori Mal College, University of Delhi, New Delhi, India
Akshay Kumar, Department of Remote Sensing, Birla Institute of Technology, Mesra, India

The authors, in this study, have quantified the changes in the geomorphic activity of the Ganga River between Rishikesh and Haridwar by using Survey of India (SOI) 1:5000 scale topographic sheets, satellite data, digital elevation model (DEM), and hydrological data from published resources. They have also correlated the change in settlement area in Haridwar and Rishikesh and connected it to the changing geomorphic features in the downstream reach of the Tehri dam site. The study finds that the change in geomorphology of the river in the studied reach of the Ganga River is attributable to reduced water and sediment supply. The study area being in the active tectonic zone, bound between MCT in the north and HFT in the south, may have experienced offsetting of the reduction in sediment supply by dam in the post dam era (after 2006) due to massive earthquakes delivering sediment directly delivered to streams through enhanced landslide and other mass wasting processes.

Compilation of References 288

About the Contributors 348

Index 354

Preface

Natural resources are the basic element for human survival and development. The rapid population growth, wide scale tree felling and devoid of regeneration consumed the natural resources to a level, which created a problem for the mankind. So, there is an urgent requirement of a coordinated approach for sustainable development and management of natural resources to keep a balance of the environment, ecology and human livelihood. This approach demands an interdisciplinary approach and sound knowledge on each specific resource, as well as on the ecological and socio-economic perspectives related to their use. It is important to conserve the natural resource bases and improve its economic viability to achieve the goal of livelihood security.

Spatial Information Science consisting of Remote Sensing (RS), Geographical Information System (GIS), and Global Positioning System (GPS) is a path breaking and advance technology for acquiring information needed for environmental protection, natural resource management and addressing the concerns related to sustainable development and planning. It offers a powerful and proficient tool for mapping, monitoring, modeling, and management of natural resources etc. It is also capable to make use of recent advancements in the imaging sensors at different observational scales for concentrate understanding of earth's surface and its dynamics. Application of spatial information science includes land use mapping and planning, crop yield estimation, water resource management, forest fire mapping, forest inventory, glacier monitoring, soil degradation, pest and disease management, and many more. Therefore, this subject plays an increasing role and offer additional role in natural resource management across the globe. Several authors have carried out their research and operational applications of direct relevance especially to natural resource management with the launch of Landsat series satellites in the 1970s. However, there is a fundamental gap between the theoretical concepts and the operational use of these advances tools. This could be resolved and explained by providing broad range of applications and prospective of this technology to the scientific and research community in the field of spatial information science. The target audience and potential readers of this book includes Researchers, Academicians, Urban Planners, Environmentalists, Policy Makers, Scientists, and Ecologists, etc. working in the field of natural resource management.

In Chapter 1, a GIS based multi-criteria decision analysis was used for habitat suitability study in Sariska Wildlife Reserve, India. Crisp and fuzzy AHP methods were used to propose optimal habitat suitability model. The Chi-square test reveals relationship between the degree of uncertainty and the difference between the maps. The method adopted in this study can be used effectively for conservation and management of biodiversity.

In Chapter 2, the delineation of ground water potential zone was performed in Jodhpur district, Rajasthan, India using geospatial approach. In this chapter an integrated approach of Remote Sensing, GIS and multi-criteria analysis was used for ground water potential zone into very poor, poor, good and very good categories. The groundwater potential zone has been validated with the information of siphoning wells. This study can provide information of fresh water zones for the drinking purpose.

In Chapter 3, multi-temporal Sentinel-1 dual-polarized data acquired from January 2018 to April 2018 in Interferometric Wide swath mode were used for discriminating crop types, such as, wheat, gram and mustard over Ashok Nagar district, Madhya Pradesh, India. The class separability analysis using Bhattacharyya Distance (BD) has been performed for multi-temporal backscatter, log-ratio and RVI to quantify the ability to distinguish temporal profiles of crops. The classification of crop types is an essential requirement for ensuring food security, crop monitoring and to understand the environmental effects of cultivated ecosystems.

In Chapter 4, natural resources management issues were discussed in Nigeria. In current time, a sustainable use of land, water, solid minerals, forest, fishery, and wild flora and fauna is difficult and overwhelming to local efforts in Nigeria. The demographic pressure and related requirement for the development and societal livability have a tendency to overuse land resources and impact negatively on their present quality and future regeneration. With the advent of Geo-spatial technology, it is possible to address these challenges and to fill the current gaps in knowledge of natural resources management.

In Chapter 5, in developing countries the cities have been expanded without a rationale and the life conditions are constantly unbearable. This chapter aims mainly at comparing NDVI values with the parameters derived from tasseled cap transformation to investigate the status of urban environment in Kharagpur city of West Bengal, India. Most of the area in the North West and south east part of city was found with healthy vegetation cover. This study will be helpful to provide valuable information for sustainable urban planning and management.

In Chapter 6, a detailed description of sustainable management of natural forest resources was presented. Two developed techniques such as hyperspectral remote sensing (HRS) and LIDAR (Light detection and ranging) remote sensing (LRS) has the capabilities to supply enhanced and accurate measurements in comparison to optical remote sensing. In this chapter, a literature review is presented to understand the position of HRS, LRS, and their integration with the machine and deep learning algorithms for the effective monitoring and management of natural forest resources.

In Chapter 7, a study was conducted to assess the drought condition and corresponding yield of paddy during Kharif season of year 2018 over Jharkhand state, India. Several indices were used to assess the paddy health condition during drought year (2018) and non-drought year (2017). Afterward, the correlation analysis was performed between the DES-based yield data and derived drought indicesto develop the district-level paddy yield model for the drought year 2018. This study can be used to provide detailed and timely information on agricultural drought in order to make mitigation strategies and appropriate decision

In Chapter 8, a case study is presented to examine and analyze the flood events occurred in July-August, 2019 over the Northern Bihar region situated in Kosi and Gandak river basins. In addition, a comparative study was performed to map the near real time flood inundation based on multi-temporal Sentinel–1A (SAR) satellite and MODIS NRT Flood data (Optical and 3-day composite). The results showed overestimated flood area due to varying spatial resolutions. This type of study can be used to provide temporal mapping of flood extent.

In Chapter 9, the distributional pattern of landslides can reveal the relative significance of tectonic and climate. A study was presented in this chapter to test this hypothesis, ~ 55 landslides of the Tons River valley, Himalaya along with the tectonic and climate proxies. The study revealed that the possible existence of a positive feedback system in the Higher Himalaya Crystalline and the simultaneous role of tectonic-climate in the Lesser Himalaya Crystalline. The despites of landslides are the surficial impression of climate-tectonic-erosion linkage which is rarely explored in the context of Himalaya

In Chapter 10, a study was performed to estimate the Above Ground Biomass (AGB) to assess the total Carbon Stock of forest cover present in Ramgarh district, Jharkhand, India. In this study, AGB is estimated using Sentinel-1A SAR data and in-situ field measurements which is conducted in 20 different sites within the forest area. Both the polarized images are correlated with field data in which cross-polarized backscatter value shown a stronger correlation of 0.75 (R2 Value). C-band is proved to be the best band for the estimation of biomass and carbon stock in tropical mixed forests. This study can be useful to assess the quantity and the components of the forest in term of biomass.

In Chapter 11, The North East Region (NER) of India has very rich forest resource having numerous unique, endangered, endemic, medicinal plant, bamboo etc. This chapter presented the uniqueness of the forests of NER. Remote sensing and GIS techniques have an important role assessment, management and conservation of forest resources. Geospatial technology is capable to provide information of ongoing forest degradation and deforestation taking because of developmental activity and economic benefits. There is continuous improvement in the forest estimates with the emergence of Unmanned Aerial vehicles (UAV) technique.

In Chapter 12, this case study presented a statistical analysis to identify the major flood events recorded in the Middle Ganga Plain (MGP) over the period of 1980-2015. The flood data were collected from Dartmouth Flood Observatory (DFO) and the Centre for Research on the Epidemiology of Disasters (CRED) Emergency Events Database (EM-DAT) and compared with the Water Level (WL) data. The statistical analysis was performed for Severity class, Flood Duration in Days, Affected flood area, Flood Magnitude, Total number of death and Total count of displaced people. This type of study can be useful in preparedness and mitigation strategies during flooding event.

In Chapter 13, the wetland ecosystems are one of the highly productive ecosystems in the world. A detailed description of Ramsar Convention is presented is to protect, maintain and restore these ecosystems. Currently, in the fourth phases saline wet landscapes can play vital role to attain nineteen targets of this plans. This chapter presents a case study of Sambhar Salt Lake, a Ramsar site of India. The Normalized Difference Water Index has been calculated for three decades in 1992, 2009 and 2019 revealing the declining phases of the lake.

In Chapter 14, a work is presented to quantify the changes in the geomorphic activity of the Ganga River between Rishikesh and Haridwar. Satellite images, digital elevation model, topographic sheets, and hydrological data were used in this study. The study shows that the changes in geomorphology of Ganga River are attributable to reduced water and sediment supply. The study area being in the active tectonic zone may have experienced offsetting of the reduction in sediment supply by dam after 2006 because of massive earthquakes.

This edited book entitled *Spatial Information Science for Natural Resource Management* contains chapters written by prominent researchers and experts. The primary focus of this book is to replenish the gap in the available literature on the subject by bringing jointly the concepts, theories and experiences of the specialists and professionals in this field. This book will be of great interest to researchers, policymakers and administrative authorities working in the field of natural resource monitoring and management.

Acknowledgment

First of all, the editors would like to pay obeisance to almighty God for keeping us in good health during the completion of edited book entitled *Spatial Information Science for Natural Resource Management.*

The words cannot express our indebtedness to Hon'ble Sunil Sharma, Chairperson, Suresh Gyan Vihar University, Jaipur for his valuable suggestions and inspiration throughout the preparation of this edited book. We want to express our heartfelt and indebted gratitude to Hon'ble Dr. Sudhanshu, Chief Mentor, Suresh Gyan Vihar University, Jaipur for his time to time guidance, expert suggestions and motivation during the completion of this edited book.

The editors would like to express sincere appreciation to all the members of editorial advisory board for their continuous support and valuable instructions provided at all stages of the preparation of this edited book. We also express our gratitude to all the reviewers for their kind and timely support during the review process. Thanks are also due to all the colleagues and friends for their constant support and help. The editors wish to gratefully acknowledge IGI Global for giving us opportunity to publish with them.

Chapter 1
Crisp and Fuzzy AHP in GIS-MCDA for Wildlife Habitat Suitability Analysis

Suman Sinha
Amity University, Kolkata, India

ABSTRACT

Geographic information system-based multi-criteria decision analysis (GIS-MCDA) is a process of decision making where geographical data and value judgments are integrated. Analytic hierarchy process (AHP) is a useful technique in MCDA for determining weights. This study focuses on the evaluation of GIS-MCDA using different uncertainty levels in AHP. Best suitable sites for tiger habitats are located and analyzed in Sariska Wildlife Reserve, India using crisp and fuzzy AHP in GIS-MCDA, and thereafter, an optimal habitat suitability model is proposed. The percentage deviation over the uncertainty levels ranges slightly over 5%. The relative difference between CAHP and FAHP is nearly 2.7%. Chi-square test reveals relationship between the degree of uncertainty and the difference between the maps. For real-world situations with increased variability, fuzzification is preferred and shows the best results. The worldwide declining status of the tigers is a serious threat to the overall biodiversity, and the methods adopted in this study thus target their conservation and management.

DOI: 10.4018/978-1-7998-5027-4.ch001

INTRODUCTION

Geographic information system-based multi-criteria decision analysis (GIS-MCDA) is a process of decision making in which geographical data and value judgments are integrated to obtain information for decision makers. GIS-MCDA is used in a wide range of decision and management situations like environment planning and ecology management, urban and regional planning, hydrology and water resources, forestry, transportation, agriculture, natural hazard management, health-care resource allocation, etc. (Vahidnia et al. 2008, Sinha 2014). Analytic hierarchy process (AHP) (Saaty 1980, 1988) is an important and useful technique in decision making that has been widely used as a useful MCDA tool or a weight estimation technique in different cases (Vaidya and Kumar 2006, Sinha 2015, 2016). AHP is a method for ranking decision alternatives and selecting the best one when the decision maker has multiple criteria for analysis (Taylor 2004). The traditional AHP requires crisp judgments. However, AHP is often criticized for its inability to effectively handle the inherent uncertainty and imprecision associated with the mapping of the decision maker's perception to exact numbers (Deng 1999). AHP applications are used for suitability analysis in several cases (Banai-Kashani 1989, Eastman et al. 1992, 1993, Xiang and Whitley 1994). Although AHP method is a powerful tool in MCDA for spatial problems (Anselin et al. 1989, Kangas 1992, Correa-Berger 2006, Sinha et al. 2011, 2012, Sinha 2015, 2016), some researchers believe that Saaty's AHP method has some weaknesses (Yang and Chen 2004).

Decision maker obviously makes an uncertain estimation rather than a precise value in real-world cases. Humans are ineffective in making quantitative predictions, whereas they are comparatively efficient in qualitative assessment. In complex systems, the judgments of humans are symbolized by linguistic and vague patterns. Hence, better representation of this linguistics can be developed as quantitative data, which then can be refined by fuzzy set theory evaluation methods (Leung and Chao 2000, Kulak and Kahraman 2005, Ozdagoglu and Ozdagoglu 2007). Due to the complexity and uncertainty involved in real-world decision problems, fuzzy judgments provide better and more realistic decisions than crisp judgments. Since uncertainty and fuzziness are common characteristics in many decision-making problems, a fuzzy AHP (FAHP) method should be able to tolerate vagueness or ambiguity (Mikhailov and Tsvetinov 2004). In other words, the conventional AHP approach may not fully reflect a style of human thinking because the decision makers usually feel more confident to give interval judgments rather than expressing their judgments in the form of single numeric values, and so, FAHP can capture the appraisal of human for ambiguity when complex multi-attribute decision-making problems are considered (Erensal et al. 2006, Sinha 2014). Uncertainty lies within the parameters for the real-world problems like tiger habitat suitability, as with changing time due to its dynamic nature, the landscape also changes resulting from changes in the nature, human activities and socioeconomic conditions. Fuzzy numbers are normal convex fuzzy sets, first introduced by Zadeh (1973). The mathematics of fuzzy set theory is described in Fuzzy Sets (Zadeh 1965) and the methods explained and detailed in Patil (2018) and El-Din et al. (2019). This theory, which is a generalization of classic set theory, allows the membership functions to operate over the range of real numbers [0, 1]. The main characteristic of fuzziness is the grouping of individuals into classes that do not have sharply defined boundaries (Hansen 2005). The uncertain comparison judgment can thus be represented by the fuzzy number.

The uncertainty associated with the mapping of decision makers' judgment to number is not taken into account by the AHP, and also the preference and personal judgment of decision maker have important effects on the AHP result (Yang and Chen 2004). In order to overcome these problems, researchers modified and fuzzified the Saaty's AHP to formulate and control the uncertainty. Hence, trapezoidal membership function for comparison ratios in AHP has been formulated (Buckley 1985), and a new approach for triangular case has been developed (Chang 1996). Having a defined uncertainty in the form of fuzzy numbers in comparison ratios and then obtaining the weights in fuzzy forms help the decision makers to get a better understanding of the final importance of the factors and the uncertainty lying within them. However, Saaty is against fuzzification of his method. He states that some uncertainty lies in the nature of the method, i.e. the ratios in the method are not absolute or crisp numbers, in fact, they are fuzzy numbers; so making the AHP fuzzier not only does not guarantee better results but also could make it worse (Saaty and Tran 2007). Some of the adequate applications of the FAHP are the flexible manufacturing systems (Chutima and Suwanfuji 1998); critical decisions in new product development (Buyukozkam and Feyzioglu 2004); evaluation of success factors in e-commerce (Kong and Liu 2005); assessment of water management plans (Sredjevic and Medeiros 2008); safety management in production (Dagdeviren and Yuksel 2008); energy sources for space heating (Jaber et al. 2008); personnel selection (Gungor et al. 2009); selection of enterprise resource planning systems (Cebeci 2009); and weapon selection (Dagdeviren et al. 2009). Use of FAHP in GIS is well explained (Vahidnia et al. 2008). Recently, Fuzzy AHP-based MCDA has been applied for several such decision making researches (Patil 2018, El-Din et al. 2019) that includes waste management (Khoshand et al. 2019), route planning (Singh et al. 2019), facility planning (Lin and Wang 2019), watershed management (Wijitkosum and Sriburi 2019), ecodesign practices (Singh and Sarkar 2019), carbon technology (Promentilla et al. 2016), disaster management (Sharma et al. 2012), sustainable manufacturing (Ocampo 2019), etc. Some spatial planning or spatial problems like site selection can be considered as a multiple criteria decision-making problem involving a set of alternatives that are evaluated on the basis of conflicting and incommensurate criteria (Malczewski 1999).

So far, the attempts and works for comparing the CAHP and FAHP were mostly done in theory. Practical examples of GIS-based decision making problem using comparative studies of CAHP and FAHP are limited. A comparative assessment of the above methods was done for the selection of the best site for constructing a dam (Kordi 2008), and FAHP was also used for the selection of wastewater treatment process (Karimi et al. 2011). Fuzzy technology in AHP has been used in real-world suitability analyses (Sinha et al. 2011a, 2012, Sinha 2014). The main goal of this study is to practically compare the results of CAHP and FAHP methods by applying them in a GIS-based real-world application of locating and analysing the best sites for tiger habitat for their relocation in Sariska Wildlife Reserve, Rajasthan, India, considering different criteria/rules. Worldwide declining status of the tigers is a serious threat to the overall biodiversity. This study also targets at the conservation of the endangered tigers using the above said methods. The comparison between the results of CAHP and FAHP in the work is done in two approaches. The remainder of this article is structured as follows: The background and theory of the methods adopted is given in Section 2. Section 3 deals with the detailed method and data used, introducing the problem and the evaluation criteria considered for the study. Section 4 describes the final results, while overall discussion and concluding remarks are summarized in Section 5.

Background

Crisp AHP (CAHP)

The AHP, which is a powerful tool in applying MCDA, was introduced and developed by Saaty in 1980. In AHP, obtaining the weights or priority vector of the alternatives or the criteria is required. For this purpose, the pairwise comparison method (PCM) is developed (Saaty 1980). For creating the pairwise comparison matrix in the PCM, Saaty (1980) employed a system of numbers to indicate how much one criterion is more important than the other. If the elements of the pairwise comparison matrix are shown with cij, which indicates the importance of *i*th criterion over *j*th, then c_{ji} could be calculated as 1/ c_{ji} (Boroushaki and Malczewski 2008). CAHP employs different techniques to determine the final weights; two of them are explained and used in his article. The first is Lambda Max (λ_{max}) technique and the other is geometric mean. A reciprocal or pairwise comparison matrix (Table 1) is developed by the authors along with expert decisions to establish the relative importance between the major parameters using PCM. A method of calculating the weights of criteria in the PCM is geometric mean method (Buckley 1985). However,

Saaty (1980) has used the λ_{max} technique to obtain the weights of the criteria in the PCM. Every matrix has a set of eigenvalues, and for every eigenvalue, there is a corresponding eigenvector. In Saaty's λ_{max} technique, a vector of weights is defined as the normalized eigenvector corresponding to the largest eigenvalue λ_{max}. The value of λ_{max} is required in calculating the consistency ratio (CR) and consistency index (CI) (Han and Tsay 1998, Malczewski 1999). The CR shows a reasonable level of consistency in the pairwise comparisons if C.R.<0.10 whereas C.R.>0.10 indicate inconsistent judgments.

Fuzzy Weights

A fuzzy number is a set described as (α, β, γ, δ), where $0 < \alpha$ £ β £ γ £ δ, and at which α, β, γ and δ are the four parameters of the fuzzy number. Therefore, instead of assigning a single numeric value, a range or set of values are assigned. It is specifically done when the decision maker is not sure of assigning a single, definite value for a particular factor. Geometric mean technique acts as a method that can easily be applied in the process of fuzzifying the AHP (Buckley 1985, Sinha et al. 2011b, Sinha 2014). If $\beta_{ij} = \gamma_{ij}$, it results in a triangular membership function which results in FAHP Triangular method. According to the last equality, the highest density of the possibility of the fuzzy number is collected at this point. By separating these two variables, it is possible to spread the distribution of the fuzzy number in order to add to uncertainty. As soon as $\gamma_{ij} - \beta_{ij}$ becomes greater than zero, it gives more uncertainty. Larger $\gamma_{ij} - \beta_{ij}$ results in higher uncertainty. Therefore, it gives opportunity to analyze the difference between CAHP and FAHP with three more options; one is low $\gamma_{ij} - \beta_{ij}$ with value of 1, while the second is a higher $\gamma_{ij} - \beta_{ij}$ with value equal to 2, giving narrow trapezoidal ratios and wide trapezoidal ratios, respectively. Simultaneously, these result in formulating the FAHP Narrow Trapezoidal and FAHP Wide Trapezoidal methods. The fuzzy number (w i) has to be defuzzified to attain a singular crisp value. In this study, we have used a simple centroid method (Opricovic and Tzeng 2003, Chang and Wang 2009).

METHODS

Coupling GIS and MCDA for Tiger Relocation

Problem Description

GIS-based decision problems usually comprising different criteria and alternatives are generally analyzed by MCDA. AHP being a powerful method of MCDA can organize spatial problems and decides most suitable alternatives for the defined problems. However, owing to some intrinsic uncertainties in the method, several authors suggest fuzzifying the AHP (Buckley 1985, Chang 1996, Bozdag et al. 2003, Kahraman et al. 2004, Sinha et al. 2011b), while others are against (Saaty and Tran 2007). The dispute over fuzzifying AHP is still unsolved, and the attempt for finding that is a practical comparison between the AHP and fuzzified AHP using real-world problem like tiger relocation. Global declining of wild tigers has made them endangered and needs improved conservation strategies for their proliferation. In Sariska Wildlife Reserve, India, there were no tigers left during 2005 on the account of extensive poaching (Jhala et al. 2008, Project Tiger Report 2009). The study aims for a geospatial solution using crisp and FAHP based GIS-MCDA methods and comparing the same for best suitable sites for tiger habitat. It can result to further relocation of tigers in order to conserve them under the prevailing conditions

Study Area and Dataset

The study area includes Sariska Wildlife Reserve, Rajasthan, India, which covers an area of nearly 1183 km^2 (Figure 1). It is located among the Aravalli hill ranges in the Alwar district of Rajasthan, with spatial extent of 27°13’ to 27°31’ N latitude and 76°15’ to 76°33’ E longitude. Along with the Survey of India toposheets, IRS P–6 LISS III 2006 data of Sariska, India, having a spatial resolution of 23.5 m is used for the study.

Evaluation Criteria in GIS-MCDA

Rules/criteria and map overlay methods were applied for Goral habitat evaluation and modelling in Chilla Sanctuary of Rajaji Wildlife Reserve (Kushwaha et al. 2000). Rules/criteria method is the assigning of appropriate scores or weights for the parameters according to the nature of suitability and the preference of the species under consideration for their habitat. The major important parameters governing tiger habitat suitability are determined from literature survey, expert knowledge, reconnaissance information and field study. The most important parameters identified were (1) land use/land cover (forest type and density, water bodies, settlements, road), (2) prey availability, (3) nature of the terrain (slope) and (4) anthropogenic factors. The land cover, mainly the vegetation cover and topography and their temporal variations are important causal factors for any ecosystem-based analysis or services (Sinha et al. 2020, Sinha & Santra 2019). In our study, each parameter occurs in the form of output map derived from source layers and are grouped into five categories (suitability or priority classes) in descending order by the authors; namely, highly suitable (class 5), very suitable (class 4), moderately suitable (class 3), poorly suitable

Figure 1. Location of the study area

(class 2) and least suitable or unsuitable (class 1). The methodology (Figure 2) includes deriving various input parameters from remote sensing and ancillary sources along with field verification and evaluating them for tiger habitat suitability using multi-criteria approach. Field work was carried out in Sariska to collect information on habitat use by tigers, along with forest type and density. Supervised classification of the satellite image using Maximum Likelihood was done in ERDAS Imagine (version 9.1) software. Thematic maps of forest type and density, road, settlement, water bodies and slope were created using ArcGIS (version 9.3) software taking digital (spectral indices) and visual interpretations into consideration. In our study, it has been found that prey is available in plenty in Sariska, which is revealed by high prey density. As for other predators too, the area is large and rich enough to support the predators without much competition. To combine MCDA and GIS, each of the criteria are represented as a map in the GIS database (Malczewski 1999, Baban & wan-Yusof 2003). Appropriate weights are assigned using knowledge-based AHP approach. The idea of multi-criteria techniques has been implemented with different uncertainty levels of AHP, i.e. CAHP and FAHP (triangular, narrow trapezoidal and wide trapezoidal). Habitat suitability map by weighted mean overlay analysis using AHP is thus obtained in GIS mode using MCDA.

Comparison Strategy

The first comparison approach is done on weights obtained from the CAHP and FAHP methods. Analysis is carried out first in normal and non-fuzzy numbers, and then fuzzified AHP is applied on fuzzy comparison ratios, which yield weights in fuzzy format. In the FAHP, pairwise comparison matrix is considered with different degrees of uncertainty for ratios which have not been considered in previous works in this field.

The second comparison approach is done in order to analyze the probable difference in the resulted maps from GIS software after performing the CAHP and FAHP methods. Also, this study, for checking the effects of fuzzification on result maps, suggest and perform the chi-square (χ^2) test as a suitable tool. The χ^2 test statistically checks the effects of fuzzification on resulted maps in terms of the degree of uncertainty in fuzzy weights. Such a comparison between the CAHP and the FAHP, which is based on maps, has not been done in this format before. For applying MCDA, a tool is needed that takes the ranking and comparison data to process them and calculates the weights of different alternatives or criteria. The AHP can serve as a powerful tool in calculating weights and pursuing MCDA procedure. Table 2 shows the weights for each factor resulting from AHP λ_{max} method and the resulting normalized and defuzzified weights obtained from FAHP triangular, FAHP narrow trapezoidal and FAHP wide trapezoidal methods.

The hypotheses for χ^2 test are designed. Null hypothesis (H_0) states that the level of uncertainty has no effect on difference between the map in experiment and the non-fuzzified CAHP map. Alternative hypothesis (H_a) states that the level of uncertainly can have some effects on difference between the map in experiment and the non-fuzzified CAHP map. The χ^2 statistic value is calculated using the following formula (Dowdy et al. 2004).

Figure 2. Methodology for deriving tiger habitat suitability maps

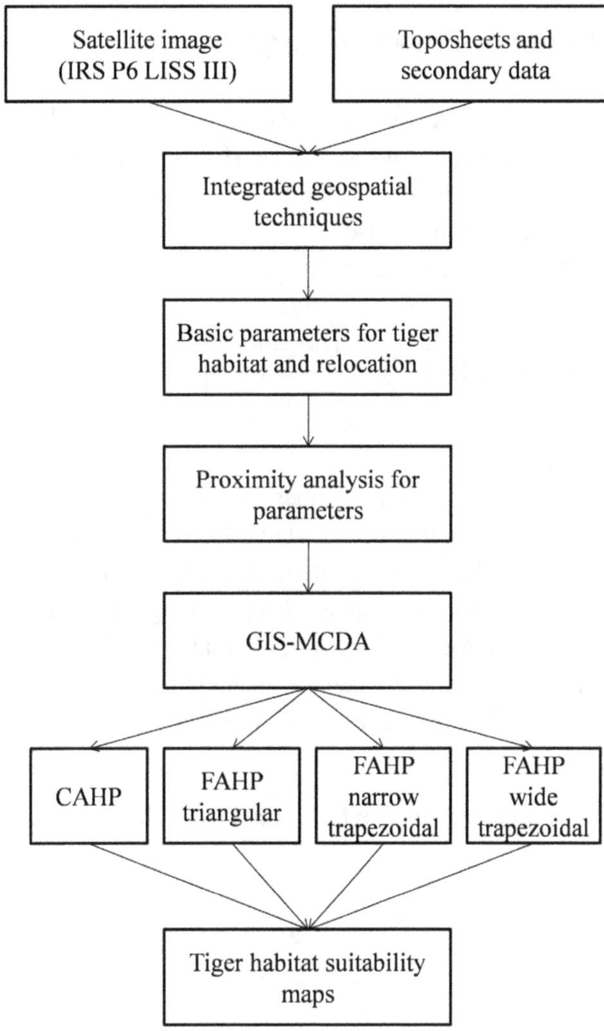

$$\chi^2 = \sum_i \left[\frac{(O_i - E_i)^2}{E_i} \right] \quad (1)$$

where, O and E are the observed values of number of pixels. In this case, O is the vector indicating the number of pixels in each class of the examining map, and E is the vector indicating the number of pixels in each class of the map related to CAHP λ_{max} method.

Results

Percentage Difference Between Weights

The results show that there are some differences between the obtained weights of different fuzzy pairwise comparison matrices. If differences between the weights obtained from non-fuzzy λ_{max} method and the weights obtained from each fuzzified method are calculated, a graph of difference (Figure 3) from the non-fuzzy method is obtained over various ranges of uncertainties.

$$Percentage\ difference = 100 \times \sum_i \left[\frac{\left(\left| w_i' - w_i \right| \right)}{w_i} \right] \tag{2}$$

where w'_i is the defuzzified weight number i which has been calculated through FAHP, and w_i is the weight number i calculated from non-FAHP λ_{max} method. The graph in Figure 3 shows that the percentage deviation over the uncertainty levels ranges slightly over 5%. Difference between FAHP triangular and narrow trapezoidal is nearly 3.9% and that between narrow trapezoidal and wide trapezoidal is 1.3%. This proves a marked difference between the different fuzzy methods. Table 2 shows the differences in the weights assigned by all the methods.

Figure 3. Deviations (in percentage in the y-axis) due to different FAHP uncertainty levels

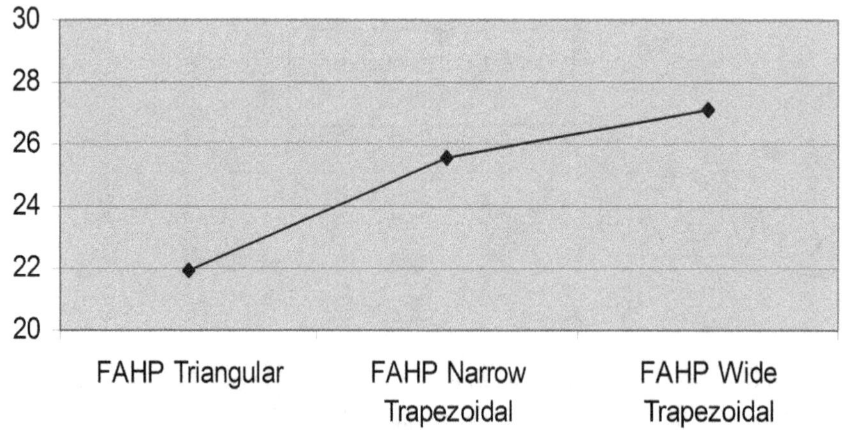

Chi-Square (χ^2) Test for Independence

Habitat suitability map of tiger is derived by multi-criteria approach integrating various input maps in weighted overlay model using AHP for assigning weights. Weighted overlay which considers both the weights and the relative importance factors (scores) as required in index modelling is considered for the number of pixels for each suitability or priority classes. Number of pixels in each priority class can be calculated from the resulting map outputs obtained and documented in Table 3. In all the maps, as per the weights determined from the AHP methods, no pixels are classified for priority class 1. As the value '0' should be omitted from χ^2 test, hence priority class 1 could not occupy a place in the test. The number of pixels in each priority class of the maps (Table 3) was required for performing the χ^2 test. The χ^2 test values, using Equation (1), for different levels of uncertainty in FAHP and CAHP λ_{max} method are provided in Table 4. As it is shown in Table 4, two of the calculated values are smaller than 7.815 ($\alpha = 0.05$, df = 3), which is the critical value in the study, i.e. FAHP triangular and FAHP narrow trapezoidal. So for them, the observed values were similar to the expected values, and the null hypothesis is accepted. On the other hand, in FAHP wide trapezoidal, the value is extremely higher than the critical value, which means that the null hypothesis must be rejected and the alternative hypothesis should be accepted. The test in this case shows that there is a relationship between degree of uncertainty and difference between the maps in context to spatial extent of each suitability classes. The best probable result is obtained with FAHP wide trapezoidal method in MCDA, and the resulting output in the form of map is represented in Figure 4. The figure also highlights the differences in area obtained as a result of the fuzzy and non-fuzzy methods applied in the study. Extensive ground truth along with expert opinion is taken into account for the verification of the results in the form of map outputs.

Effect of Fuzziness

Figures 5(a)–(c) show the membership function values (weight intervals) of all parameters for FAHP triangular, narrow trapezoidal and wide trapezoidal methods, respectively. Classes obtained from each method are compared in the form of matrix. The matrixes include the number of classes classified by each method in a pairwise manner. Results show that the methods applied for assigning weights do have significant impacts on the outputs. CAHP is more similar to FAHP narrow trapezoidal method in the classification. FAHP triangular method is more similar to FAHP narrow trapezoidal method than compared to FAHP wide trapezoidal method. This reveals probable difference in the resulting maps from GIS software after performing the AHP methods. Increase in the levels of uncertainty leads to

Figure 4. Tiger habitat suitability map outputs using CAHP (left) and FAHP wide trapezoidal (right) methods through GIS-MCDA showing differences in the output (encircled)

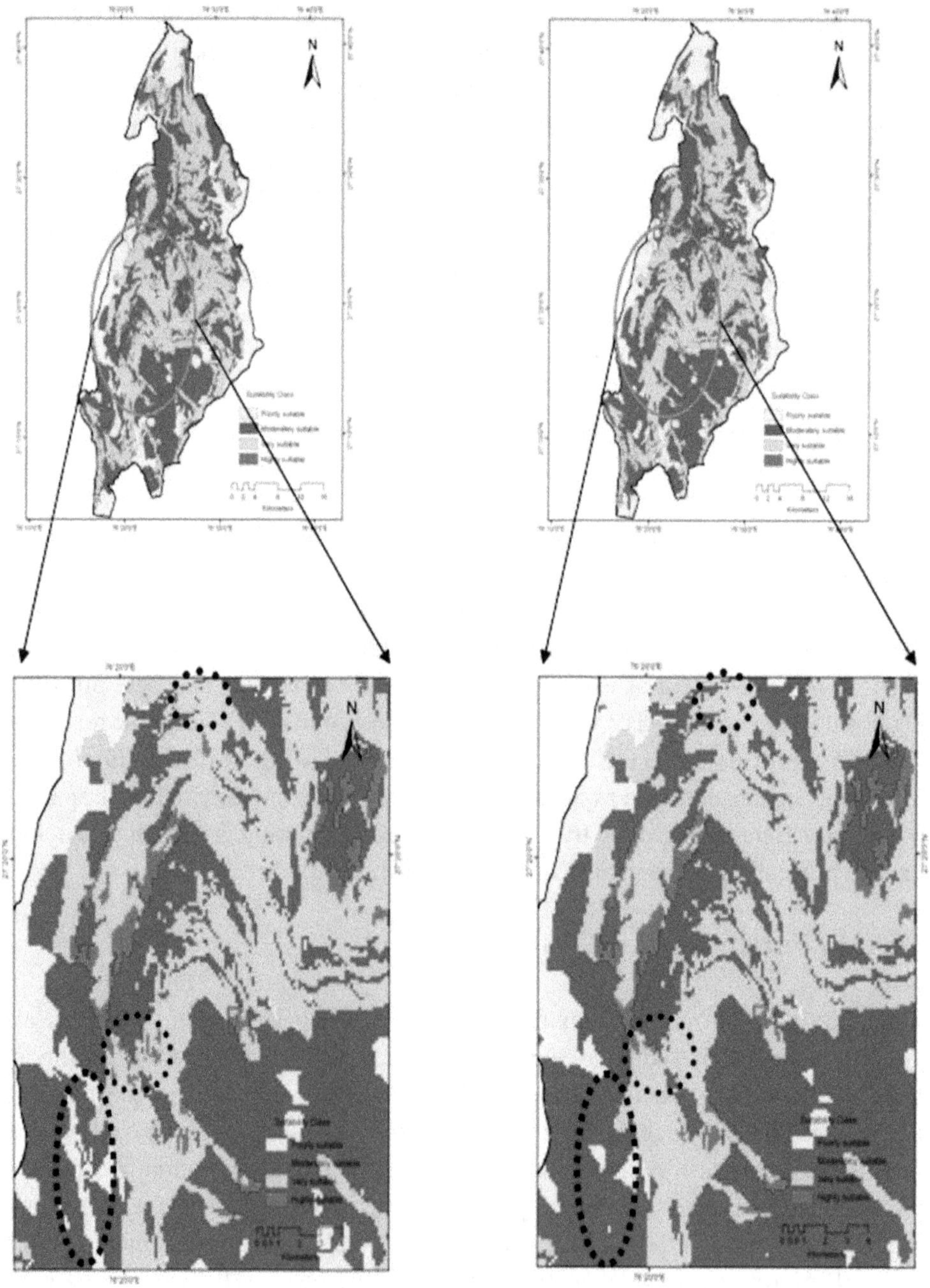

more difference in the map outputs revealed in Table 5. The results mean that the differences are prominent using varied uncertainty levels. For real-world problems, increased fuzziness is preferred due to the involvement of a number of constraints and factors. Analysis of variance (ANOVA) test showed relatively better value of significance for FAHP wide trapezoidal (= 0.118) method as compared to other methods at the significance level of 0.05. The values were 0.139 and 0.136 for FAHP triangular and narrow trapezoidal methods, respectively, while it was 0.164 for CAHP.

DISCUSSIONS AND CONCLUSION

MCDA has the potential for integrating geospatial data and expert knowledge in GIS domain to be implemented in site suitability studies for wildlife habitat evaluation. AHP provides an efficient method for determining the importance of the factors responsible. Weights generated through AHP are integrated in GIS, and overlay analysis produces effective results for site suitability study. The results pointed out that Sariska is found to be highly suitable for tiger habitat and relocation. For real-world situations, like the case of habitat suitability considered here where a number of variables are conjugated, the weights assigned by FAHP methods are found to be extremely effective. Uncertainty levels in AHP have significant effects on the map outputs, and a prominent relationship exists between the degree of uncertainty and the difference between the maps as shown by χ^2 test. In general, the more the uncertainty level, the more is the difference, i.e. deviation is more prominent in case of FAHP wide trapezoidal method. It means that with increase in the difference between bij and gij, the overall fuzziness increases. There is a marked difference between CAHP and FAHP methods and also among different uncertainty levels in FAHP and that is revealed in the study. Percentage deviation over the uncertainty levels ranges slightly over 5%, when weights are quantified using AHP methods. In GIS mode, the relative difference between CAHP and FAHP is observed to be nearly 2.7%. Thus, differences in uncertainty levels provide different results. The utility of the methods will depend on the nature of the real-world spatial decision-making problem that the decision makers need to deal with; otherwise this might end up giving erroneous results.

The aforementioned statement that AHP is already a fuzzy process is acceptable. It is observed that the resulting weights from different AHP methods (λ_{max} and geometric mean) are different, revealing some differences between the methods (Table 2). However, interval judgments are rather preferred than single numeric values and so FAHP can be considered for complex multi-attribute decision-making problems. Decision makers prefer to use greater degrees of fuzziness or uncertainty in real-world applications, and no doubt, the results will vary depending upon the

Figure 5. The membership function values (weight intervals) of all parameters for (a) FAHP triangular, (b) narrow trapezoidal and (c) wide trapezoidal methods respectively

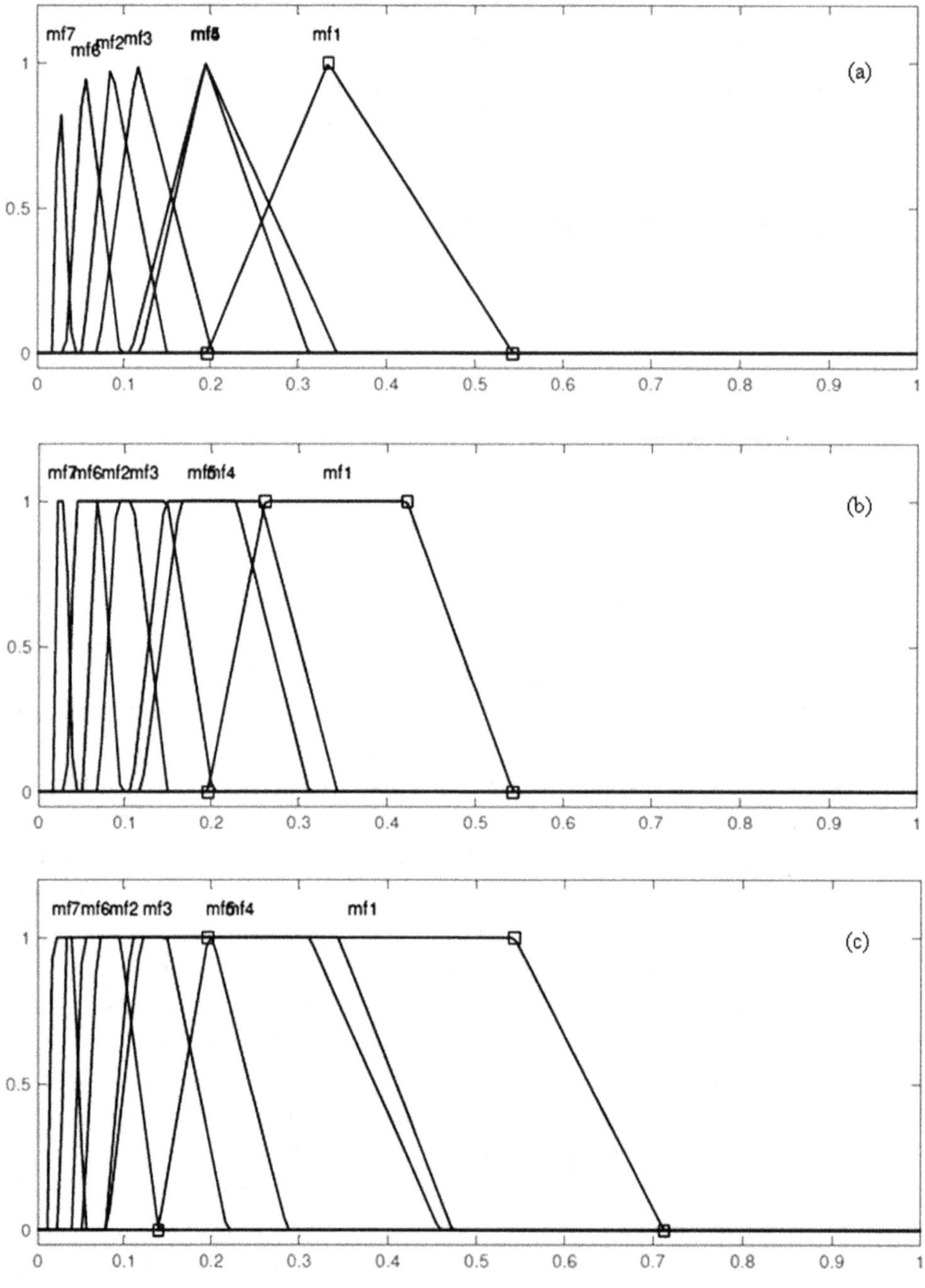

degree of uncertainty level used. Nowadays, especially in complex conditions, many decisions are made in such an environment where decision-making process is uncertain; fuzzy numbers are used for evaluation pertaining to some deviations of decision makers. Thus, FAHP should be used in spite of its complex calculations. Uncertainty and imprecision involved in expert knowledge can be well addressed using fuzzy logic. In real-world problems, we can not be very sure of assigning a definite numeric value for any factor as they might be subjected to a number of variables. In our opinion, it would be safe and more realistic if a range of values be assigned for a particular factor in the problem. ANOVA performed in the study also confirms the observations.

As for the case discussed here, tigers are counted among the endangered umbrella species, and special care should be taken for their conservation as tigers are found only in some scattered pockets of Asia, mainly in India. The worldwide declining status of the tigers is a serious threat to the overall biodiversity, and the methods (crisp and FAHP–MCDA) adopted in this study shows positive response towards their relocation in suitable sites, thus favouring their conservation and proper management. Parameters considered in this study are somewhat uncertain, like all other real-world situations, and hence, the authors suggest the addition of fuzziness to the classical AHP for assigning weights. The study of suitable habitat sites for tiger relocation shows good results using CAHP and even better using FAHP. The study proves the utility of FAHP wide trapezoidal method as the most effective method for assigning weights rather than simple crisp or other FAHP methods. All such real-world multivariate complex problems should be addressed with fuzzy approaches. The study would be of significant use if considered with management and policy-making strategies, such as forest management, land use planning and sustainable management of natural resources. It can help the decision makers to decide upon the target problems and appropriate method used to solve the same. If the problem does not deal with factors that are not uncertain and vague and numerical judgments can be made, then CAHP can be applied; otherwise, it is advisable to take fuzzy interval judgments. Here lies the role of the expert decision makers.

REFERENCES

Anselin, A., Meire, P. M., & Anselin, L. (1989). Multicriteria techniques in ecological evaluation: An example using the analytical hierarchy process. *Biological Conservation*, *49*, 215–229.

Baban, S. J., & wan-Yusof, K. (2003). Modelling optimum sites for locating reservoirs in tropical environments. *Water Resources Management*, *17*, 1–17.

Banai-Kashani, R. (1989). A new method for site suitability analysis: The analytic hierarchy process. *Environmental Management, 13*, 685–693.

Boroushaki, S., & Malczewski, J. (2008). Implementing an extension of the analytical hierarchy process using ordered weighted averaging operators with fuzzy quantifiers in ArcGIS. *Computers & Geosciences, 34*, 399–410.

Bozdag, C. E., Kahraman, C., & Ruan, D. (2003). Fuzzy group decision making for selection among computer integrated manufacturing systems. *Computers in Industry, 51*, 13–29.

Buckley, J. J. (1985). Fuzzy hierarchical analysis. *Fuzzy Sets and Systems, 17*, 233–247.

Buyukozkam, G., & Feyzioglu, O. (2004). A fuzzy-logic-based decision-making approach for new product development. *International Journal of Production Economics, 90*, 27–45.

Cebeci, U. (2009). Fuzzy AHP-based decision support system for selecting ERP systems in textile industry by using balanced scorecard. *Expert Systems with Applications, 36*, 8900–8909.

Chang, D. Y. (1996). Application of the extent analysis method on fuzzy AHP. *European Journal of Operational Research*,, 649–655.

Chang, T.-H., & Wang, T.-C. (2009). Using the fuzzy multi-criteria decision making approach for measuring the possibility of successful knowledge management. *Information Sciences, 179*, 355–370.

Chutima, P., & Suwanfuji, P. (1998). Fuzzy analytical hierarchy process part routing in FMS. *Thammasat International Journal of Science and Technology, 3*(2), 29–47.

Correa-Berger, B. P. (2006). *Developing a habitat suitability model for the spotted turtle using a hybrid-deductive approach* (MSc dissertation). New York: Rochester Institute of Technology.

Dagdeviren, M., Yavuz, S., & Kilinc̦, N. (2009). Weapon selection using the AHP and TOPSIS methods under fuzzy environment. *Expert Systems with Applications, 36*, 8143–8151.

Dagdeviren, M., & Yuksel, I. (2008). Developing a fuzzy analytic hierarchy process (AHP) model for behavior-based safety management. *Information Sciences, 178*, 1717–1733.

Deng, H. (1999). Multicriteria analysis with fuzzy pairwise comparisons. *International Journal of Approximate Reasoning, 21*, 215–231.

Dowdy, S. H., Wearden, S., & Chilko, D. (2004). *Statistics for Research* (3rd ed.). John Wiley and Sons Inc.

Eastman, J. R., Kyem, P., & Toledano, J. (1993). A procedure for multi-objective decision making in GIS under conditions of conflicting objectives. EGIS '93.

Eastman, J. R., Toledano, J., Jin, W., & Kyem, P. A. K. (1992). Participatory procedures for multi-criteria evaluations in GIS. In GIS'92, Buffalo, NY.

El-Din, H. K., Abd El Munim, H. E., & Mahdi, H. Decision-making in fuzzy environment: a survey. In *Application of Decision Science in Business and Management*. IntechOpen.

Erensal, Y. C., Oncan, T., & Demircan, M. L. (2006). Determining key capabilities in technology management using fuzzy analytic hierarchy process: A case study of Turkey. *Information Sciences*, *176*, 2755–2770.

Gungor, Z., Serhadlioglu, G., & Kesen, E. (2009). A fuzzy AHP approach to personnel selection problem. *Applied Soft Computing*, *9*, 641–646.

Han, W. J., & Tsay, W. D. (1998). *Formulation of quality strategy using analytic hierarchy process*. In 27th annual meeting of the Western Decision Science Institute, University of Northern Colorado.

Hansen, H. S. (2005). GIS-based multi-criteria analysis of wind farm development. In H. Hauska, & H. Tveite (Ed.), *Proceedings of the 10th Scandinavian research conference on geographical information science (ScanGIS'2005)* (pp. 75–87). Stockholm, Sweden: Royal Institute of Technology.

Jaber, J. O., Jaber, Q. M., Sawalha, S. A., & Mohsen, M. S. (2008). Evaluation of conventional and renewable energy sources for space heating in the household sector. *Renewable & Sustainable Energy Reviews*, *12*, 278–289.

Jhala, Y. V., Gopal, R., & Qureshi, Q. (Eds.). (2008). Status of the tigers, co-predators, and prey in India. National Tiger Conservation Authority, Govt. of India, New Delhi, and Wildlife Institute of India, Dehradun. TR 08/001, 151.

Kahraman, C., Cebeci, U., & Ruan, D. (2004). Multi-attribute comparison of catering service companies using fuzzy AHP: The case of Turkey. *International Journal of Production Economics*, *87*, 171–184.

Kangas, J. (1992). Multiple-use planning of forest resources by using the analytic hierarchy process. *Scandinavian Journal of Forest Research*, *7*, 259–268.

Karimi, A. R., Mehrdadi, N., Hashemian, S. J., & Nabi-Bidhendi, Gh. R., & Tavakkloi-Moghaddam, R. (2011). Using of the fuzzy TOPSIS and fuzzy AHP methods for wastewater treatment process selection. *International Journal of Academic Research*, *3*, 737–745.

Khoshand, A., Rahimi, K., Ehteshami, M., & Gharaei, S. (2019). Fuzzy AHP approach for prioritizing electronic waste management options: A case study of Tehran, Iran. *Environmental Science and Pollution Research International*, *26*(10), 9649–9660. PMID:30729431

Kong, F., & Liu, H. (2005). Applying fuzzy analytic hierarchy process to evaluate success factors of e-commerce. *International Journal of Information and Systems Sciences*, *1*(3–4), 406–412.

Kordi, M. (2008). *Comparison of fuzzy and crisp analytic hierarchy process (AHP) methods for spatial multicriteria decision analysis in GIS* (MSc dissertation). University of Gavle.

Kulak, O., & Kahraman, C. (2005). Fuzzy multi-criterion selection among transportation companies using axiomatic design and analytic hierarchy process. *Information Sciences*, *170*, 191–210.

Kushwaha, S. P. S., Munkhtuya, S., & Roy, P. S. (2000). Mountain goat habitat evaluation in Rajaji National Park using remote sensing and GIS. *Photonirvachak (Dehra Dun)*, *28*, 293–303.

Leung, L. C., & Chao, D. (2000). On consistency and ranking of alternatives in fuzzy AHP. *European Journal of Operational Research*, *124*, 102–113.

Lin, Q. L., & Wang, D. (2019). Facility layout planning with SHELL and fuzzy AHP method based on human reliability for operating theatre. *Journal of Healthcare Engineering*, *2019*, 1–12. PMID:30792832

Malczewski, J. (1999). *GIS and multicriteria decision analysis*. John Willey and Sons Inc.

Mikhailov, L., & Tsvetinov, P. (2004). Evaluation of services using a fuzzy analytic hierarchy process. *Applied Soft Computing*, *5*, 23–33.

Ocampo, L. A. (2019). Decision modeling for manufacturing sustainability with fuzzy Analytical Hierarchy Process. *Global Business Review*, *20*(1), 25–41.

Opricovic, S., & Tzeng, G.-H. (2003). Defuzzification within a multicriteria decision model. *International Journal of Uncertainty, Fuzziness and Knowledge-based Systems*, *11*(5), 635–652.

Ozdagoglu, A., & Ozdagoglu, G. (2007). Comparison of AHP and fuzzy AHP for the multicriteria decision-making processes with linguistic evaluation. *Istanbul Ticaret Universitesi Fen Bilimleri Dergisi*, *6*, 65–85.

Patil, A. N. (2018). Fuzzy AHP methodology and its sole applications. *International Journal of Management Research & Review*, *8*(5), 24–32.

Project Tiger Report–National Tiger Conservation Authority, Government of India. (2009). *Project Tiger. Stripes*, *1*(1), 1–20.

Promentilla, M. A. B., Antonio, M. R. S., Chuaunsu, R. M. M. II, & Serra, A. J. (2016). *A calibrated fuzzy AHP approach to derive priorities in a decision model for low carbon technologies*. In *DLSU Research Congress*.

Saaty, T. L. (1980). *The analytic hierarchy process*. McGraw-Hill International.

Saaty, T. L. (1988). *Multicriteria decision making: the analytic hierarchy process*. RWS Publications.

Saaty, T. L., & Tran, L. T. (2007). On the invalidity of fuzzifying numerical judgments in the analytic hierarchy process. *Mathematical and Computer Modelling*, *46*, 962–975.

Sharma, L. K., Kanga, S., Nathawat, M. S., Sinha, S., & Pandey, P. C. (2012). Fuzzy AHP for forest fire risk modeling. *Disaster Prevention and Management*, *21*(2), 160–171.

Singh, M. P., Singh, P., & Singh, P. (2019). Fuzzy AHP-based multi-criteria decision-making analysis for route alignment planning using geographic information system (GIS). *Journal of Geographical Systems*, *21*, 395–432.

Singh, P. K., & Sarkar, P. (2019). A framework based on fuzzy AHP-TOPSIS for prioritizing solutions to overcome the barriers in the implementation of ecodesign practices in SMEs. *International Journal of Sustainable Development and World Ecology*, *26*(6), 506–521.

Sinha, S. (2014). Integrating fuzziness to wildlife relocation and habitat analysis in Rajasthan, India. *International Journal of Earth Science and Engineering*, *7*(1), 280–288.

Sinha, S. (2015). Bland-Altman analysis for evaluating AHP-based wildlife habitat suitability models. *Research & Reviews: Journal of Space Science & Technology*, *4*(2), 11–18.

Sinha, S. (2016). Integrated GIS and Multi Criteria approach for the wildlife habitat suitability. *Journal of Advanced Research in Geo Sciences & Remote Sensing*, *3*, 1–16.

Sinha, S., Mohan, S., Das, A. K., Sharma, L. K., Jeganathan, C., Santra, A., Mitra, S. S., & Nathawat, M. S. (2020). Multi-sensor approach integrating optical and multi-frequency synthetic aperture radar for carbon stock estimation over a tropical deciduous forest in India. *Carbon Management*. 11(1), 39-55. doi:10.1080/17583004.2019.1686931

Sinha, S., & Santra, A. (2019). Estimation of change in forest aboveground carbon in Bhimbandh Wildlife Sanctuary, Bihar, India between 2007 and 2016. *Current Science*, *117*, 1090–1094.

Sinha, S., Sharma, L. K., & Nathawat, M. S. (2011b). Retrieving tiger habitats: Conserving wildlife geospatially. *Applied Remote Sensing Journal*, *2*(1), 1–5.

Sinha, S., Sharma, L. K., & Nathawat, M. S. (2012). Tigers losing grounds: Impact of anthropogenic occupancy on tiger habitat suitability using integrated geospatial-fuzzy techniques. *The Ecoscan*, *1*, 259–263.

Sinha, S., Sharma, L. K., Pandey, P. C., Nathawat, M. S., & Kanga, S. (2011a). Impact of human intrusion on tiger habitat and conservation using integrated geospatial techniques. *International Journal of Earth Sciences and Engineering*, *4*(3), 39–45.

Sredjevic, B., & Medeiros, Y. D. P. (2008). Fuzzy AHP assessment of water management plans. *Water Resources Management*, *22*, 877–894.

Taylor, B. W. (2004). *Introduction to management science*. Pearson Education Inc.

Vahidnia, M. H., Alesheikh, A., Alimohammadi, A., & Bassiri, A. (2008). Fuzzy analytical hierarchy process in GIS application. *The International Archives of the Photogrammetry, Remote Sensing and Spatial Information Sciences*, *37*, 593–596.

Vaidya, O. S., & Kumar, S. (2006). Analytic hierarchy process: An overview of applications. *European Journal of Operational Research*, *169*, 1–29.

Wijitkosum, S., & Sriburi, T. (2019). Fuzzy AHP Integrated with GIS Analyses for Drought Risk Assessment: A Case Study from Upper Phetchaburi River Basin, Thailand. *Water (Basel)*, *11*(5), 939.

Xiang, W., & Whitley, D. L. (1994). Weighting land suitability factors by the PLUS method. *Environment and Planning. B, Planning & Design*, *21*, 273–304.

Yang, Ch.-Ch., & Chen, B.-Sh. (2004). Key quality performance evaluation using fuzzy AHP. *Journal of the Chinese Institute of Industrial Engineers*, *21*(6), 543–550.

Zadeh, L. A. (1965). Fuzzy sets. *Information and Control*, *8*(3), 338–353.

Zadeh, L. A. (1973). The concept of linguistic variable and application to approximate reasoning. Berkeley, CA: ERL-M411.

APPENDIX

Table 1. Reciprocal matrix for associated parameters

	Forest type	Forest density	Settlement	Water hole	Drainage	Road	Slope
Forest type	1	3	3	3	3	5	7
Forest density	1/3	1	1/3	1/3	1/3	3	5
Settlement	1/3	3	1	1/3	1/3	3	5
Water hole	1/3	3	3	1	1	3	7
Drainage	1/3	3	3	1	1	3	7
Road	1/5	1/3	1/3	1/3	1/3	1	3
Slope	1/7	1/5	1/5	1/7	1/7	1/3	1

Table 2. Weightage assigned by AHP methods

Sl.no.	Parameters	Forest density	Settlement	Water hole	Drainage	Road	Slope
1	Forest type	3	3	3	3	5	7
2	Forest density	1	1/3	1/3	1/3	3	5
3	Settlement	3	1	1/3	1/3	3	5
4	Water hole	3	3	1	1	3	7
5	Drainage	3	3	1	1	3	7
6	Road	1/3	1/3	1/3	1/3	1	3
7	Slope	1/5	1/5	1/7	1/7	1/3	1

Table 3. Number of pixels in each priority or suitability class

Methods	No. of pixels					
	Priority Class 1	Priority Class 2	Priority Class 3	Priority Class 4	Priority Class 5	Total
CAHP λmax	0	12457	31758	24783	1974	70972
FAHP Triangular	0	12326	32018	24622	2006	70972
FAHP Narrow Trapezoidal	0	12537	31943	24580	1912	70972
FAHP Wide Trapezoidal	0	11065	33244	24847	1816	70972

Table 4. χ^2 test values for different levels of uncertainty in FAHP and CAHP λ_{max} method

Method	FAHP Triangular	FAHP Narrow Trapezoidal	FAHP Wide Trapezoidal
χ^2 test value	5.070879	5.201557	237.8918

Table 5. Relative difference (in percent) among different AHP methods from mapped outputs

	FAHP Triangular	FAHP Narrow Trapezoidal	FAHP Wide Trapezoidal
CAHP	1.4	1.3	2.7
FAHP Triangular		1.4	2.3
FAHP Narrow Trapezoidal			2.6

Chapter 2
GIS-based Multi-Criteria Analysis for Delineation of Groundwater Potential Zones:
A Case Study from Jodhpur District, Rajasthan, India

Raghib Raza
Quantum Asia Pvt. Ltd., India

Gajendra Kumar Chawla
Watershed Development and Soil Conservation Department, Jodhpur, India

Chandra Shekhar Dwivedi
Central University of Jharkhand, India

ABSTRACT

In the present study, delineation of ground water potential in Jodhpur district, Rajasthan is framed by using geospatial technique (i.e., remote sensing and GIS) and with multi-criteria leadership (MCDM) procedure. The analytical network process (ANP) is one of the suitable strategies that makes the research workable for delineation methodically, and corporate the analytic hierarchy process (AHP) as an uncommon case. For proper value to recognize the ground water potential zone in Jodhpur, Rajasthan, the AHP and ANP methods are utilized to decide the values of different parameters and their classes. The AHP values are then connected in a direct mix into raster calculator to get five distinctive groundwater potential zones in the investigation region, to be specific as 'very poor' (2052.0 km2), 'poor' (4225.9 km2), 'moderate' (6355.1 km2), 'good' (6451.2 km2), and 'very good' (3301.7 km2). It has been presumed that about 9752.9 km^2 area of Jodhpur district has very good to good groundwater potential, which is about 43.56% of the complete study area.

DOI: 10.4018/978-1-7998-5027-4.ch002

INTRODUCTION

Water scarcity is a major issue in all over the countries of world, mostly in developing countries. As per day to day expansion of human population water utilization has significantly expanded as seen throughout the most recent couple of decades. From the survey of food and agriculture organization (FAO 2015), the rate of growth of human population is as much that it cans twice the world's requirement for water. India is the second largest populated country in the world and the fresh water for many purposes is increasing day by day and also having facing water crisis in some part of India due to developing country. The people present in the field of water assets the board are confronting about various problems and issues including water quality, environmental changes, and water supply demand increments. Dry and semiarid districts are mainly facing the water scarcity and soil moisture declination, which hamper the agricultural fields and fail to fulfill the required amount of water supply to the demand raised. The spatial– temporal precipitation data convict the real difficulties and erratic precipitation designs prompts aridity and climate disaster conditions. These issues influence the yield generation and expanding sustainable hazards. One most possible way for relieving this water crisis can be done by the water harvesting structure. All the water boards must make a feasible by the agribusiness part which is crucial for sustenance emergency and also for financial improvement of the nation. For the battle with water scarcity in the dry and semidry districts issues for a huge number of years is a promising elective wellspring or water harvesting structure for enhancing water. Recharge water harvesting (RWH) can be frame for the proper rationing water overflow for horticulture, and incredibly significant particularly in dry region where it all fulfill the water shortage issues and make a good to very good groundwater extraction, trimming danger and increment in crop yield. RWH enhance groundwater for saves and builds water accessibility and improve socio-economic problems facing by the districts (Ammar et.al. 2016). Many techniques and devices are utilized in past for the research of groundwater potential in a area but at present Geographic information system and remote sensing program are advance version for the delineation of groundwater potential zone. Adaption of remote sensing with GIS technology has made easier to identify the hydrological potential in a region. GIS based water models are very suggesting techniques and additionally devices. MCA (AHP) model offers high potential to the GIS technique for RWH sites zoning for the information of poor groundwater zone and good quality of groundwater zone (Ziadat et.al, 2012; Bulcock and Jewitt 2013).

The physical and financial qualities were considered best due to GIS process. GIS and remote sensing access the remote area where on field is difficult to reach but at present due to modern technique it is easy to develop the remotely area, and the water scarcity and fulfilling of water demand properly made easy with GIS based modeling and tools. (Sayl et. al. 2017) has utilized GIS-RS with multi criteria choice methods, for example: Geographically, hydrology, and precipitation, socio-economically, and the various conditions to analysis and calculate the reasonable water harvesting structure for the proper groundwater recharge and groundwater arability. Researchers applying remote sensing and GIS for the hydrological potential sites for water harvesting structure to maintain the ground water potentiality such check dams, storage tank, farm pond, Talab, gully plug, trench etc. RS-GIS hence used by the researchers with many parameters such as the rainfall, soil texture, geology, geomorphology, drainage, land use& land cover, slope for the more accurate result to identify the site for water harvesting structure and remove the problems of groundwater scarcity.

Enormous numbers of people are migrating from the rural area to urban in search of water, due to crop failure and water quality. Governments are facing various challenges in the field of hydrological sector. Water consumption is increased as the urban sprawl rate is increased and one of the key issue of water management. However water management is unsustainable. Furthermore for proper sites of water harvesting structure based on GIS is necessary in India, mainly in arid and semiarid region. The Jodhpur district is selected is selected for this study because it is one of the governorates that have been exposed to the most disaster water shortages. This kind of research has not been carried out in this area yet. The study aims for delineation of groundwater fresh water zone which is good for all the purpose to be fulfill as required in the Jodhpur district.

Study Area

The study area covers the Jodhpur district, Rajasthan, India which is in the northwestern part of India. Total area of Jodhpur district is 22667.51 km^2 and perimeter is 1055.60 km (Fig.1). Jodhpur is situated at 26.28°N latitude and 73.02 °E longitude, Jodhpur is a 2nd largest city of Rajasthan. The study area is popular for its tourist destination, featuring many palaces, forts and temples and also having Thar Desert. Total population is about 36, 85,681 according to census, 2011, rate of male population is 52.62% and rate of female population is 47.38%. Average rate of literacy is 81.56%.

Jodhpur geological significance led to bring the district declared as a national geological monument. The upper boundary of volcanic is mainly eroded and sandstone of Jodhpur forms the basal portion of the Marwar super group. At present there are 16 blocks in Jodhpur district namely Bap, Phalodi, Luni, Shergaerh, Shekhala, Bhopalgarh, Balesar, Bapini, Baori, Pipar City, Mandore, Lohawat, Dechoo, Tinwari, Bilara, Osian. From 17 november, 2019 total block of Jodhpur is 21 and they are divided from "Luni to Dhawa", "Bap to Ghantiyali", "Shekhala to Chamu", "Mandore to Keru", "Bapini to Aau".

Pahlodi block of Jodhpur is also called as the salt city due to salt industry in Rin. Phalodi is under the buffer zone of Thar Desert and having extreme temperature conditions due to arid climate. Jodhpur is well surrounded by many of districts of Rajasthan i.e., Nagaur, Ajmer, Pali, and Jaisalmer. Major portion of northern and western in Jodhpur district is covered by Thar Desert.

The average precipitation is 280 mm in 2018. Maximum and minimum temperature within the district is having a good variation, the average maximum temperature is about 48°c, and the minimum is about 3°c. The winter season starts in November last month till the February, followed summer which lasting up to June.

Figure 1. Location Map

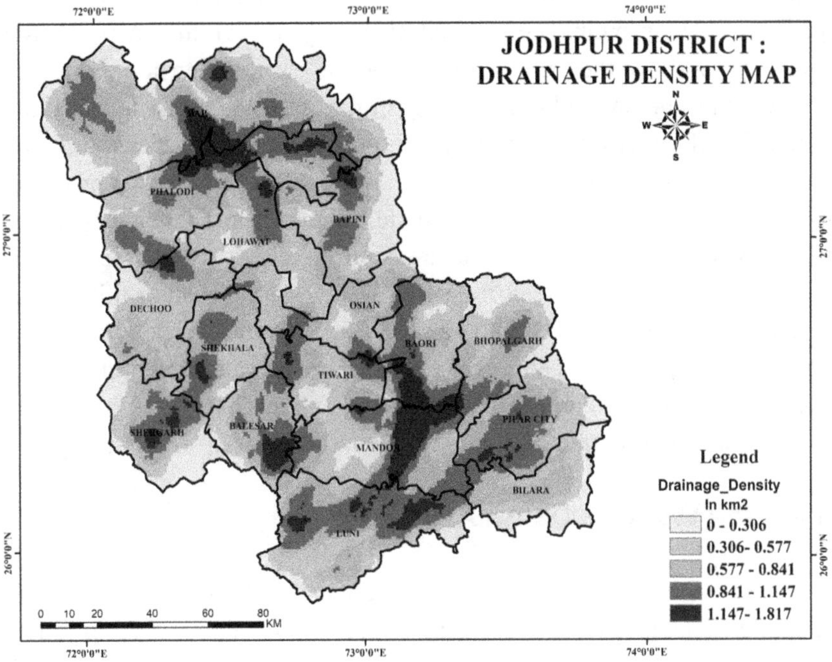

Topographically Jodhpur can be divided into three units, i.e. Sand dunes, Alluvium plain and Escarpment. The maximum slope degree is about 63.93° and Bilara has topography with 260m of maximum elevation zone. There are two river flows in district which is Luni River which enters the district near Jarkand flows southwesterly direction which cover mainy block of Jodhpur like Bilara, Mandor and Luni blocks and leaving the Jodhpur district at Dhundara to enter in Barmer district. Also Mithri River flow in Jodhpur district. Groundwater level of Jodhpur district is very good but most of the areas are under salt affected and mineral affected.

Criteria and Methodology

In this case following three steps are used for the conclusion of groundwater potential zone.

Analytical hierarchy process is a multi-criteria methodology and a modern technique like a network process which chose the most feasible technique which recognize appropriate points or location for suitable groundwater potential zone with geographic information system processes. AHP is used as wide utilization in many research works for the recognizable proof for groundwater potential zoning (krois and Schulte, 2014). In AHP strategy the mind-boggling choices are composed and dissected in an organized manner dependent on information of mathematic and specialists. The principle code of AHP is to symbolize the parts of any issue progressively to show the connections between each dimension. The primary objective ought to be on the highest dimension for settling an issue and the lower level comprise of increasingly point by point criteria that impact the fundamental target. For the most part, in AHP technique a network of pair wise correlations is connected to decide the loads for every model. The appropriateness appraisals of a given target which include two criteria are controlled by their relative significance with the assistance of pair wise correlations. The 9 - point scale is utilized to look at and rate the two criteria. The odd values 1, 3, 5, 7, and 9 relate individually to similarly, modestly, unequivocally, in all respects firmly, what's more, critical criteria when contrasted with one another, and the even qualities 2, 4, 6, and 8 are moderate qualities. Figure 2 shows the following flow chart of overall water management.

Deriving the Weights Using AHP/ANP

ANP is an extension of AHP for decision making in which a problem is divided into various parameters, arranging the min a hierarchical structure, making judgments on the relative importance of pairs of elements and synthesizing the results (Saaty 1999). The methodology for deriving the weights to the thematic layers and their corresponding classes using ANP and AHP respectively, involves following steps:

Generation of pair wise comparison matrices: The relative importance values are determined with Saaty's 1–9 scale (table 1), where as core of 1 represents equal importance between the two themes, and as core of 9 indicates the extreme importance of one theme compared to the other one (Saaty1980). Table 1 shows a matrix for comparing the classes in order to achieve the priority. A pair wise comparison matrix derived using Saaty's nine-point importance scale based on thematic layers used for delineation of ground water potential. Table 2 shows the pair wise matrix which is important for the AHP calculation. The AHP captures the idea of uncertainty in judgment through the principal Eigen value and the consistency index (Saaty, 2004). Saaty gave a measure of consistency, called Consistency Index (CI) as deviation or degree of consistency using the following equation(1):

$$CI = \lambda max - n / n - 1 \tag{1}$$

where λmax is the largest Eigen value of the pair wise comparison matrix and ninth number of classes. Consistency Ratio (CR) of the study area is 0.0006 and it is a measure of consistency of pair wise comprises on matrix and is given by equation..(2)

$$CR = CI / RI \tag{2}$$

where RI is the Ratio Index. The value of RI for different n values is given in table 3.

Figure 2.

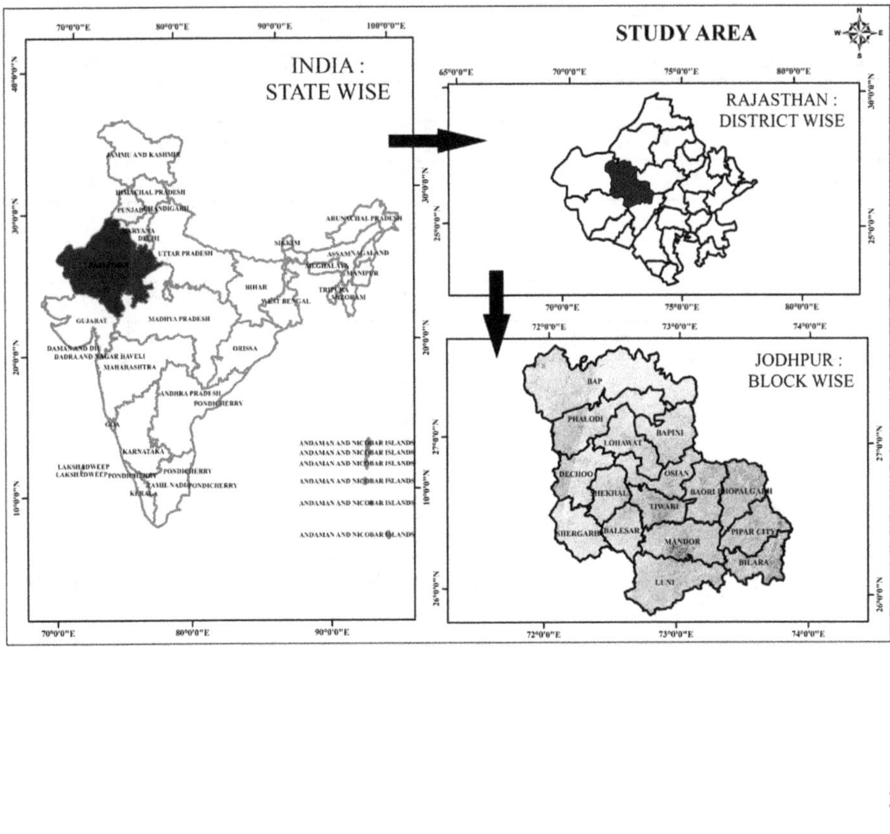

Overlay analysis to find groundwater potential zone is calculated in Arc GIS 10.4 on raster calculator to calculate with each parameter multiplied by assigned value and each values are then added with each other. The assigned normalized value of AHP is mentioned in table 3. Table 4 shows the assigned value and normalized value with AHP. Table 5 shows the normalized value of each parameter themes and area covered by each themes used for the calculation of AHP.

Table 1. Saaty's 1–9 scale of relative importance

Scale	1	2	3	4	5	6	7	8	9
Importance	Equal importance	Weak	Moderate importance	Moderate plus	Strong importance	Strong plus	Very strong importance	Very, very Strong	Extreme importance

Table 2. Pair wise comparison matrix

Parameter	Ground Water Depth	Aquifer	Chloride conc.	Geomorphology	Soil	Slop e	LU/L C	Rainfall	Drainag e Density
Ground Water Depth	9/9	9/8	9/7	9/6	9/5	9/4	9/3	9/2	9/1
Aquifer	8/9	8/8	8/7	8/6	8/5	8/4	8/3	8/2	8/1
Chloride conc.	7/9	7/8	7/7	7/6	7/5	7/4	7/3	7/2	7/1
Geomorphology	6/9	6/8	6/7	6/6	6/5	6/4	6/3	6/2	6
Soil	5/9	5/8	5/7	5/6	5/5	5/4	5/3	5/2	5
Slope	4/9	4/8	4/7	4/6	4/5	4/4	4/3	4/2	4
LU/LC	3/9	3/8	3/7	3/6	3/5	¾	3/3	3/2	3
Rainfall	2/9	2/8	2/7	2/6	2/5	2/4	2/3	2/2	2
Drainage Density	1/9	1/8	1/7	1/6	1/5	¼	1/3	½	1

Table 3. Saaty's ratio index for different values of n

n	1	2	3	4	5	6	7	8	9	10
RI	0	0	0.58	0.89	1.12	1.24	1.32	1.41	1.45	1.49

Table 4. Weights of the thematic maps of the potential groundwater

Themes	Assigned weight	Normalized Weight
Ground Water Depth	9	0.196
Aquifer	8	0.178
Chloride conc.	7	0.152
Geomorphology	6	0.1303
Soil	5	0.108
Slope	4	0.089
LU/LC	3	0.065
Rainfall	2	0.043
Drainage Density	1	0.036

Table 5. Assigned and normalized weights of the features of themes for the delineation of the groundwater potentials in the study area

Categories	Criterion	Weight	Normalized weight	Rank	Area (km2)	Percentage area %
Ground Water Depth	0-5	5	0.22	0.196	6624.61	29.21
	5-10	4	0.19		9971.89	43.96
	10-15	3	0.16		4463.75	19.68
	15-20	2	0.14		1058.64	4.66
	20>	1	0.11		565.415	2.49
Aquifer	Bilara Limestone	7	0.14	0.178	4562.77	20.06
	Jalor Granite	4	0.19		1571.88	06.91
	Rhyolite	1	0.28		589.89	02.59
	Schist	5	0.095		311.47	1.37
	Nagaur / Jodhpur sandstone	6	0.24		11494.86	50.53
	Older Aluvium	3	0.047		3268.25	14.36
	Younger Aluvium	2	0.21		949.82	04.18
Chloride Concentration	>1000	1	0.143	0.152	5513.83	24.29
	250 - 1000	2	0.071		2590.83	11.42
	<250	3	0.214		14593.3	64.29

continues on following page

Table 5. Continued

Categories	Criterion	Weight	Normalized weight	Rank	Area (km2)	Percentage area %
Geomorphology	Aeolian Origin	5	0.33	0.196	20566.3	90.61
	Denudational Origin	4	0.26		1482.78	46.216.53
	Fluvial Origin	3	0.2		343.437	1.52
	Hills	1	0.06		151.373	0.67
	Water Body	2	0.13		152.959	0.67
Soil	Sandy	3	0.5	0.37	9066.237	39.94
	Clayey	1	0.2		1155.308	05.09
	Loamy	2	0.4		12475.41	54.97
Slope	0 - 1.75	5	0.33	0.196	7739.218	34.09
	1.75 - 4.01	4	0.26		10612.4	46.76
	4.01 - 7.02	3	0.2		3372	14.86
	7.02 - 12.28	2	0.13		863.24	03.81
	12.28 - 63.93	1	0.06		109.312	0.48
LU/LC	Sandy Land	1	0.02	0.095	3847.36	16.46
	Settlement	2	0.03		518.17	2.22
	Barren land	3	0.046		2094.36	8.96
	Arable Land	4	0.07		2521.34	10.78
	Crop land	5	0.08		8304.49	35.52
	Grazing Land	6	0.09		121.23	0.52
	Shrub Land	7	0.11		191.83	0.82
	Water Body	8	0.12		160.867	0.69
	Agricultural Plantation	9	0.14		4003.10	17.12
	Spars Forest	10	0.15		1014.90	4.34
	Intense Forest	11	0.17		600.79	2.57
Rainfall	103.04 – 169.26	1	0.066	0.043	5927.06	33.19
	169.26 – 217.03	2	0.133		2127.08	12.01
	217.03 – 249.60	3	0.200		7698.35	41.44
	249.60 – 301.70	4	0.266		1804.07	10.18
	301.70 – 379.86	5	0.333		563.52	3.18

continues on following page

Table 5. Continued

Categories	Criterion	Weight	Normalized weight	Rank	Area (km2)	Percentage area %
Drainage Density	0 - 0.306	5	0.333	0.036	1452.95	06.41
	0.306 - 0.577	4	0.266		4912.22	21.66
	0.577 - 0.841	3	0.200		6661.07	29.38
	0.841 - 1.147	2	0.133		6571.38	28.98
	1.147 - 1.817	1	0.066		3076.53	13.57

RESULT AND DISCUSSION

Ranking Classes of Various Parameters

Ranking for the zoning and movement of groundwater in an area is analyzed by the several layers, such as chloride concentration data, Aquifer of the area, land use & land cover, soil, Slope, drainage density, groundwater depth, Geomorphology, Rainfall and the interrelationship between these layers (Jaiswal et al. 2003).

Land Use and Land Cover

Land use and Land cover are under the natural activity and man-made activity. Land use and land cover are interpreted by Landsat 8 (30m spatial resolution) satellite image of April, 2018 (Fig.3). Various types of land use & land cover pattern are identified in the study area which includes Spars Forest, Intense Forest, Grazing land, Barren land, water body, Crop Land, Settlement, Arable land, Sandy land, Shrub land, Agriculture plantation. Spars Forest & Intense Forest are excellent sites for groundwater exploration, and hence given as the highest rank. Agricultural and Crop Land is considered to have good groundwater prospects and Arable land are in moderate condition, while the water bodies, Barren land and Settlement area have poor groundwater potential zone in the study area (fig.4).

Drainage Density

The inverse property of permeability is known as drainage density and that's the main reason for using drainage density parameter as important for the study of the groundwater zone. The stream order of the study area is calculated through Aster DEM data (30m resolution) shown fig. 5 and then the drainage density is calculated

Figure 3.

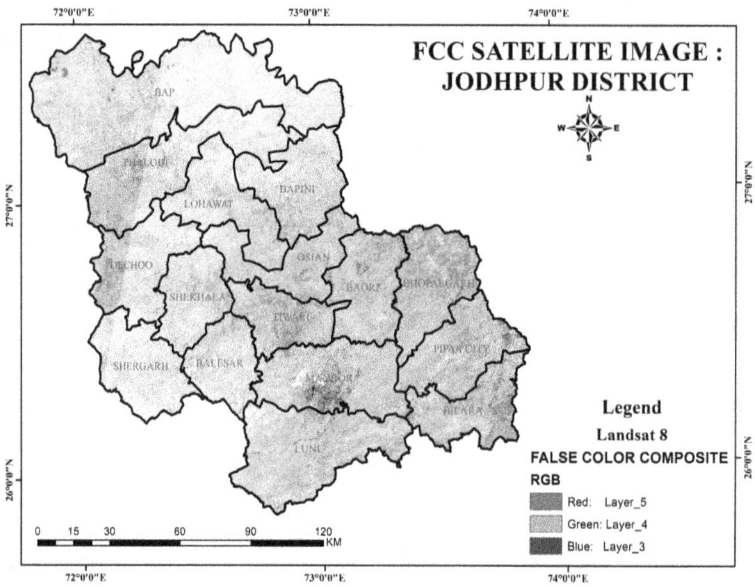

SOURCE : Earthexplorer.usgs.gov

Figure 4.

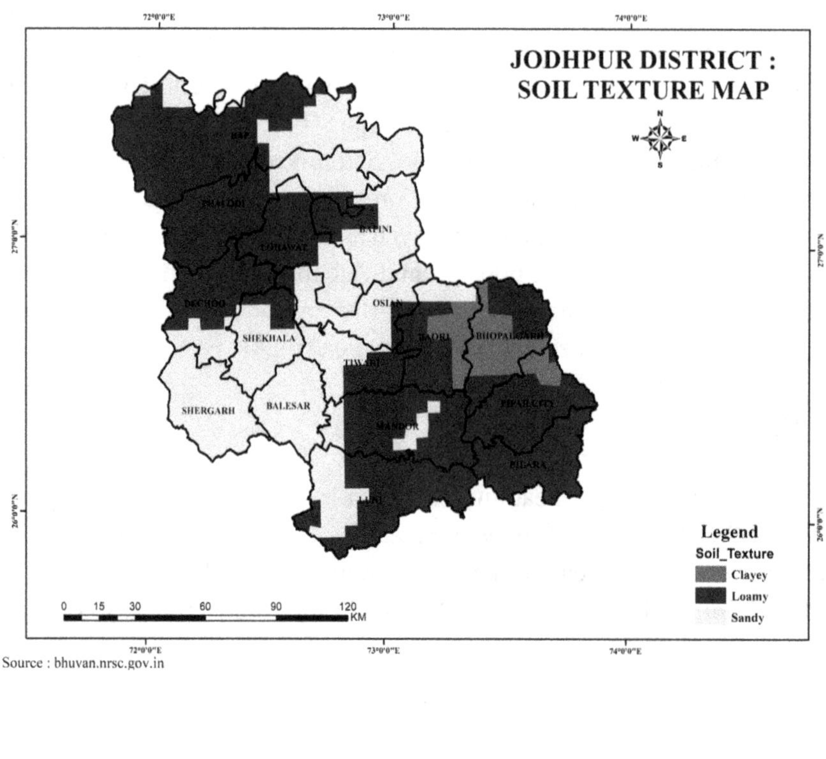

Source : bhuvan.nrsc.gov.in

Figure 5.

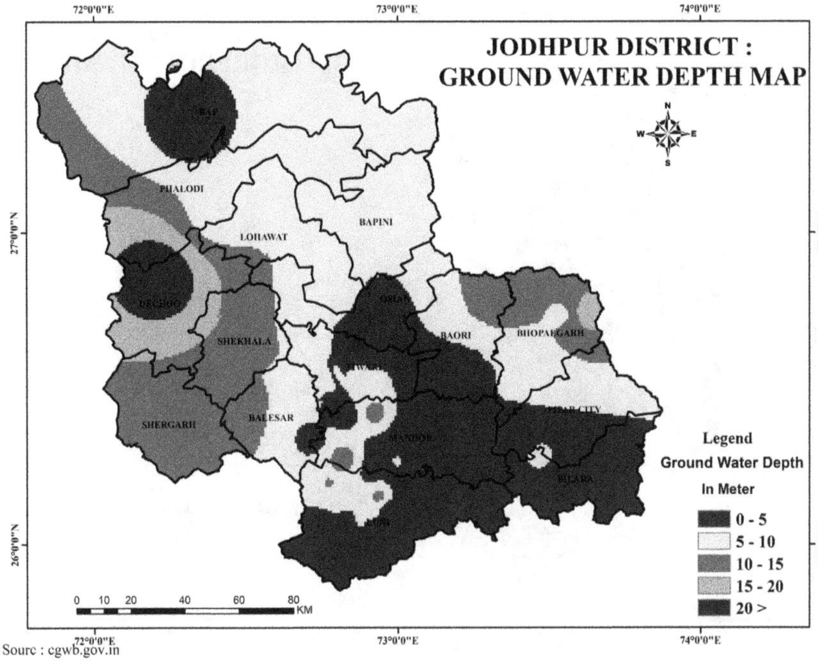

Sourc : cgwb.gov.in

through line density. High drainage density values are more prone for run-off, and hence indicates low groundwater potential zone. High ranks are assigning to low drainage density area and vice versa (fig.6).

Slope

The slope (in degree) of the study area mainly varies between 0° and 63.93°. Classes having less slope value are assigned higher rank due to almost flat terrain while the class having maximum value is categorized as lower rank due to relatively high run-off (fig.7).

Soil

The soil texture present in the study area is mainly having three main soil categories namely sandy, loamy and clayey soil. Rank of soil has been assigned on the basis of their infiltration rate. Sandy soil has high infiltration rate, hence given higher priority, while the clayey soil has least infiltration rate hence assigned low priority (fig. 8).

Figure 6.

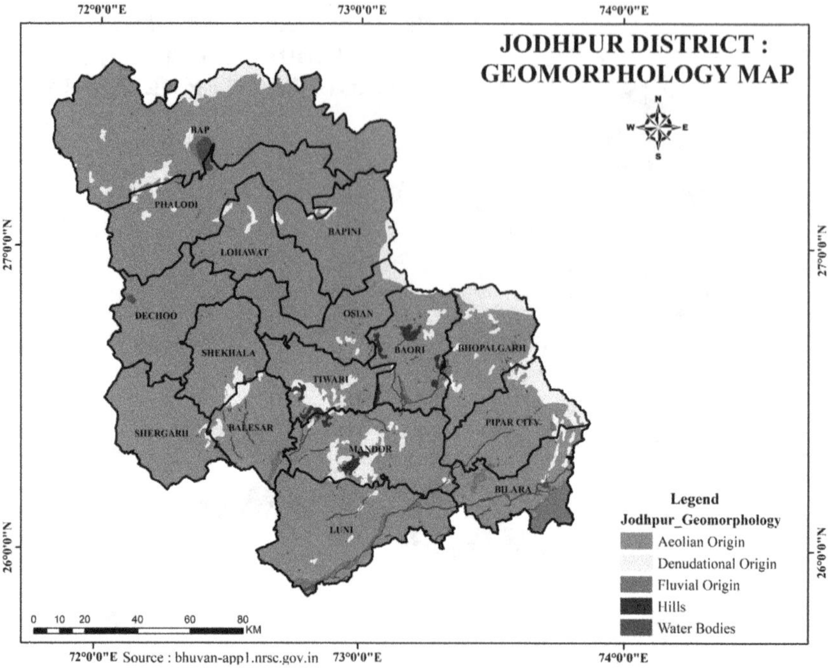

Figure 7.

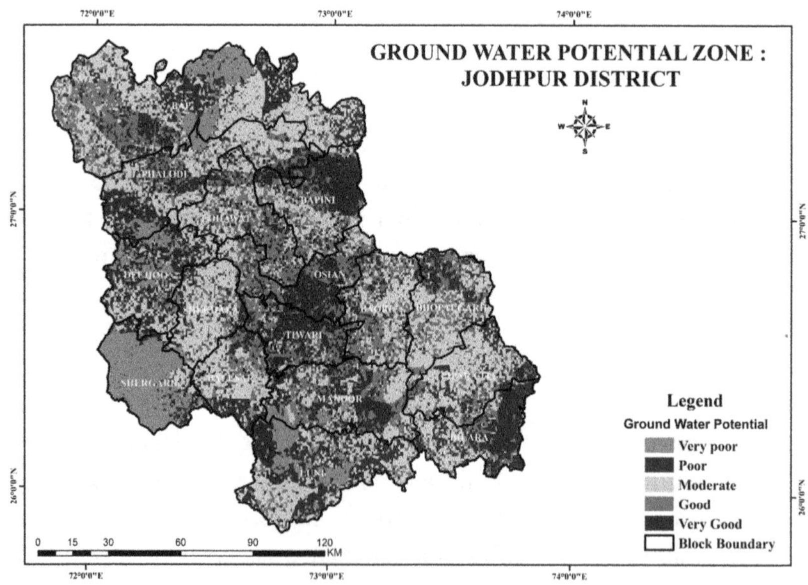

Figure 8.

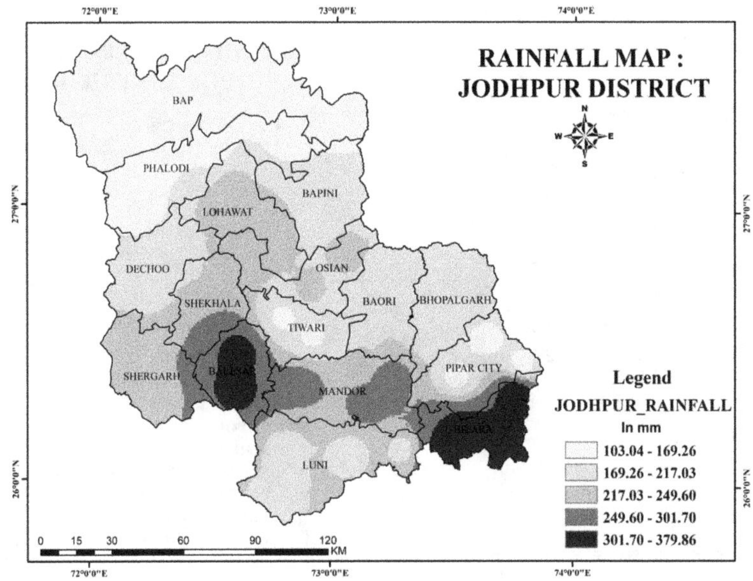

Average Post-Monsoon Groundwater Depth

Contour maps of Ravi season have been prepared from groundwater level data, collected from CGWB. In Rabi season, average groundwater depth generally ranges from 8m to 12m, with a major portion of the area having 5m - 10m depth. The developed groundwater level maps are classified into five classes (Fig. 9). On the basis of Interpolation analysis, we found that which area of groundwater depth has more suitable for analysis.

Geomorphology

The variable landform of study area is a fundamental event for the analysis of dispersion of groundwater. Five sorts of geomorphic units are distinguished in the examination zone. Aeolian and Denudational Origin has higher water level surface, hence it is the best landform for high groundwater potential then the structural origin (fig. 10).

Figure 9.

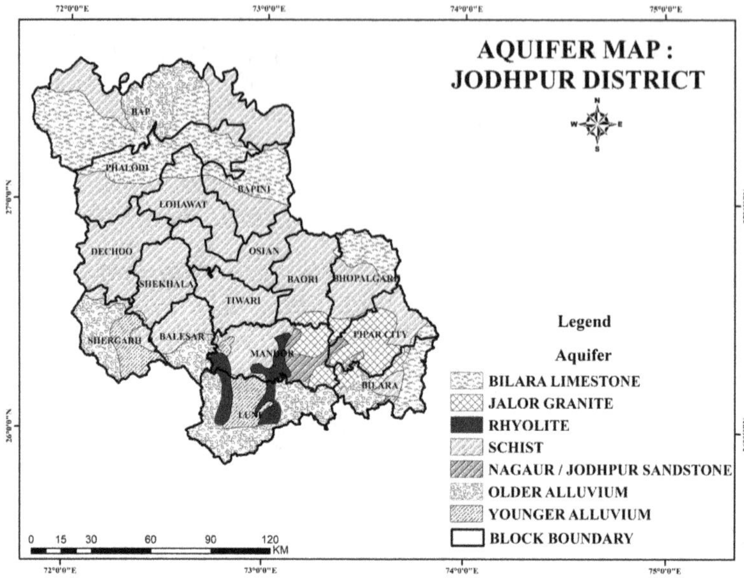

Source : cgwb.gov.in

Figure 10.

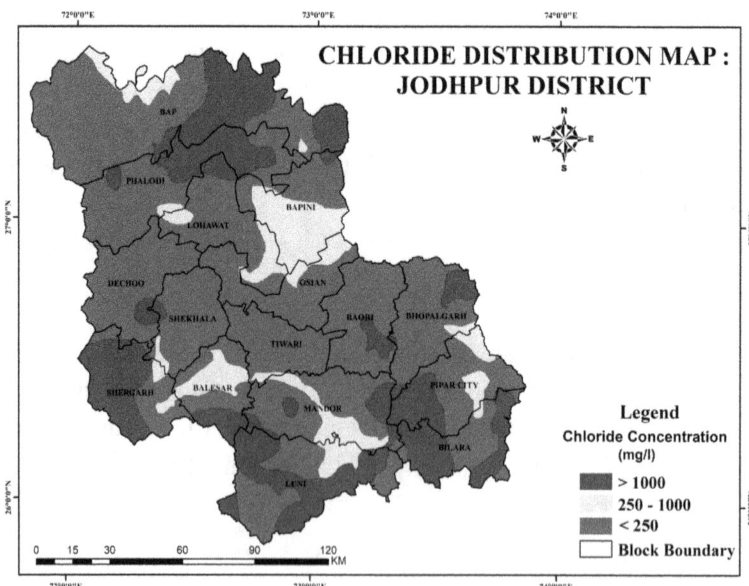

Source : cgwb.gov.in

Chloride Concentration

The important concern is a Very high chloride concentration present in ground water and their erratic effect. In Jodhpur district, the high concentrations of chloride are mainly present in the north and in southern portion of district namely Baap and Phalodi block. The Moderate concentration are in the lower portion of the central part of Jodhpur district as found south of Bapini, around Balesar and to the western part of Mandor Block. Large area cover the low chloride concentration are found in the eastern (east of Bhopalgarh and Pipar cities), in middle (Tinwari, Shekhala block) and in the north (north of Bap and Dechoo). Very high concentrations (unsuitable for drinking and domestic purposes) are seen in southern parts of the district and in the Northern parts. In between these the areas have moderate concentrations suitable for drinking and domestic purposes (fig. 11).

Figure 11.

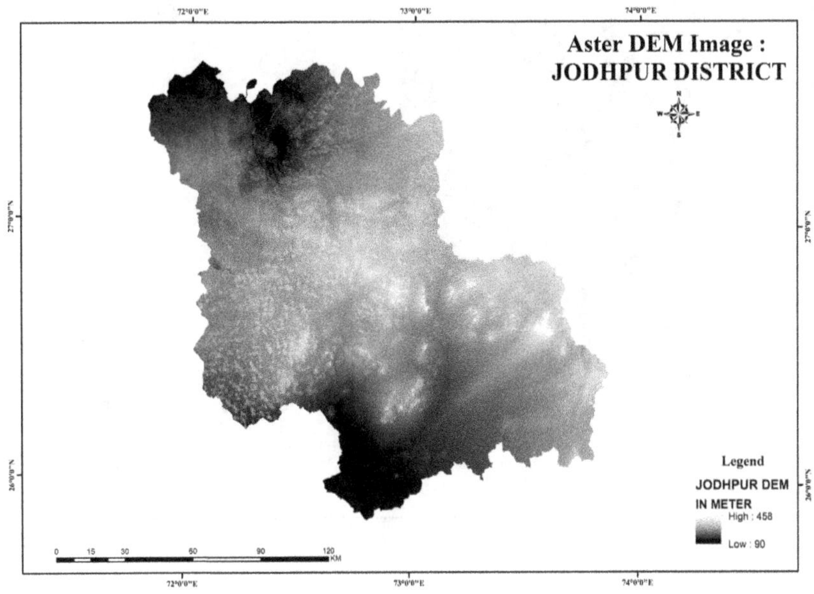

Aquifer

Semi consolidated to unconsolidated clay, pebble, sand gravel are mainly present in the younger and older alluvium and about 18% of the district is covered with alluvium. Nagaur/Jodhpur sandstone covers about 50.53% area of Jodhpur district; it belongs to Marwar super group. Jodhpur sandstone is the main source for ground water which mainly occurs under confined to semi confined conditions. The Bilara Limestone is also the main source for groundwater, and occupies extensive area of Bap and Phalodi blocks and some parts of Bilara, Bhopalgarh, and Osian block (Fig.12).

Figure 12.

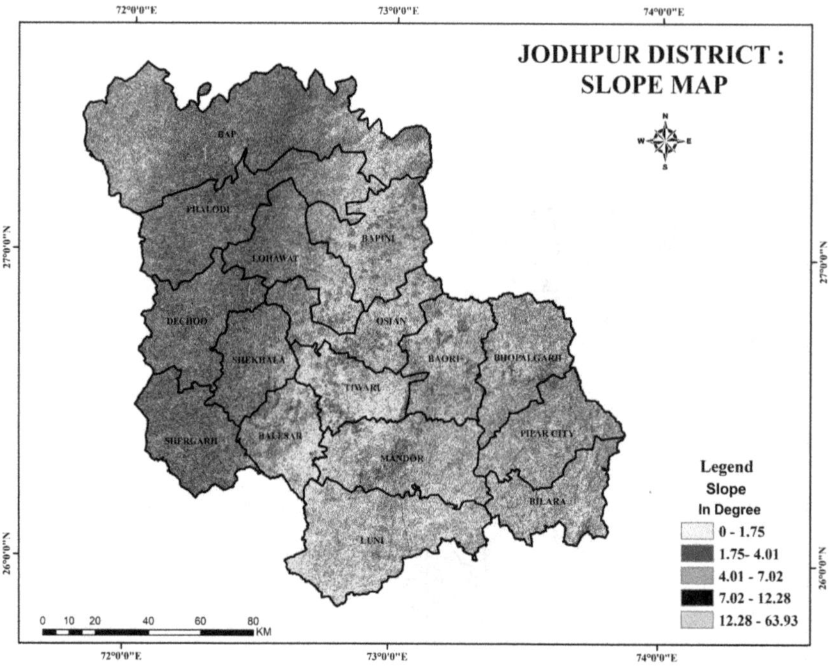

Rainfall

The Jodhpur district receives very low rainfall of an average of 280 mm in 2018. The point data collected from the all blocks of Jodhpur by PHD department and with the interpolation in GIS we calculated an average rainfall the study area. By the analysis we can say that Bap and Phalodi receives very low rainfall and that's why the area is given low priority in comparison with the area like Bilara and Balesar which receives high amount of Rainfall and given higher priority for the analysis of groundwater potential (Fig.13).

Figure 13.

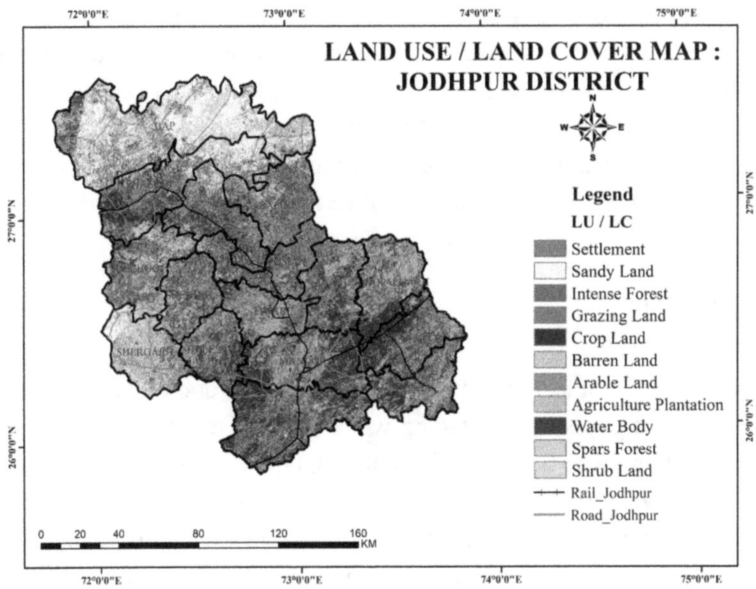

Ground Water Potential

In the study area, the methodology is based on AHP process which supports the relative importance of various thematic layers and their corresponding classes which are affecting the groundwater and calculating groundwater potential zone. Table 5 above shows the ranking of each thematic layer and weight of each sub classes of all the different parameters and the weights of each parameter is calculated by using

AHP method. The ground water potential which is evaluated by the weighted linear combination is shown in fig. 14.

Result shows that about 28% of the total area is under the poor and very poor groundwater due to various reasons present in that reason which is discussed above and the region which comes under poor zone is Shergarh, Luni, Dechu and middle portion of Bap block. 28.39% of the area is under the moderate zone of ground water potential which is good for the utilizing for domestic purpose or for agriculture purpose. 28.82% of the study area is under the Good and 14.74% area is under very good ground water prospect zone mainly present in the middle portion like Osian, Tiwari, and southeast portion of Bilara block and the water quality is mainly good for all the purpose and extraction of fresh water from these region will be beneficial to maintain water harvesting.

Thus the above study shows the capabilities of remote sensing and GIS and the different types of modeling used in GIS for calculating close accuracy in the site selection of groundwater potential. The Jodhpur district has a good ground water depth but due to presence of sandy land, low rainfall and also various contaminant minerals like chloride concentration, fluoride concentration, arsenic concentration etc. which effect the ground water in large amount. Therefore all area in the district having good groundwater depth but cannot be used by the local people.

CONCLUSION

In this study, we have selected Jodhpur district because due to various undulations present in area and population is facing water scarcity every year. Hence we have used successfully remote Sensing, GIS and MCDM techniques and demonstrated for evaluation of groundwater potential zone. A three-step methodology was used that includes development of thematic layers, deriving the weights using ANP/AHP and overlay analysis to find groundwater potential zone. Groundwater potential zoning was important for the fresh water estimation and where to recharge fresh water artificially. Remotely sensed satellite image data and digitization of existing maps using GIS were used for the preparation of thematic layers. The ANP and the AHP were used to provide utility weights for the alternatives. The overlay analysis of various thematic maps and assign weights are used for the delineation of zone of groundwater potential. In this study area, five categories of groundwater potential zone have been delineated based on remote sensing, GIS and AHP/ANP techniques i.e. Very Poor, Poor, Moderate, Good, and Very Good. Final result is shown in Fig. 14 which will help many organizations and Government for the future development and proper growth.

REFERENCES

Ahmed, Shahid, bin Harun, Ismail, Nawaz, & Shamsudin (2014). Assessment of groundwater potential zones in an arid regionbased on catastrophe theory. *Earth Science Informatics*. Advance online publication. doi:10.100712145-014-0173-3

Ammar, A., Michel, R., Ouessar, M., & Ritsema, C. (2016). Identification of suitable sites for rainwater harvesting structures in arid and semi-arid regions: A review. *ISWCR*, *4*, 108–120.

District Census 2011. (2011). Census2011.co.in

Jaiswal, R., Mukherjee, S., Krishnamurthy, J., & Saxena, R. (2003). Role of remote sensing and GIS techniques for generation of groundwater prospect zone towards rural development – an approach. *International Journal of Remote Sensing*, *24*(5), 993–108. doi:10.1080/01431160210144543

Khamis, Afan, Nur Muhammad, & ElShafie. (2017). *Development of a spatial hydrologic soil map using spectral reflectance band recognition and a multiple – output artificial neural network model*. doi:10.5194/hess-2017-13

Saaty, T. L. (1980). *The Analytic Hierarchy Process*. McGraw- Hill.

Saaty, T. L. (1996). *Decision making with dependence and feedback, The Analytic Network Process*. RWS Publications.

Saaty, T. L. (1999). Fundamentals of the analytic network process. *International Symposium of the Analytic Hierarchy Process (ISAHP)*, Kobe, Japan.

Saaty, T. L. (2004). Fundamentals of the analytic network process – multiple networks with benefits, costs, opportunities and risks. *Journal of Systems Science and Systems Engineering*, *13*(3), 348–379. doi:10.100711518-006-0171-1

Science Learning Hub. (2016). *What is clay*. University of Waikato.

Ziadat, F., Bruggeman, A., Oweis, T., Haddad, N., Mazahreh, S., Sartawi, W., & Syuof, M. (2012). A participatory GIS approach for assessing land suitability for rainwater harvesting in an arid rangeland environment. *Arid Land Research and Management*, *26*(4), 297–311. doi:10.1080/15324982.2012.709214

Chapter 3
Evaluation of Multi-Temporal Sentinel-1 Dual Polarization SAR Data for Crop Type Classification

Thota Sivasankar
https://orcid.org/0000-0003-2422-8731
NIIT University, India

Pavan Kumar Sharma
Amnex Infotechnologies Pvt. Ltd., India

M. N. S. Ramya
Independent Researcher, USA

Pithani Venkatesh
Skymet Weather Services Pvt. Ltd., India

G. D. Bairagi
M.P. Council of Science and Technology, India

ABSTRACT

India is one of the highly populated countries, and its economy mainly depends on agriculture. The crop type classification is an essential requirement for ensuring food security, crop monitoring, and to understand the environmental consequences of cultivated ecosystems. This study exploits freely available multi-temporal SAR data for discriminating crop types, such as wheat, gram, and mustard, over Ashok Nagar district, Madhya Pradesh, India. Nine Sentinel-1 dual-polarized data acquired from

DOI: 10.4018/978-1-7998-5027-4.ch003

January 2018 to April 2018 in interferometric wide swath mode are used. Class separability analysis using Bhattacharyya Distance (BD) has been performed for multi-temporal VV and VH backscatter, log-ratio, and Radar Vegetation Index (RVI) to quantify the ability to distinguish temporal profiles of crops. RVI has shown the significant result in class separability analysis in comparison with other parameters. Crop type classification map has been generated using a support vector machine classifier with overall accuracy and Kappa coefficient of 96.32% and 0.95, respectively.

INTRODUCTION

According to the United Nations (United Nations, 2019), India is one of the top ten countries whose global population proliferate more than half of the projected growth between 2019 and 2050. It is also forecasted that India may be the highest populated country by 2027 based on the current growth rate. The steady increase in population will contemporaneously increase demand for food, which influence the future actions of mankind towards food security and nature conservation (FAO, 2009). Several studies, such as Yu et al. (2015) and Pingali et al. (2019) discussed the importance of agriculture and its need for a strategy in India to ensure future food security.

The spatial distribution of crop types and timely health status are significantly important for decision makers at regional, national and even global level (Brisco et al., 1998; Wu et al., 2015). In this regard, satellite Earth Observation (EO) data has become an essential data source. With the increase in EO satellites' operating in optical and microwave regions of the electromagnetic spectrum enable continuous land monitoring/observations. Particularly, free and open satellite data provided from Landsat and Sentinel missions have been extensively used for these applications (Wulder et al., 2012; Veloso et al., 2017; Kamilaris et al., 2017). Although, several studies (Thenkabail et al., 2005; Dheeravath et al., 2010) have shown the potential capabilities of optical multi-spectral data for crop classification. As the data is highly sensitive to the weather conditions, these approaches may not provide acceptable results over areas having frequent cloud cover or foggy conditions. Since radar remote sensing can provide data during almost all-weather conditions, this may be effectively used for un-interrupted crop growth monitoring than optical. In addition, its unique sensitivity towards structural, geometrical and dielectric properties of various components of the crop makes possible for crop identification and biophysical parameters retrieval, concurrently assess the health status (Patel and Srivastava, 2013; Gao et al., 2018; Sivasankar et al. 2018). Furthermore, the ability of radar signal to penetrate through vegetation and interact with underneath soil enables to estimate soil characteristics like moisture (Srivastava et al., 2006; Baghdadi et al., 2006; Sharma et al., 2019), roughness (Baghdadi et al., 2002; Srivastava et al., 2008) and texture (Singh and Kathpalia, 2007). However, previous studies have also demonstrated the synergistic use of optical and SAR datasets for agricultural crop applications (Moran et al., 1997; McNairn et al., 2009; Gibril et al., 2017; Sun et al., 2019).

The distinct advantage of radar remote sensing techniques and the increasing number of space-borne SAR satellites has made researchers to widely use the data for these applications. The backscatter signal from the agricultural field is a complex function of both target (crop as well as soil) parameters and SAR sensor configuration (Henderson and Lewis, 1998). Thus, a proper chosen of sensor parameters is vital to enhance the sensitivity of SAR data for a specific application. The SAR sensor parameters such are frequency, polarization and incidence angle have a significant influence on scattering mechanisms in vegetation canopies. For more details, a brief review on radar remote sensing for agricultural applications is given in Sivasankar et al. (2018). Previous studies (Jia et al., 2012; Foody et al., 1989; Freeman et al., 1994; Patel and Srivastava, 2007; Tan et al., 2011; Kothapalli Venkata et al., 2017) have demonstrated that multi-configuration SAR datasets like multi-temporal, multi-polarization, multi-frequency and multi-incidence angle provide better accuracy for crop identification than single SAR data acquired at a given frequency, polarization and incidence angle. Among these, multi-temporal SAR data is a particularly attractive data source for crop type classification.

Several studies (Whelen and Siqueira, 2018; Lasko et al., 2018; McNairn et al., 2009; Wang et al., 2010) have demonstrated the effectiveness of multi-temporal SAR data for crop classification. For instance, Lasko et al., (2018) investigated the performance of multi-temporal dual polarized Sentinel-1 SAR data for mapping double and single crop rice in Hanoi, Vietnam. The study observed the highest overall accuracy of 93.5%,±1.33%using both VV and VH backscatter from the random forest classifier. Xu et al., (2019) proposed a new classification approach for crop classification using temporal backscatter signatures. Skriver et al., (2011) analyzed multi-temporal SAR datasets acquired in C- and L-bands for crop classification. The study results identified that temporal SAR backscatter signature is an essential input for accurate crop mapping and particularly, cross-polarized backscatter shown the best results, with errors down to 3% - 6% at the two frequencies. As the temporal information from SAR data is very important for crop identification, it is still an active research area for developing an appropriate approach for this purpose.

Machine learning techniques are receiving more and more attention for crop classification from SAR data (Sonobe et al., 2014a; Tan et al., 2007; Xu et al., 2014). Among these, support vector machine has shown better performance in several studies (Sonobe et al., 2014b; Navale and Haldar, 2019). Therefore, the temporal signatures of VV & VH backscatter, VV/VH (log-ratio) as well as radar vegetation index (RVI) generated from C-band dual-polarized Sentinel-1 SAR data are analyzed for various crops. Based on the performance of each parameter's ability to discriminate various crops, SVM has been utilized for mapping crops.

MATERIALS

The study site is located in Ashok nagar district, the northern part of Madhya Pradesh, India. It consists of total area about 4700 km^2 whose average elevation is about 507 m above the sea level (Figure 1). The study area has sufficient water for irrigation as it is situated between two major rivers Sindh on western and Betwa on eastern sides. The soil composition of this region is mainly three types such as black, brown and bhatori (stony). The climate of Ashok nagar is sub-tropical and its economy essentially depends on agriculture. The temperature varies drastically, 47 °C in summer and even drops to 4 °C in the winter. Winter is the longest duration of the other seasons, extending from October to mid-March where its daily temperature ranges from 15 °C to 20 °C, though on some nights it can fall as low as 5 °C.

The Sentinel-1 (S-1) mission, developed by the European Space Agency (ESA), provides C-band SAR data acquired in Interferometric Wide swath (IW) mode globally for free of cost to the research community. The S-1 mission is based on a constellation of two satellites, S-1A and S-1B launched on April 3, 2014 and April 25, 2016 respectively with a temporal revisit of 12 days (about 6 days by combining both ascending and descending passes). The IW mode is operated as TOPS (Terrain Observation with Progressive Scans in azimuth) which offers single look data at a large swath width of 250 km at a nominal ground resolution (range × azimuth) of 5 × 20 m. Nine Sentinel-1 IW mode dual-polarized Level-1 ground range detected (GRD) images acquired from January 10, 2018 to April 20, 2018 are used in this chapter. Figure 1 shows the location and color composite image of the experimental data, where the VH-backscatter observed on Jan 22, 2018, Feb 27, 2018 and Mar 23, 2018 represent the RGB channels respectively.

METHODOLOGY

In this chapter, multi-temporal C-band dual-polarization (VV and VH) Sentintel-1 Level-1 ground range detected (GRD) SAR images have been used. The pre-processing which involves thermal noise removal, radiometric calibration and terrain correction has been done using the Sentinel-1 toolbox. All the images were co-registered to extract correct temporal profiles of different crop types. In addition

Figure 1. Location of the study area

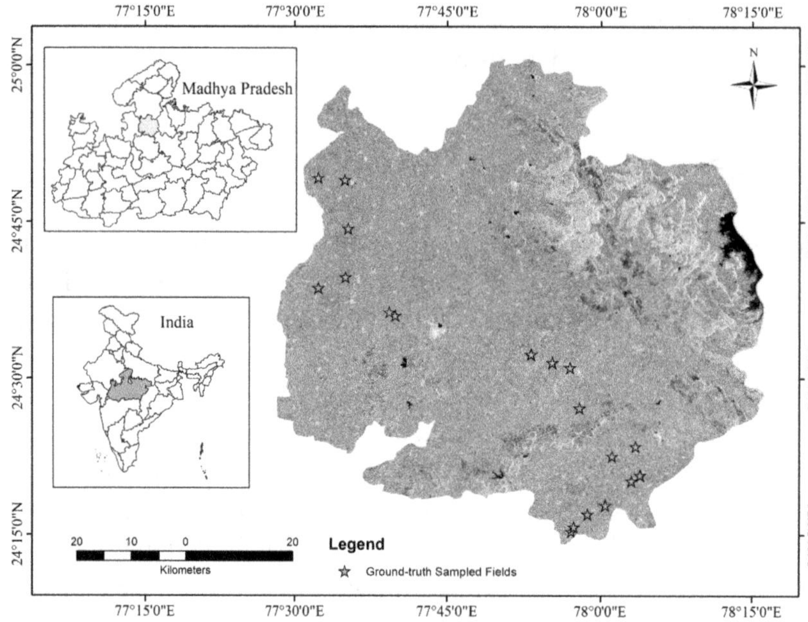

to the backscattering coefficients (VV and VH), VV/VH (log-ratio) and RVI were generated for each data. Actually, RVI was developed for full polarization to assess the vegetation growth using time series data. This has been modified by Charbonneau et al. (2005) for dual polarization SAR data, assuming that $\sigma^\circ_{HH} \approx \sigma^\circ_{VV}$. In recent,

Table 1. Sentinel-1 SAR acquisitions over the study site

Dates	Polarization	Pass direction	Acquisition mode	Relative orbit
Jan 10, 2018	VV and VH	Descending	IW	63
Jan 22, 2018	VV and VH	Descending	IW	63
Feb 03, 2018	VV and VH	Descending	IW	63
Feb 27, 2018	VV and VH	Descending	IW	63
Mar 11, 2018	VV and VH	Descending	IW	63
Mar 23, 2018	VV and VH	Descending	IW	63
Apr 04, 2018	VV and VH	Descending	IW	63
Apr 16, 2018	VV and VH	Descending	IW	63
Apr 28, 2018	VV and VH	Descending	IW	63

few studies have considered for dual polarization (VV and VH) Sentinel-1 SAR data, following Charbonneau et al. (2005) as given in equation 1 to assess the vegetation cover (Nasirzadehdizaji et al., 2019).

$$RVI = \frac{4\sigma_{VH}^{0}}{\sigma_{VV}^{0} + \sigma_{VH}^{0}} \quad (1)$$

The co-registered multi-temporal Sentinel-1 SAR derived parameters i.e., VV & VH backscatter, VV/VH and RVI have been layer stacked. Vector representing farmers' field boundaries collected during ground-truth campaigns using handheld GPS were overlaid on the layer stacked file to extract the temporal signatures. These profiles have been analyzed and assessed its ability to discriminate crop types. In this regard, the Bhattacharyya or Jeffries-Matusita Distance (BD) developed by (Bhattacharyya, 1943) has been used to quantify the ability to distinguish temporal profiles of various crop types i.e., wheat, mustard and gram. The BD is a commonly used measure that is bounded between values of 0 and 2. It calculates covariance weighted distance between the class mean of each class. The highest value states that the classes have good separability and help to classify crops. The expression to measure BD is given in the following equation 2:

$$BD(i,j) = 2\left[1 - e^{-a(i,j)}\right] \quad (2)$$

where $BD(i,j)$ = Bhattacharyya Distance between class i and j

$$a(i,j) = 0.125 \times T\left[M(i) - M(j)\right] \times Inv\left[A(i,j)\right] \times \left[M(i) - M(j)\right] + 0.5 \times \ln\left\{\frac{\det\left(A(i,j)\right)}{\sqrt{\left[\det\left(S(i)\right) \times \det\left(S(j)\right)\right]}}\right\} \quad (3)$$

where:

T, M, S, and Inv are defined as transpose of matrix, mean vector and covariance matrix respectively

A(i,j) = 0.5*[S(i)+S(j)]

det() = determinant of a matrix

The better separability shown by temporal profiles of parameters are chosen for supervised classification using support vector machine (SVM). SVM is a non-parametric statistical learning technique widely used and tested for land cover mapping from satellite dataset. It fits an optimal hyper plane through n-dimensional spectral space distinguishing classes based on a user defined training samples (Mather and Brandt, 2016). The potential of SVM is its intelligence to classify data even with small training dataset successfully known as support vectors (Foody and Mathur, 2004). The accuracy of classification mainly depends on the quality of support vectors that lie at the edge of the class distributions for discriminating class boundaries. SVM consists of different kernels, such as Linear, polynomial, radial basis function and sigmoid kernels for building hyperplane. Yang (2011) stated that among four kernels, radial basis function kernel gave prominent results for classification. Therefore, Radial basis function kernel (RBF) is chosen further for the analysis. Figure 2 shows the detailed flow chart of our methodology.

Figure 2. Flow chart of the methodology for crop type classification

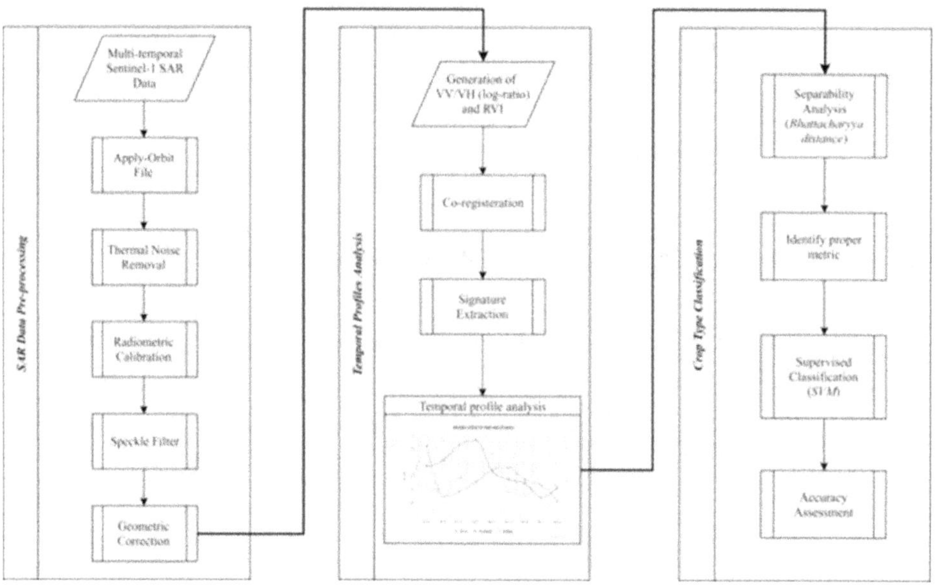

RESULTS AND DISCUSSIONS

The temporal dynamics of VV & VH backscatter, VV/VH (log ratio) and RVI for the three crop types, such as, wheat, mustard and gram are analyzed (see Figure 3-6). The study results shown an increasing behavior of backscatter from initial stage to mature stage of the crop and then decreased during ripening stage. Similar trend was also observed in most of the previous studies carried over various SAR datasets (Patel and Srivastava, 2013; Lasko et al., 2018). However, few studies have also reported a decreasing trend of backscatter with crop growth (Sivasankar et al., 2020; Fontanelli et al., 2013). Analysis has shown that increasing trend of VV & VH backscatter as well as RVI with crop growth whereas decreasing trend of VV/VH (log-ratio).

Figure 3. Time-series VH-backscatter for various crops

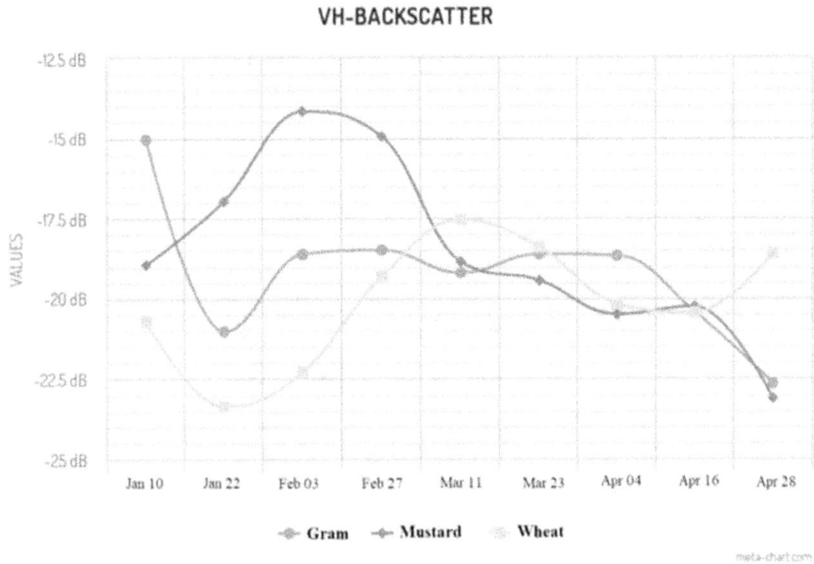

The VH backscattering is the contribution of multiple reflections of the SAR signal, which depends mainly on the crop characteristics. As the crop grows, the canopy increases imply increasing multiple reflections. Therefore, VH backscatter has shown increasing trend with the crop growth. VV backscattering is sensitive to both soil and crop characteristics due to which it shows uneven variations. It gives very less information about the crops but structural variations are well identified. This helps in classifying crops of different structures such as mustard from gram or wheat, can be observed in Figure 4. The VV backscatter of wheat increases with time whereas different pattern for gram and mustard.

Further, these temporal signatures were quantified, based on Bhattacharyya Distance (BD) its ability to distinguish crop types. The BD values for VH-backscatter, VV-backscatter, VV/VH (log-ratio) and RVI are given in Table 2, 3,

Figure 4. Time-series VV-backscatter for various crops

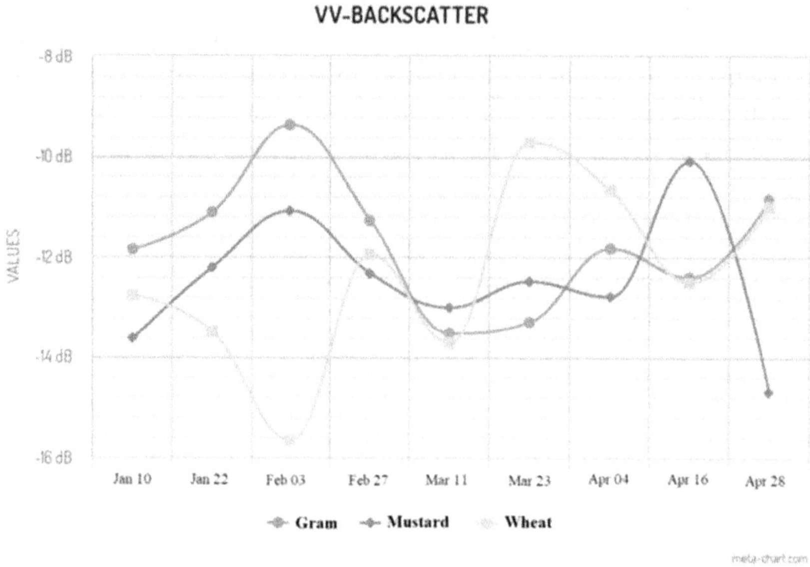

Figure 5. Time-series VV/VH (Log-ratio) for various crops

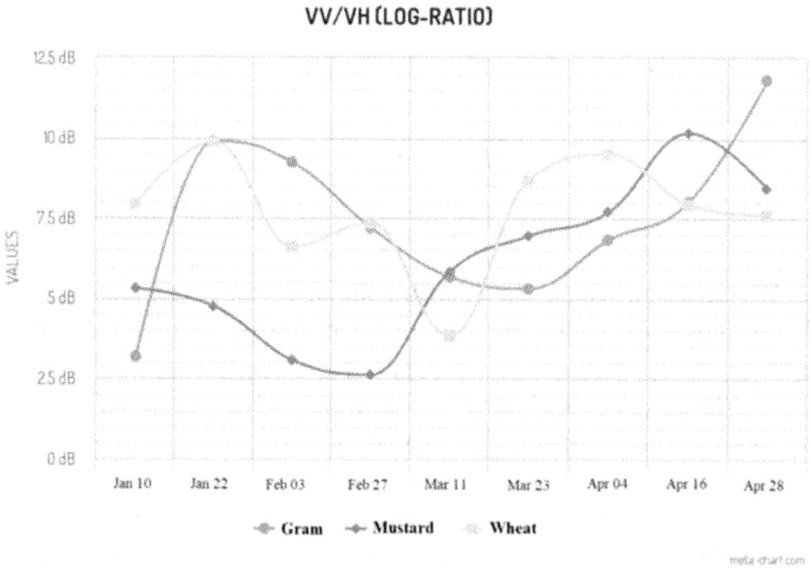

Figure 6. Time-series Radar Vegetation Index (RVI) for various crops

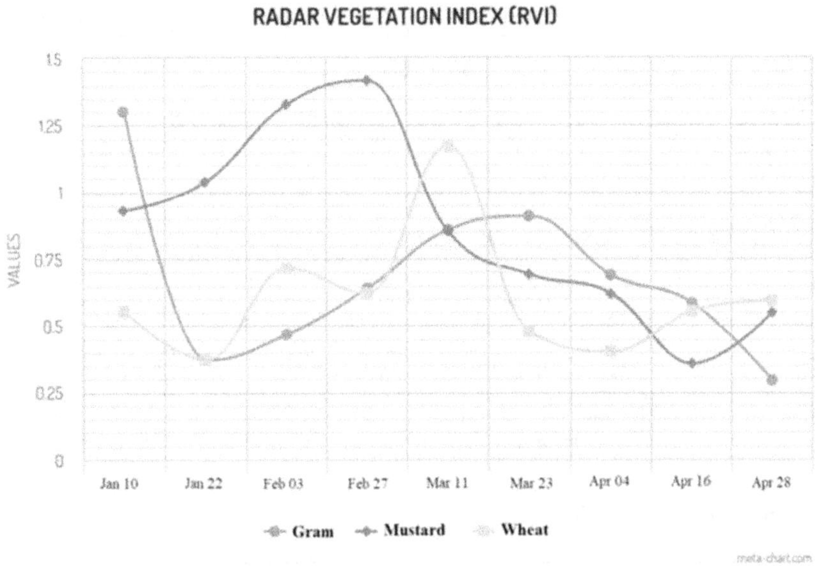

4 and 5 respectively. From table-2, the BD value is high for wheat and gram with mustard is 1.91 and 1.95 respectively whereas low value for gram with wheat is 1.72. The mustard crop is a medium level biomass while wheat and gram are lower level biomass. Therefore, mustard is well differentiated with gram and wheat using VH backscatter. VV backscatter shows low BD values in comparison with VH backscatter for mustard with wheat and gram. The BD values for gram and wheat is 1.82 in VV backscatter while 1.72 for VH backscatter. In VV backscatter, gram and mustard are not well separated show low value 1.67.

Table 2. BD values for VH-backscatter

	Wheat	Mustard	Gram
Wheat	0		
Mustard	1.91	0	
Gram	1.72	1.95	0

Table 3. BD values for VV-backscatter

	Wheat	Mustard	Gram
Wheat	0		
Mustard	1.88	0	
Gram	1.82	1.67	0

Table 4. BD values for VV/VH (log-ratio)

	Wheat	Mustard	Gram
Wheat	0		
Mustard	1.79	0	
Gram	1.75	1.73	0

Table 5. BD values for RVI

	Wheat	Mustard	Gram
Wheat	0		
Mustard	1.94	0	
Gram	1.89	1.98	0

To improve the potentiality of identifying crops, log-ratio and RVI parameters were also investigated further in this study. The log-ratio of VV/VH shown low BD values for all crops in comparison with VV and VH backscatter can be observed from table 4. This implies that log ratio may not give a significant result. The separability of crops are well improved in RVI. The advantages of both VH and VV backscatter can be observed in RVI. The BD values for RVI are represented in table 5. The BD for mustard with wheat and gram is 1.94 and 1.98 respectively. The separability between gram and wheat is also high in RVI nearly 1.89 compared to the other backscatters. Therefore, RVI is considered for the final classification (see Figure 7). SVM classifier with RVI is used further for crop type classification. The Overall Classification Accuracy and Kappa Coefficient of the crop classification is 96.32% and 0.95 respectively.

Figure 7. Crop type classification using Support Vector Machine (SVM)

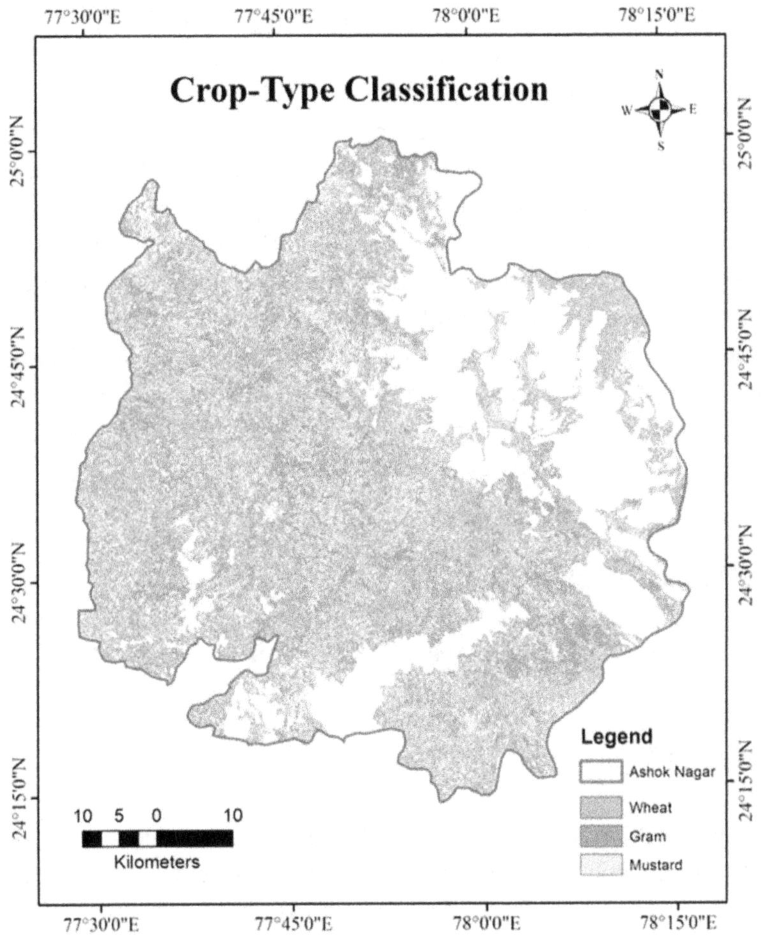

CONCLUSION

The study investigated multi-temporal Sentinel-1 dual polarized SAR dataset for crop type classification. VV and VH backscattering have been generated for the crops from January, 2018 to April, 2018. The VH backscattering shown an increasing trend with the time period while VV backscattering shown irregular variations. Class separability analysis was performed based on the Bhattacharyya Distance (BD) for wheat, mustard and gram. The BD value is high for wheat and gram with mustard approximately 1.91 and 1.95 respectively with VH backscatter whereas gram and wheat have shown good separability with VV backscatter nearly 1.82. To improve the separability of classes, VV and VH backscatter are combined and generated log ratio and RVI for further analysis. The lower separability between crops is observed with log-ratio images. RVI gave promising results in separability of the crop types. The advantages of VV and VH backscatter are incorporated in RVI for decreasing the misclassification of the crops. The BD for mustard with wheat and gram is 1.94 and 1.98 respectively. The separability between gram and wheat is also high in RVI nearly 1.89 compared to the other backscatters. The class separability analysis helped to assess the quality of training samples for classifying crops. The outcomes of this separability analysis suggest that RVI has shown adequate results and chosen for the classification of crops. The high separbility between classes resembles the existence of pure pixels for the classification. The SVM classifier is applied to multi-temporal SAR data for assisting crop type classification. Three crops are classified with an overall accuracy of 96.32% and a kappa coefficient of 0.95. The combination of multi-temporal dataset led to abundant information and outperformed the classification procedure with high accuracy.

ACKNOWLEDGMENT

We acknowledge European Space Agency (ESA) for Sentinel-1 SAR data provision.

REFERENCES

Baghdadi, N., Gaultier, S., & King, C. (2002). Retrieving surface roughness and soil moisture from synthetic aperture radar (SAR) data using neural networks. *Canadian Journal of Remote Sensing*, *28*(5), 701–711. doi:10.5589/m02-066

Baghdadi, N., Holah, N., & Zribi, M. (2006). Soil moisture estimation using multi-incidence and multi-polarization ASAR data. *International Journal of Remote Sensing*, *27*(10), 1907–1920. doi:10.1080/01431160500239032

Bhattacharyya, A. (1943). On a measure of divergence between two statistical populations defined by their probability distributions. *Bulletin of the Calcutta Mathematical Society*, *35*, 99–109.

Brisco, B., Brown, R. J., Hirose, T., McNairn, H., & Staenz, K. (1998). Precision agriculture and the role of remote sensing: A review. *Canadian Journal of Remote Sensing*, *24*(3), 315–327. doi:10.1080/07038992.1998.10855254

Charbonneau, F., Trudel, M., & Fernandes, R. (2005). Use of Dual Polarization and Multi-Incidence SAR for soil permeability mapping. *Advanced Synthetic Aperture Radar (ASAR) 2005.*

Dheeravath, V., Thenkabail, P. S., Chandrakantha, G., Noojipady, P., Reddy, G. P. O., Biradar, C. M., Gumma, M. K., & Velpuri, M. (2010). Irrigated areas of India derived using MODIS 500 m time series for the years 2001–2003. *ISPRS Journal of Photogrammetry and Remote Sensing*, *65*(1), 42–59. doi:10.1016/j.isprsjprs.2009.08.004

Fontanelli, G., Paloscia, S., Zribi, M., & Chahbi, A. (2013). Sensitivity analysis of X-band SAR to wheat and barley leaf area index in the Merguellil Basin. *Remote Sensing Letters*, *4*(11), 1107–1116. doi:10.1080/2150704X.2013.842285

Foody, G. M., Curran, P. J., Groom, G. B., & Munro, D. C. (1989). Multi-temporal airborne synthetic aperture radar data for crop classification. *Geocarto International*, *4*(3), 19–29. doi:10.1080/10106048909354218

Foody, G. M., & Mathur, A. (2004). Toward intelligent training of supervised image classifications: Directing training data acquisition for SVM classification. *Remote Sensing of Environment*, *93*(1-2), 107–117. doi:10.1016/j.rse.2004.06.017

Freeman, A., Villasenor, J., Klein, J. D., Hoogeboom, P., & Groot, J. (1994). On the use of multi-frequency and polarimetric radar backscatter features for classification of agricultural crops. *International Journal of Remote Sensing*, *15*(9), 1799–1812. doi:10.1080/01431169408954210

Gao, Q., Zribi, M., Escorihuela, M., Baghdadi, N., & Segui, P. (2018). Irrigation mapping using Sentinel-1 time series at field scale. *Remote Sensing*, *10*(9), 1495. doi:10.3390/rs10091495

Gibril, M. B. A., Bakar, S. A., Yao, K., Idrees, M. O., & Pradhan, B. (2017). Fusion of RADARSAT-2 and multispectral optical remote sensing data for LULC extraction in a tropical agricultural area. *Geocarto International*, *32*(7), 735–748. doi:10.1080/10106049.2016.1170893

Henderson, F. M., & Lewis, A. J. (1998). Principles and applications of imaging radar. *Manual of Remote Sensing, 2.*

Jia, K., Li, Q., Tian, Y., Wu, B., Zhang, F., & Meng, J. (2012). Crop classification using multi-configuration SAR data in the North China Plain. *International Journal of Remote Sensing*, *33*(1), 170–183. doi:10.1080/01431161.2011.587844

Kamilaris, A., Kartakoullis, A., & Prenafeta-Boldú, F. X. (2017). A review on the practice of big data analysis in agriculture. *Computers and Electronics in Agriculture*, *143*, 23–37. doi:10.1016/j.compag.2017.09.037

Kothapalli Venkata, R., Poloju, S., Mullapudi Venkata Rama, S. S., Gogineni, A., Prabir Kumar, D., Allakki Venkata, R., & ... & KRP, S. (. (2017). Multi-incidence angle RISAT-1 Hybrid Polarimetric SAR data for large area mapping of maize crop–a case study in Khagaria district, Bihar, India. *International Journal of Remote Sensing*, *38*(20), 5487–5501. doi:10.1080/01431161.2017.1338783

Lasko, K., Vadrevu, K. P., Tran, V. T., & Justice, C. (2018). Mapping double and single crop paddy rice with Sentinel-1A at varying spatial scales and polarizations in Hanoi, Vietnam. *IEEE Journal of Selected Topics in Applied Earth Observations and Remote Sensing*, *11*(2), 498–512. doi:10.1109/JSTARS.2017.2784784 PMID:30151066

Mather, P., & Tso, B. (2016). *Classification methods for remotely sensed data*. CRC press. doi:10.1201/9781420090741

McNairn, H., Champagne, C., Shang, J., Holmstrom, D., & Reichert, G. (2009). Integration of optical and Synthetic Aperture Radar (SAR) imagery for delivering operational annual crop inventories. *ISPRS Journal of Photogrammetry and Remote Sensing*, *64*(5), 434–449. doi:10.1016/j.isprsjprs.2008.07.006

McNairn, H., Shang, J., Champagne, C., & Jiao, X. (2009, July). TerraSAR-X and RADARSAT-2 for crop classification and acreage estimation. In *2009 IEEE International Geoscience and Remote Sensing Symposium* (Vol. 2, pp. II-898). IEEE. 10.1109/IGARSS.2009.5418243

Moran, M. S., Vidal, A., Troufleau, D., Qi, J., Clarke, T. R., Pinter, P. J. Jr, ... Neale, C. M. U. (1997). Combining multifrequency microwave and optical data for crop management. *Remote Sensing of Environment*, *61*(1), 96–109. doi:10.1016/S0034-4257(96)00243-X

Nasirzadehdizaji, R., Balik Sanli, F., Abdikan, S., Cakir, Z., Sekertekin, A., & Ustuner, M. (2019). Sensitivity Analysis of Multi-Temporal Sentinel-1 SAR Parameters to Crop Height and Canopy Coverage. *Applied Sciences (Basel, Switzerland)*, *9*(4), 655. doi:10.3390/app9040655

Navale, A., & Haldar, D. (n.d.). Evaluation of machine learning algorithms to Sentinel SAR data. *Spatial Information Research*, 1-11. Doi:10.100741324-019-00296-8

Patel, P., & Srivastava, H. S. (2007, May). Polarimetric SAR classification using physical based scattering mechanism: comparative evaluation of L and P bands. In Conference of Joint Experiment Project towards microwave Remote Sensing Data Utilisation, Ahmedabad (pp. 15-16). Academic Press.

Patel, P., & Srivastava, H. S. (2013). RADARSAT-2 announcement of opportunity project on soil moisture, surface roughness and vegetation parameter retrieval using SAR polarimetry. SAC/EPSA/MPSG/CVD/TDP R&D/01/13, SOAR International Closing and Reporting–2013.

Pingali, P., Aiyar, A., Abraham, M., & Rahman, A. (2019). *Transforming food systems for a rising India*. Springer. doi:10.1007/978-3-030-14409-8

Sharma, P. K., Kumar, D., Srivastava, H. S., Patel, P., & Sivasankar, T. (2019). Soil moisture retrieval under wheat crop using RISAT-1 hybrid polarimetric SAR data. *Journal of Agrometeorology*, *21*(1), 58–62.

Singh, D., & Kathpalia, A. (2007). An efficient modeling with GA approach to retrieve soil texture, moisture and roughness from ERS-2 SAR data. *Progress In Electromagnetics Research*, *77*, 121–136. doi:10.2528/PIER07071803

Sivasankar, T., Kumar, D., Srivastava, H. S., & Patel, P. (2018). Advances in radar remote sensing of agricultural crops: A review. *International Journal on Advanced Science. Engineering and Information Technology*, *8*(4), 1126–1137.

Sivasankar, T., Kumar, D., Srivastava, H. S., & Patel, P. (2020). Wheat leaf area index retrieval using RISAT-1 hybrid polarized SAR data. *Geocarto International*, *35*(8), 905–915. doi:10.1080/10106049.2019.1566404

Skriver, H., Mattia, F., Satalino, G., Balenzano, A., Pauwels, V. R., Verhoest, N. E., & Davidson, M. (2011). Crop classification using short-revisit multitemporal SAR data. *IEEE Journal of Selected Topics in Applied Earth Observations and Remote Sensing*, *4*(2), 423–431. doi:10.1109/JSTARS.2011.2106198

Sonobe, R., Tani, H., Wang, X., Kobayashi, N., & Shimamura, H. (2014a). Random forest classification of crop type using multi-temporal TerraSAR-X dual-polarimetric data. *Remote Sensing Letters*, *5*(2), 157–164.

Sonobe, R., Tani, H., Wang, X., Kobayashi, N., & Shimamura, H. (2014b). Parameter tuning in the support vector machine and random forest and their performances in cross-and same-year crop classification using TerraSAR-X. *International Journal of Remote Sensing*, *35*(23), 7898–7909. doi:10.1080/01431161.2014.978038

Srivastava, H. S., Patel, P., & Navalgund, R. R. (2006). How far SAR has fulfilled its expectation for soil moisture retrieval. In *Microwave Remote Sensing of the Atmosphere and Environment V* (Vol. 6410, p. 641001). International Society for Optics and Photonics.

Srivastava, H. S., Patel, P., Navalgund, R. R., & Sharma, Y. (2008). Retrieval of surface roughness using multi-polarized Envisat-1 ASAR data. *Geocarto International*, *23*(1), 67–77. doi:10.1080/10106040701538157

Sun, Y., Luo, J., Wu, T., Zhou, Y. N., Liu, H., Gao, L., ... Wang, L. (2019). Synchronous Response Analysis of Features for Remote Sensing Crop Classification Based on Optical and SAR Time-Series Data. *Sensors (Basel)*, *19*(19), 4227. doi:10.339019194227 PMID:31569430

Tan, C. P., Ewe, H. T., & Chuah, H. T. (2011). Agricultural crop-type classification of multi-polarization SAR images using a hybrid entropy decomposition and support vector machine technique. *International Journal of Remote Sensing*, *32*(22), 7057–7071. doi:10.1080/01431161.2011.613414

Tan, C. P., Koay, J. Y., Lim, K. S., Ewe, H. T., & Chuah, H. T. (2007). Classification of multi-temporal SAR images for rice crops using combined entropy decomposition and support vector machine technique. *Progress In Electromagnetics Research*, *71*, 19–39.

Thenkabail, P. S., Schull, M., & Turral, H. (2005). Ganges and Indus river basin land use/land cover (LULC) and irrigated area mapping using continuous streams of MODIS data. *Remote Sensing of Environment*, *95*(3), 317–341.

United Nations, Department of Economics and Social Affairs, Population Division. (2019). *World Population Prospects 2019: Highlights (ST/ESA/SER.A/423)*. Author.

Veloso, A., Mermoz, S., Bouvet, A., Le Toan, T., Planells, M., Dejoux, J. F., & Ceschia, E. (2017). Understanding the temporal behavior of crops using Sentinel-1 and Sentinel-2-like data for agricultural applications. *Remote Sensing of Environment*, *199*, 415–426. doi:10.1016/j.rse.2017.07.015

Wang, D., Lin, H., Chen, J., Zhang, Y., & Zeng, Q. (2010). Application of multi-temporal ENVISAT ASAR data to agricultural area mapping in the Pearl River Delta. *International Journal of Remote Sensing*, *31*(6), 1555–1572. doi:10.1080/01431160903475258

Whelen, T., & Siqueira, P. (2018). Time-series classification of Sentinel-1 agricultural data over North Dakota. *Remote Sensing Letters*, *9*(5), 411–420. doi:10.1080/2150704X.2018.1430393

Wu, B., Gommes, R., Zhang, M., Zeng, H., Yan, N., Zou, W., Zheng, Y., Zhang, N., Chang, S., Xing, Q., & van Heijden, A. (2015). Global crop monitoring: A satellite-based hierarchical approach. *Remote Sensing*, *7*(4), 3907–3933. doi:10.3390/rs70403907

Wulder, M. A., Masek, J. G., Cohen, W. B., Loveland, T. R., & Woodcock, C. E. (2012). Opening the archive: How free data has enabled the science and monitoring promise of Landsat. *Remote Sensing of Environment*, *122*, 2–10. doi:10.1016/j.rse.2012.01.010

Xu, J., Li, Z., Tian, B., Huang, L., Chen, Q., & Fu, S. (2014). Polarimetric analysis of multi-temporal RADARSAT-2 SAR images for wheat monitoring and mapping. *International Journal of Remote Sensing*, *35*(10), 3840–3858. doi:10.1080/01431161.2014.919679

Xu, L., Zhang, H., Wang, C., Zhang, B., & Liu, M. (2019). Crop Classification Based on Temporal Information Using Sentinel-1 SAR Time-Series Data. *Remote Sensing*, *11*(1), 53. doi:10.3390/rs11010053

Yang, X. (2011). Parameterizing support vector machines for land cover classification. *Photogrammetric Engineering and Remote Sensing*, *77*(1), 27–37. doi:10.14358/PERS.77.1.27

Yu, W., Elleby, C., & Zobbe, H. (2015). Food security policies in India and China: Implications for national and global food security. *Food Security*, *7*(2), 405–414. doi:10.100712571-015-0432-2

Chapter 4
Geo-Spatial Technology for Land Resources Management in Nigeria

Ugonna Chimnonyerem Nkwunonwo
https://orcid.org/0000-0002-6944-0675
University of Nigeria, Enugu, Nigeria

ABSTRACT

There is little discussion regarding natural resources management issues in Nigeria, unlike many other places around the world, where such issues have made significant contributions to research growth. This is a troubling situation that complicates Nigeria's present need to address her increasing sequence of aggravations related to land cover modifications. Sustainable use of land, water, solid minerals, and forest is difficult and overwhelms local efforts. Demographic pressures and the corresponding need for developments and societal livability have a proclivity to overuse land resources and impact negatively on their present quality and future regeneration. Traditional and indigenous approaches are still the bases of natural resources management. The cumulative challenges of these issues with other prevalent anomalies are increasingly compromising Nigeria's land resources base. However, geo-spatial technology with its potential for policy and decision-support is being set forth to address these challenges and to fill the current gaps in knowledge of natural resources management.

DOI: 10.4018/978-1-7998-5027-4.ch004

INTRODUCTION

Of the natural resources in Nigeria, land, specifically soil, is over-stressed by several factors—human, climatic, ecological and biological. In the south-eastern part of Nigeria, gully erosion threatens the soil, leading to a loss of major nutrients for agricultural purposes. For the northern region, desert encroachment is fast taking over the vegetative characteristics and changing local hydrological design (Akpodiogaga-a & Odjugo, 2010). Within Lagos state, and most places within the south-western and north-central regions of Nigeria, flooding is a major problem creating ever more vulnerable areas and more water-logged soil which are so overwhelming for human liveability and other purposes. The cost of oil exploration on the land resources within the Niger delta cannot be more obvious. The vast majority of the land resources within the region are no longer useful for farming, fishing, and other socio-economic activities. Therefore, debates have increased over land and environmental management in the area. Nkwunonwo *et al.* (2019) argued that the soil is a fundamental factor in the Fulani-herdsmen-farmers conflicts that have become perennial and intractable in the northern part of Nigeria. Although there is no validation of the argument as yet, it is an undeniable fact that soil in the northern part of Nigeria is fast losing its potential to support animal grazing, forcing herdsmen within the region to roam about in search of grazing lands. This condition has caused and still causes tension for agriculture and other development purposes, but it is also a major issue of dialogue for research and environmental governance.

More than ever, we need research to come up with ways to assuage the difficulties with soil in Nigeria and to increase its availability for food security and other agro-based activities. Researchers in pedometry and soil science have done ample work, for example, Vanlauwe et al. (2005), Nkwunonwo & Okeke (2013) and Mashi & Shaibu (2018). However, those works lack the foundation of current natural resources management, which is sustainable development and environmental protection (example: Sharma & Shivakoti, 2017). A realistic management of land resources should highlight a culture that affords land resources the opportunity to contribute towards achieving man's existential purpose. This culture also embraces the ability to address environmental issues such as erosion, flooding, gas flaring, pollution, urban heat and noise, whilst maximizing the usefulness of land resources. It is an integrated culture in the sense that it finds expression in many other fields of knowledge involving the soil.

The UNDP-Nigeria collaboration (*Agenda 21*), which focused on the institutional aspects of sustainable development based on five core areas: (1) Integrated decision making, (2) major groups, (3) science, (4) information and (5) international law, has made some strides at advancing the course of natural resources management in Nigeria. In view of land resources management, one critical achievement of this collaboration is the enactment of a national policy on the environment. Land use, soil conservations, wildlife are among the fourteen vital areas of Nigeria's economy where the achievement of sustainable development goals (SDGs) is of critical importance. Despite its robustness and dynamic nature, inadequate financing constrains institutional framework within the overall gambit of this collaboration, while the mode of its operation is weak since it is mostly a top-down approach with little participation from the various target communities (Kolstad & Søreide, 2009; Ene, 2011; Sala-i-Martin & Subramanian, 2013). Evidence from extant research shows that little attention is being given to scientific and technical expertise, which is a major propeller in aligning local land management objectives to the general best practice of natural resources management (Areola & Ofomata, 1978; Mmom & Arokoyu, 2010; Thi *et al.*, 2015). So far, incorporating modern scientific approaches and methodologies into natural resources management in Nigeria has been far from being fully actualized. This creates a major loophole in practice of land resources management in Nigeria and raises a major concern regarding how to raise the awareness of current techniques of managing natural resources in Nigeria and how to optimize the increasing geo-spatial technology for sustainable management of land resources within the country.

In view of these concerns, current research within Nigeria's context should primarily focus on three key points: (1) to investigate and highlight the strength and weaknesses of traditional natural resources management, and to suggest ways of improving the current practices, (2) to identify the specific gaps in knowledge of land resources management in Nigeria and to suggest ways to bridge those gaps, (3) to introduce geo-spatial technology in such a way to ease a swift migration to best practices in land resources management. To address these points, this chapter presents a systematic review of natural resources management in Nigeria. It draws from best practices in natural resources management in a way to identify research direction for effective land resources management in Nigeria. Finally, it discusses how the potentials and prospects of emerging geo-spatial technology can substantially benefit the sustainable management of Nigeria's land resources.

BACKGROUND

Geo-spatial technology is an emerging scientific concept that embodies a set of tools to access, manipulate, store, distribute, model and visualize geographic information or spatially referenced data (or spatial data). This is just the data that have location attributes (longitude, latitudes or any geocentric properties) with examples that include satellite images, land surveying parcels, coordinate positions of points on and above the earth's surface. With geo-spatial technology, rapid resolution of critical issues and informed decisions on several earth-fixed concepts are quick and simple. The key features of geo-spatial technology are conversion of spatial data to useful information, mapping (design of logically relevant, and aesthetically viable static and interactive maps that will serve a myriad of human applications), database creation, decision-support and policy design, and modelling and analyses of spatial data. Geo-spatial technology finds expression in everyday life as it brings to bear its various tools to query events, actions and unexpected life's outcomes for spatial components such as '*Location*', '*pattern*', '*Trend*', '*Condition*', and '*Implications*'. For example, in an accident situation (shown in Figure 1 below), geo-spatial technology can notify the police or an emergency team where the accident happened (*location*), the frequency of accidents occurring at this location (*pattern*), the number of such accidents within the last six months (*trend*), how far is the location from the nearest response team (let's say within three minutes) (*condition*), and given the current traffic, how long it will take for one response team to arrive this location (*implication*). Regarding land resources management, these geo-spatial identities may take a different form and context, although the relevance of geo-spatial technology in exploring and assessing the whole spectrum of land resources—dimension, topological relationships, present and future valuations, developmental impacts, risk factors, etc., — may be key to effective management of land resources.

Within the academy of geo-spatial technology, there are three major components: geographic information system (GIS), remote sensing (RS) and global positioning system (GPS). Some literature might consider software as a service (SaaS) and several database infrastructures including spatial data geo-portals and open data cube (ODC), which specifically supports analysis-ready earth observation datasets. These developments are crucial responses to the growing advancement in software and hardware development and the need to meet the challenges of lack of quality geo-spatial data in many localities. There are now several research developments which show the increasing availability of this database infrastructure in such places as Australia, Colombia and Switzerland (refer to: Ariza-Porras *et al.,* 2017; Giuliani *et al.,* 2017; Lewis *et al.,* 2017; Dong *et al.,* 2019). However, this chapter will devote its discussions to the three major components of geo-spatial technology and how they can influence land resources management within the context of Nigeria.

1. Geographic Information System (GIS)

In fact, GIS which is the most fundamental and inclusive part of geo-spatial technology, comprises a set of hardware, software, people, and methods for access, storage, analysis and display of spatial data. Roger Tomlinson (1933 - 2014) in 1968 became the first person to mention GIS when he proposed a solution based on extensive computerized spatial data, to design a land map covering one million

Figure 1. A conceptualised accident scene around Elm Groove, Southsea, Portsmouth, UK
Source: Drawn by author

square miles, intended to manage farmland, wildlife, forestry, and to explore areas suitable for tourism development. Since then, GIS has grown into a major industry that deals with education, government, urbanism, tourism and several important business initiatives. One of the persistent ambiguities of GIS which has embroiled the scientific communities within the past few decades, is the conclusion on whether GIS is a more of a tool or mere science. Many front-line studies, especially those of Wright *et al.* (1997), Schuurman (2000), Pavloskaya (2006) and Martin & Wing (2007) have attempted to clarify this issue by presenting GIS in a way that makes it relevant in both paradigms—tool and science, but with a caveat that enables the GIS technology to ease freely into the world of practice and academy of science.

Amidst these discussions, it is increasingly being acknowledged that GIS has assumed an almost irreplaceable position in everyday life and so might well debunk any debates pertaining to its scientific relevance and disposition. In their highly insightful book, *'Introduction to GIS'*, Ian *et al.* (2011) argue that, although it is not easily perceptible, people engage with GIS every day, especially when asking and answering local questions which define their routine and daily experiences. This is a fun feature in GIS as everything within the context of human activities seems serendipitously organised in such a way that follows a special pathway of spatial thinking and behaviour. Thus, GIS can serve a key purpose in land resources management. As discussed later on, several land use and land cover analyses, change detection and mapping in Nigeria have sufficiently applied GIS, but scantly in studies involving restoration of soil fertility and land ownership debates. This is a major gap in knowledge of land resources management in Nigeria.

2. Remote Sensing

Remote sensing, as the name implies, is the acquisition of information about an earth feature without contacting the feature. This definition, by every means, represents one of the most simplified ways to communicate the object of remote sensing to a layman. But from grounded theoretical perspectives that suffice academic purpose, the layman's illustrative idea of remote sensing is abjectly insufficient, since it does not capture the entire paraphernalia of remote sensing technology including its payload which is the key component that determines the geometric and radiometric characteristic of the imaged earth feature. Invariably, a small chapter of a book cannot exhaustively discuss the vast theory of remote sensing. Refer to Congalton & Greem (2002); Sabins (2007); Lillesand *et al.* (2015); Islam *et al.* (2018) for detailed discussions and a full suite of satellites available for remote sensing data capture. To wade through this epistemological limitation in a cursory description of the subject of remote sensing and to foster a more relevant theoretical premise that affords a sound linking of geo-spatial technology with land resources management, this chapter will consider adopting a favourable definition. Campbell & Wynne (2011) summarized remote sensing:

remote sensing is the practice of deriving information about the earth and water surfaces using images gained from an overhead perspective using electromagnetic radiation in one or more regions of the electromagnetic spectrum, reflected or emitted from the earth's surface.

The authors' concept of remote sensing in the foregoing proposition potentially offers a depth of understanding into the place of remote sensing in the overall built-up of geo-spatial technology. From this concept, remote sensing traverses through three inalienable contact points—physical object, sensor data, information extraction—before its application to a specific earth-based activity. This is the basic operational concept of remote sensing which explains the behaviour of electromagnetic radiation from sunlight as it passes through various layers of the atmosphere, interacts with earth's features and then received by sensors on board a remote sensing platform (refer to figure 2). The concepts of electromagnetic radiation (EMR) and electromagnetic spectrum (EMS) are very useful to remote sensing as they determine the nature and quality of geo-spatial data being imaged. The variety of geo-spatial data as of today resonates well with the full potentials of remote sensing systems and the EMRs they support (Villmann *et al.*, 2008 and Huang *et al.*, 2018). The remote sensing systems

Figure 2. The operational procedure of remote sensing technology
Source: Drawn by author

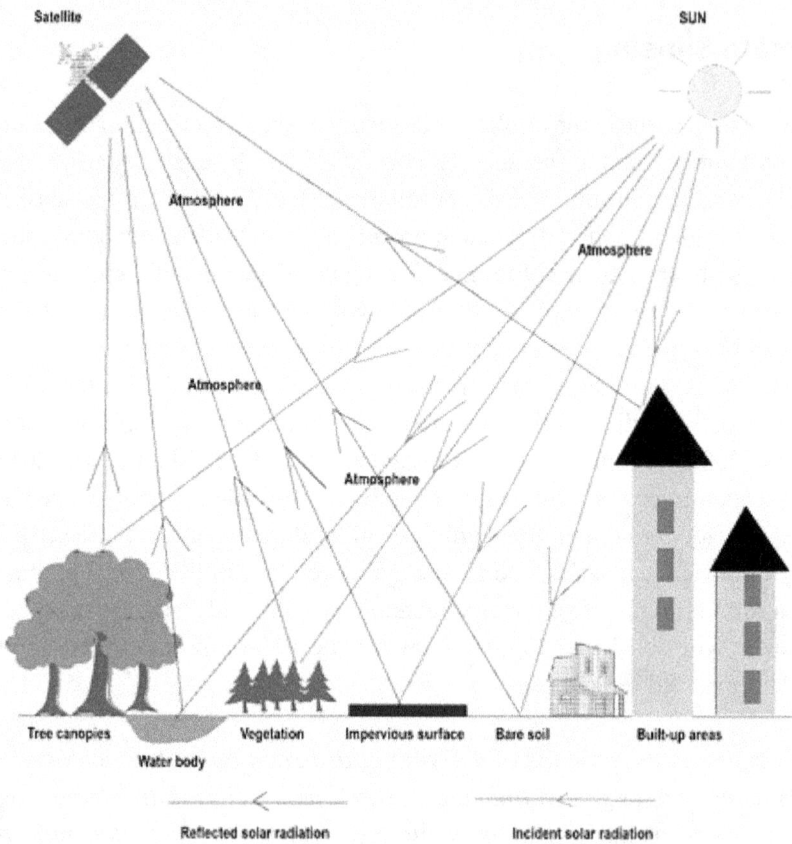

record these geo-spatial datasets as digital numbers (DNs) and they represent earth features which have temporal and spatial identities needed for monitoring, modelling and predicting changes within land and water surfaces. This is of importance to land resources management in which changes and impacts of human activities are major considerations.

3. Global Positioning System (GPS)

The global positioning system (GPS) is the world's largest and most accessible navigational satellite system. Anyone with a GPS receiver can access absolute positions (latitude and longitudes) and real-time information of anywhere on the surface of the earth where there is unobstructed coverage to four or more GPS satellites. This makes the system an everyday necessity in business, military, environmental systems, tracking and surveillance. In land resources management, the use of GPS facilitates property survey and mapping for boundary demarcation and the development of efficient cadastre towards a functional land tenure system which is a serious issue in Nigeria. GPS is the property of the US government, which it has operated through the US Air Force since 1973. The operational procedure of GPS is beyond this chapter. We can find details of such procedures and how to improve GPS measurement accuracies in Parkinson & Gilbert (1983), El-Rabbany (2002) and Wing *et al.* (2005). GPS operates based on time and position of known GPS satellites that carry stable atomic clocks that interact with one another and with the ground clock. Radio signals transmitted continuously by GPS satellites are used by the GPS receivers to solve complex equations to compute absolute ground positions. The accuracy specification of GPS signals, which is too low for most precise monitoring and modelling requiring geodetic precisions is a major impasse with the system. However, this is being improved based on other current techniques such as the use of Relative Kinematic Positioning (RKP) which can resolve GPS signals to less than 10cm accuracy (Bisnath & Langley, 2002).

4. Land Resources Management

Land resources management (LRM) represents as a strategic and organizational approach that applies to all land-related projects and initiatives to ensure the sustainability of land assets and gains. These can also accommodate a variety of academic discussions, political debates, communiqué, legislative, policy, investment, community participation and institutional issues. Current discussions highlight the inter-disciplinary nature of LRM and why it is important to use the best practice in the management of land resources. Within natural resources management and environmental science research, LRM highlights the importance of land use and land cover analyses, urban planning and renewal, and nature and biodiversity conservation (Akinyemi, 2017; Sfeir-Younis, 2019). Other procedures are assessment of ecosystem quality, coastal area management, land suitability evaluation, forest management policy and laws on land reform and tenure system, and all operations to ensure that the soil replenishes its lost nutrients (Pfister *et al.,* 2011; Barik, 2019; Chen *et al.,* 2019). Geo-spatial technology, through its components—GIS, remote sensing and GPS have been integral over the years to effective LRM, although a major persistent challenge is how to make quality geo-spatial datasets such as LiDAR (Light detection and ranging) more and more available to poorer countries. There is a flurry of overwhelming evidence to show that the potentials of geo-spatial technology is not widely available and therefore the discussion in this chapter is useful in science, not just as it advocates for the improvement and integration of geo-spatial technology to

LRM, but since it seeks for research directions and inputs towards the same for the sustainability of land resources in Nigeria. Again, these discussions aim to fill in the gaps on how to promote community involvement in LRM which is a key research topic now given the many inputs that a community can make within the context of strategic planning and management of land resources (Stoll-Kleemann *et al.*, 2010; Bowler *et al.*, 2012). In line with these discussions, this chapter devotes the next section to a brief review of LRM in Nigeria, to identify specific knowledge gaps that should underscore the importance of integrating geo-spatial technology with LRM.

REVIEW OF LAND RESOURCES AND MANAGEMNT IN NIGERIA

Land Resources in Nigeria

Areola & Ofomata (1978) is one of the most famous studies on land resources and its management in Nigeria. With ample evidence, the study provided some crucial insights into Nigeria's land resources and some key factors in its deterioration and general failures. Interestingly, Nigeria's land resources reflect a wealth of a unique character and beauty of the sub-Saharan region. Within a land area of approximately 923000 square kilometres, the land resources in the country encompass a rich agricultural land, based on classical geological backdrop underlying soils with great potential for crop production and animal farming (Gefu & Gilles, 1990; Thapa & Yila, 2012). The country is rich in solid minerals and crude oil, which has remained the mainstay of Nigeria's economy. There are mangrove swamps, lush vegetation and forestland with trees that serve a unique purpose of wild game reserve, timber production and textile manufacturing. There is an interesting landscape with hills, valleys, plateaus and extant plain lands that resonate with an excellent creative appeal. The water body is rich with many marine life-forms that make up a large part of the word's extensive biodiversity. The studies by Ewak (2010) and Ogunjinmi *et al.* (2014) recently considered protected areas or areas set aside for nature's conservation as part of the land resources in Nigeria. A few years ago, Asadu *et al.* (2004) reviewed the uses of land resources in Nigeria and found that the use of Nigeria's land resources revolve around agriculture (crop and animal production, industry: both oil and non-oil) and social needs such as the provision of infrastructure. Although Asadu's and his team's work are over a decade and a half old, it still provides a credible insight into the major operationalization of land resources in Nigeria, since no other study has presented as much comprehensive knowledge of the status of land resources in Nigeria.

Remarkably, climate change, rapid population growth and urbanization are key issues in Nigeria's land resources debates (Obioha, 2009; Madu, 2012). The general perspective from extant research on Nigeria's natural resources is that of a challenge for Nigerians, not necessarily because of any uncertainties regarding the ability of the country's natural resources to meet current needs of the people and future expectations, but from a lack of principles of altruistic and effective management strategies (Areola & Ofomata, 1978; Collier, 2010; Sala-i-Martin & Subramanian, 2013). In particular, the authors' view in Sala-i-Martin & Subramanian (2013) on natural resources in Nigeria suggests that the administration of land resources do not foster the development of principles of equality of human beings in Nigerian society. Even though Nigeria is abundantly endowed with a variety of resources, the level of poverty and lack of basic needs in a country that is moving towards a more sustainable future, explains the negative. This situation easily evokes a strong feeling of disillusionment with any incumbent political leadership, and a tendency to suspect that the myth: *'natural resources are a curse'* could be a fact and a reality that Nigerians have to grapple with, considering the many challenges and conflicts subject to land resources in particular. This is not a peculiar issue to Nigeria, but in many countries in the Middle East, Asia and the Caribbean where the economy of natural resources is a problem of justice, communal peace and social-economic development. However, the peculiarity of Nigeria's situation is that it is compounded by the ethnic and religious biases that has defined the country.

In addressing this unfortunate and extreme situation due to a lack of good resources management, it will make a good theoretical road map to evolve a new management paradigm that ensures that land resources' wealth in Nigeria serve its purpose based on equality and in consideration to individuals and places that have been largely affected by routine exploitation of land resources. So, if effective land resources management is key to a sustainable development in Nigeria, then best practices in resource management should be adopted which is why this study promotes geos-spatial technology that drives contemporary strategies in natural resources management.

Climate Change, Nigeria's Land Resources and Geo-Spatial Technology

Climate change refers to variations in the average weather pattern, caused by inevitable geophysical processes, such as tectonic movements, volcanic eruptions and El Niño oscillations effects, all of which make up the universe. The key indicators of climate change are high earth surface temperature, increased frequency and intensity of rainfall events, melting of the glaciers, and rising sea levels. Current academic debates on climate change have intensified following the fact that these indicators associate with extreme weather events such as wildfires flooding, heat waves, earthquake, erosion, landslide and droughts (Lenderink & Van Meijgaard, 2008; McMichael *et al.,* 2013). World over, the frequency of occurrence of these events, and the intensity of their impacts are increasing and natural resources are being destroyed. With Nigeria, there is a vital link between climate change and natural resources, which, if explored, will help to evolve a crucial discussion in natural resources management (Ebhuoma & Simatele, 2019). Although a chapter of this nature may not pride itself of any exhaustive discussions on the real impacts of climate change on natural resources, it is important to spotlight important issues that prevails when one considers natural resources management alongside climate change. Many studies have shown that the socio-economic impacts of climate change measured across geographical and temporal scales are significant (Adger *et al.,* 2013; Nkwunonwo, 2015; Nkwunonwo *et al.,* 2016). There is a consensus that climate change affects DCs disproportionately, even though they are not one of the largest contributors to climate change (Storma, 2019). Thus, how to adapt communities in the DCs to the unpredictable conditions of climate change through current risk reduction techniques is a key topic of increased interest in climate change research (e.g. Gillard *et al.,* 2016; Tàbara *et al.,* 2019).

The concept of climate change is rapidly changing the overall ideology of resources management. Ample research has shown that climate change is a major threat to social and environmental factors (Goklany, 2012; Tai *et al.,* 2014; Papworth *et al.,* 2015; Tripathi *et al.,* 2016). The design of natural resources management takes cognizance of the fact that extreme weather events that result from climate change have huge impacts on land, water and the broader context of biodiversity (Virkkala *et al.,* 2019). However, a climate-based approach to natural resources management can be subject to complications for some reasons. Although it can have a real global impact, undue politics with scientific, economic and juridical limitations, all of which often straddle issues of strategic and community-based management ideologies, can complicate climate-based natural resources management for the DCs. Land resources interlocked within the general framework of climate change implies that the goal of climate-based natural resources management must include building the coping capacity and resilience of environmental systems, bearing in mind the pervasiveness of climate change and the economic significance of mitigating large-scale social and environmental impacts. In view of these goals, and with particular reference to Nigeria, current research that focuses on the DCs must promote geo-spatial technology with its necessary scientific adaptations to mediate climate change effects in land resources and its management.

Land Resources Management in Nigeria

Proper management of land resources in Nigeria has existed for over four decades. Initially, it was a response to the need to explore the richness of the country's land resources. We can assume real evidence of this from the reconnaissance soil mapping project which the Department of Agricultural Land Resources (DALR) carried out in 1990. The aim of the project was to identify the soil base of Nigeria, its distribution across various regions and to prepare a map that delineates its constituent units, chemical and physical properties. Within the same period, the geological structure of Nigeria and her vegetation and physical landforms were also investigated. Refer to Vine (1949), Olorunniwo *et al.* (1987), FDALR (1990) and Nkwunonwo & Okeke (2013) for details of these investigative studies. With the fast-growing human population and emergence of cities across the country, it became important to relate the original focus of land resources management with sustainable human development and several global agenda. In view of these objectives, there is a flurry of important discussions which this chapter has given special credence to. These discussions, summarized in Table 1, highlight three important areas of land resources management: urban land use planning, land degradation and soil nutrients conservation, and land ownership and rights (which forms the basis of land use Act of 1978 and Land reform policy of 1999. I will discuss this as a separate section). It will be a non-exhaustive attempt to review the vast literature on land resources management in Nigeria within a chapter of a book. Though, many of the studies are very similar in their approach and objectives. Therefore, this chapter reviewed only the studies that are true representatives of the extant research—that considered Nigeria as a whole and presented more overarching discussions within their topical areas.

One of the key issues arising from land resources management in Nigeria is the lack of conservation culture. Areola (1984) discussed this issue extensively from the point of view of political ramifications. Conservation is a major focus of land resources management, but Nigeria's land resources management which shows a clear derail of politics within the country, represents Africa's unique perspective of natural resources management (Eze & Neba, 2006; Chigbu, 2019; Mbuyi, 2019). Nigeria particularly, political agenda is fashioned to impress the elites at the expense of obvious needful citizens. This creates an atmosphere of the biblical '*whited sepulchre*' which is a simile of the true situation in Nigeria. There are white elephant projects and policies which have continued to elude a true land management within the country and reducing the land available for sustainable economic development within the country (Ehiorobo, 2018).). As a result, there is a constant and enormous pressure on land resources leading to land use/land cover modifications, and land degradation now experienced everywhere in the country (Arowolo *et al.,* 2018). Apart from agricultural food production which takes about 65% of the land area of Nigeria (Asadu *et al.,* 2004), extensive urban growth, mineral exploitation, industrialization and the need for tourism development, Nigeria's land availability is also been threatened by natural hazards, particularly erosions, flooding, and desert encroachment—issues that have escalated too much unresolved conflicts on the land tenure regime and holding pattern in Nigeria (Famoriyo, 1984; Otubu, 2018).

Table 1 shows that over the past four decades at least, research has developed in Nigeria with increasing topical discussions which involves several land resources management approaches to address an almost perennial pressure and threats on land resources. Rapid population growth fraught with competitive wants and interest is a major challenge that has created an asymmetric situation in the land tenure system whereby the right of land becomes a legacy of the rich, leaving the poor without access to real land for any socio-economic purpose. With a view to the land management techniques in Nigeria—the key ones are shown in Table 1, there is mainly empirical and theoretical discussions, qualitative, quantitative and mixed methods research involving both secondary and primary survey data and analyses of land use and land cover modifications, urban and regional land use planning, soil management and degradation issues. Despite the prospects of these research inputs, several studies have argued that land resources management in Nigeria need further development to integrate new scientific information and strategies, and to address environmental and socio-economic objectives (Nkwunonwo & Okeke, 2013; Essien & Cyrus, 2019). It is crucial to mention the study by Babalola & Uyi (2019) which recently implemented a web-based platform for land information management for Akure Nigeria. Equally, Babalola & Hull (2019) considered the use of a domain model of social tenure to record land rights in Ekiti State, Nigeria.

These studies demonstrate the critical importance of modern science in land resources management for Nigeria. A fascinating study by Thapa & Yila (2012) which was focused on soil conservation and restoration of fertility. The authors examined a wide range of land management practices, including application of fertilizers, mulching and agroforestry, which have been adopted by Nigerian farmers especially in the Jos area to control degradation of land. They examined thirteen land management practices based on mainly qualitative primary information collected through a questionnaire survey of 150 farm households and group discussions held during to identify their relative popularity in the study area. The results of the analysis revealed four practices application of chemical fertilizers and farm yard manure, inter-cropping and crop rotation were highly popular. These practices need little technical skills, show positive short-term benefits and short establishment time, suggesting that the benefits of increased productivity can be obtained quickly. This shows the apparent lack of modern science and technology in most land management practices in Nigeria, and points to a major gap between what is obtainable in best practices and what applies to Nigeria.

Table 1. Summary of reviewed studies on land resources management in Nigeria

S/No	Author(s)	Study	Focus	Technique
1.	Famoriyo (1984).	Administration of land allocation in Nigeria.	Land ownership and rights	Theoretical discussions
2.	Thapa & Yila (2012).	Farmers' land management practices and status of agricultural land in the Jos Plateau, Nigeria.	Soil management	Qualitative primary data research
3.	Asadu *et al.* (2004).	Land use and soil management situations in Nigeria: an analytical review of changes.	Land use planning Soil management	A review of secondary data
4.	Fasona & Omojola (2009).	Land cover change and land degradation in parts of the southwest coast of Nigeria	Land use planning Soil management	Geo-spatial technology
5.	Fabusoro *et al.* (2008)	Land rights regimes in southwest Nigeria: implications for land access and livelihoods security of settled Fulani agropastoralists.	Land ownership and rights	Qualitative primary data research
6.	Mengistu & Salami (2007).	Application of remote sensing and GIS inland use/land cover mapping and change detection in a part of south western Nigeria	Land use planning Soil management	Geo-spatial technology
7.	Omotayo & Musa. (1999).	The role of indigenous land classification and management practices in sustaining land use system in the semi-arid zone of Nigeria.	Land use planning Soil management	Qualitative primary data research
8.	Osemeobo (1991).	Effects of common property resource utilization on wildlife conservation in Nigeria	Land use planning Soil management	Qualitative primary data research
9.	Odemerho. (1992).	Land degradation impacts on tropical agricultural basins and their management: the Nigerian example.	Land degradation	Review and secondary data research
10.	Olokeogun *et al.* (2014).	Application of remote sensing and GIS in land use/land cover mapping and change detection in Shasha forest reserve, Nigeria.	Land use planning Soil management	Geo-spatial technology
11.	Adedeji & Adeofun, (2014).	Spatial Pattern of Land Cover Change Using Remotely Sensed Imagery and GIS: A Case Study of Omo-Shasha-Oluwa Forest Reserve, SW Nigeria	Land use planning	Geo-spatial technology
12.	Mabogunje (2010)	Land Reform in Nigeria: progress, problems & prospects.	Land ownership and rights	Theoretical discussions
13.	Babalola & Uyi (2019)	A web-based land information system, tool for online land administration in Akure Nigeria.	Land information management	Geo-spatial technology

For example, advanced land resources management in the United States, United Kingdom and several of the European countries apply a variety of integrative approaches and intellectual investment (e.g. adaptive predictive frameworks, and advanced soil mapping modelling) that enable them to meet the environmental and socio-economic needs of the wider populations (Herrick *et al.,* 2006; Haygarth & Ritz, (2009).In comparison, Nigeria's land governance lacks many of these mechanisms, whilst management of land resources is reactive, ad hoc and arguably represents a travesty of the actual aim of land resources management as enshrined in the United Nations charter of sustainable land management goal. Amidst the conundrum of land management in Nigeria, most soil scientist and core experts in land administration, who are also appointees of political organizations often find that commitment and duty to professional standards conflicts with their loyalty to the political organizations that appointed them. In which case, a situation of he that pays the pipe dictates the tome applies so they can achieve no goal beyond what the incumbent political leadership defines in any aspects of the leadership framework, including land administration. This is a real limitation to address if land resources management in Nigeria must take a turn to the batter. Such a situation is beyond the interest of this paper, however, the purpose of integrated frameworks that incorporate the principles of geo-spatial technology into land resources management is highlighted. This framework relies on and optimizes the emerging geo-spatial infrastructure and tools including GIS, remote sensing and GPS. The benefit of this framework will come from stakeholder engagement to address the major issues in Nigeria's land resources administration and management.

Land Use Act (1978) and Land Reform (1999)

Land policy and legislation provide a viable way to understand the current knowledge and realities in the management of land resources. With Nigeria, the two legislative tools that serve this important purpose are the land use Act of 1978 and land reform policy of 1999. Mabogunje (2009; 2010) presented some important discussion regarding the implications of these two legislative tools on land resources and management in Nigeria. Professor Mabogunje who in 1999, was the chairman of the presidential technical committee on land reform has published first hand on the critical issues, dialectics and milestones in implementing Nigeria's land reform. Other studies pertinent to these issues are Chidebelu *et al.* (10980) and Babalola *et al.* (2015). Amidst the cloud of debates, land ownership policy in Nigeria from the land use Act of 1978 has been clad with problems which made it difficult to achieve a stable state of land ownership rights for all Nigerians who need them. The land use Act conferred the ownership of all lands to the state governor and stipulates certain rules and limitations on individual ownership of lands. Although the Act was a unification of the multiplicity of land ownership rights and policies in Nigeria prior to and a little longer after the British colonization of the country, it did not recognize the traditional and customary basis of land ownership which resonated well with the peace, unity and progress of the country. The Act was seen as a major clog in the when of progress of Nigeria primarily since it brought about serious delays in securing rights from governors, and according to Mabogunje (2010), the Act was an abrogation of the right of land ownership by many Nigerians especially in the southern half of the country, while its nationalization by government remained inconsistent with the democratic practices and operations of free market economic system. This was the major challenge of the Act, which over many decades, remained the clamour and agitations of many Nigerians to evolve a better and more strategic land ownership basis for the country.

Land reform policy of 1999 which came on board with the birth of Nigeria's nascent democracy, was a paradigm shift from the Land use Act to a more integrated policy on land right and ownership which recognizes the importance of customary and ancestral lands and the uniqueness of people's primordial cultures and belief systems in land ownership. Land reform, particularly within African setting, is about how to induce or encourage peasant land-owning communities to confer individual property rights, rather than communal tenure, to the present tillers of the land. The land reform is underscored by this aim of confirming the property rights on most of the population, whose land use pattern holds a vitally potential stake in Nigeria's most important natural resource development. The policy also supported geo-spatial technology and training of para-surveyors, and farmers on the applications of these modern science in land administration—not only the plans of their holdings but also the many implications of a land title for their economic advancement. According to Mabogunje (2010), Nigeria's land reform of 1999 was aimed at:

...to determine individuals' 'possessory rights' using best practices and most appropriate technology to determine the process of identification of locations and registration of title holdings," as a forerunner to a National Land Commission. To identify, demarcate, and title the landholdings of individuals, one of the committee's major challenges will be to create mechanisms that are co-ordinately implemented by state and local governments.

Besides what Mabogunje (2010) expected as a major challenge to the land reform – which was supported by Kasim & Agbola, (2018)—that there will be a slow regime in the policy's implementation in many places within Nigeria and that the practices of land use Act itself will be hard to overcome quickly, Jacobs (2002) on a post-structural premises queried the place of women in the land reform policy. This is due to the historical experience and current changes which has failed to adopt women's status on the right of ownership of Nigeria. This is a crucial point considering the struggle of rural women mostly widows whose land are often revoked arbitrarily.

GEOSPATIAL TECHNOLOGY, LAND RESOURCES MANAGEMENT IN NIGERIA FUTURE RESEARCH DIRECTIONS

The implications of LRM and land-based education in Nigeria suggest that a lot remains to be done. In all of this, integrating geo-spatial technology in the management of land resources is of such an importance that it is useful for this chapter to outline two areas of research importance and focus which will mediate the current gaps in knowledge of natural resources management in Nigeria.

1. It is beneficial to start various research that engages geospatial technology proposed in the original design of Nigeria's land reform policy. Non-implementation of this proposal is a major gap which must be addressed by ensuring that every land management activity henceforth adheres to such a rule. This will make it possible to adhere to science and technology enshrined in the best practices of natural resources management.
2. The second is the ability to make the geo-spatial technology a possibility for all land resources management operations, not just land use and land cover analyses and change detection. Researchers should evolve ways of adopting modern methodologies to Nigeria's context. The national ecological fund should be made available for researchers to access to advance research in this direction.

CONCLUSION

Land resources—which include soils, solid minerals, vegetation, urban space, geographical landforms, etc —are arguably the world's most important factors in socio-economic development. How to manage this natural endowment on man and his society raises serious issues, especially in societies with little or no advancement in science and research. Within the context of Nigerian, which is the focus of this treatise, the challenge of managing land resources is clear and shows obvious constraints in sustainable development. The current state of land right administration and land resources management is a snap of the right view of land management that is associated with best practices and experienced in other developed countries such as the United States, United Kingdom and other countries in Europe. This current situation in Nigeria is because of right wing political ideologies and poor knowledge of the land management system. Therefore, this paper is specifically about geo-spatial technology including at least geographic Information System (GIS), remote sensing and global positioning system (GPS) to approach the most effective resource management in Nigeria. It is based on the belief that the best results in land management for Nigeria can be achieved by combining geo-spatial technology and other advanced, state-of-the-art scientific techniques with the variety of land resources management solutions, not just land use land cover analyses which is the most focused aspect of Nigeria's system of land resources management.

REFERENCES

Adger, W. N., Barnett, J., Brown, K., Marshall, N., & O'brien, K. (2013). Cultural dimensions of climate change impacts and adaptation. *Nature Climate Change*, *3*(2), 112–117. doi:10.1038/nclimate1666

Agunbiade, M. E., Rajabifard, A., & Bennett, R. (2016). What scope for integrating land management policies, land administration processes and data infrastructures for housing production in Nigeria? *Journal of Housing and the Built Environment*, *31*(1), 51–68. doi:10.100710901-015-9443-8

Akinyemi, F. O. (2017). Land change in the central Albertine rift: Insights from analysis and mapping of land use-land cover change in north-western Rwanda. *Applied Geography (Sevenoaks, England)*, *87*, 127–138. doi:10.1016/j.apgeog.2017.07.016

Akpodiogaga-a, P., & Odjugo, O. (2010). General overview of climate change impacts in Nigeria. *Journal of Human Ecology (Delhi, India)*, *29*(1), 47–55. doi:1 0.1080/09709274.2010.11906248

Alao, J. S., & Shuaibu, R. B. (2013). Agroforestry practices and concepts in sustainable land use systems in Nigeria. *Journal of Horticulture and Forestry*, *5*(10), 156–159.

Areola, O. (1987). The political reality of conservation in Nigeria. *Conservation in Africa: people, policies and practice*, 277-292.

Areola, O., & Ofomata, G. E. K. (1978). Land use control and natural resource management in Nigeria. *GeoJournal*, *2*(5), 429–436. doi:10.1007/BF00156220

Ariza-Porras, C., Bravo, G., Villamizar, M., Moreno, A., Castro, H., Galindo, G., Cabera, E., Valbuena, S., & Lozano, P. (2017, September). CDCol: A geoscience data cube that meets Colombian needs. In *Advances in Computing. CCC: Colombian Conference on Computing. Communications in Computer and Information Science*, vol. 735 (pp. 87-99). Springer. 10.1007/978-3-319-66562-7_7

Arowolo, A. O., Deng, X., Olatunji, O. A., & Obayelu, A. E. (2018). Assessing changes in the value of ecosystem services in response to land-use/land-cover dynamics in Nigeria. *The Science of the Total Environment*, *636*, 597–609. doi:10.1016/j.scitotenv.2018.04.277 PMID:29723833

Asadu, C. L., Ezeaku, P. I., & Nnaji, G. U. (2004). Land use and soil management situations in Nigeria: An analytical review of changes. *Outlook on Agriculture*, *33*(1), 27–37. doi:10.5367/000000004322877755

Babalola, K. H., & Hull, S. A. (2019). Using a domain model of social tenure to record land rights: A Case Study of Itaji-Ekiti, Ekiti State, Nigeria. *South African Journal of Geomatics*, *8*(2), 221–237.

Babalola, S. O., Rahman, A. A., Choon, L. T., & Van Oosterom, P. J. M. (2015). Possibilities of land administration domain model (LADM) implementation in Nigeria. *ISPRS Annals of The Photogrammetry. Remote Sensing and Spatial Information Sciences*, *2*(2), 155.

Babalola, S. O., & Uyi, I. O. (2019). A web-based land information system, tool for online land administration in Akure Nigeria. International Archives of the Photogrammetry, Remote Sensing & Spatial Information Sciences. XLII-4/W16, 137-142. doi:10.5194/isprs-archives-XLII-4-W16-137-2019

Barik, R. K. (2019). CloudGanga: cloud computing based SDI model for ganga river basin management in India. In *Geospatial Intelligence: Concepts, Methodologies, Tools, and Applications* (pp. 278–297). IGI Global. doi:10.4018/978-1-5225-8054-6.ch013

Bisnath, S. B., & Langley, R. B. (2002). High-precision, kinematic positioning with a single GPS receiver. *Navigation*, *49*(3), 161–169. doi:10.1002/j.2161-4296.2002.tb00265.x

Bowler, D. E., Buyung-Ali, L. M., Healey, J. R., Jones, J. P., Knight, T. M., & Pullin, A. S. (2012). Does community forest management provide global environmental benefits and improve local welfare? *Frontiers in Ecology and the Environment*, *10*(1), 29–36. doi:10.1890/110040

Campbell, J. B., & Wynne, R. H. (2011). *Introduction to remote sensing*. Guilford Press.

Chen, C., Park, T., Wang, X., Piao, S., Xu, B., Chaturvedi, R. K., ... Tømmervik, H. (2019). China and India lead in greening of the world through land-use management. *Nature Sustainability*, *2*(2), 122.

Chidebelu, A. N., Ames, G. W., & Brown, E. E. (1980). Land use in Nigeria. *Food Policy*, *5*(4), 306–310. doi:10.1016/0306-9192(80)90059-7

Chigbu, U. E. (2019). Concept and Approach to Land Management Interventions for Rural Development in Africa. In *Geospatial Technologies for Effective Land Governance* (pp. 1–14). IGI Global. doi:10.4018/978-1-5225-5939-9.ch001

Collier, P. (2010). The political economy of natural resources. *Social Research*, *77*(4), 1105–1132.

Congalton, R. G., & Green, K. (2002). *Assessing the accuracy of remotely sensed data: principles and practices*. CRC Press.

Dong, J., Metternicht, G., Hostert, P., Fensholt, R., & Chowdhury, R. R. (2019). Remote sensing and geospatial technologies in support of a normative land system science: Status and prospects. *Current Opinion in Environmental Sustainability*, *38*, 44–52. doi:10.1016/j.cosust.2019.05.003

Ebhuoma, E. E., & Simatele, D. M. (2019). 'We know our Terrain': Indigenous knowledge preferred to scientific systems of weather forecasting in the Delta State of Nigeria. *Climate and Development*, *11*(2), 112–123. doi:10.1080/17565529.2017.1374239

Ehiorobo, A. O. (2018). Efficient Resource Allocation and Utilization: The Missing Link in Nigeria's Quest for Sustainable Development. *Economics and Business*, *32*(1), 264–275. doi:10.2478/eb-2018-0020

El-Rabbany, A. (2002). *Introduction to GPS: the global positioning system*. Artech house.

Enaruvbe, G. O., & Atafo, O. P. (2016). Analysis of deforestation pattern in the Niger Delta region of Nigeria. *Journal of Land Use Science*, *11*(1), 113–130. doi: 10.1080/1747423X.2014.965279

Eneh, O. C. (2011). Managing Nigeria's environment: The unresolved issues. *Journal of Environmental Science and Technology*, *4*(3), 250–263. doi:10.3923/jest.2011.250.263

Essien, E., & Cyrus, S. (2019). Detection of urban development in Uyo (Nigeria) using remote sensing. *Land (Basel)*, *8*(6), 102. doi:10.3390/land8060102

Ewak, J. O. (2010). Expanding mandate and corporate social responsibility in the management of National Parks and protected areas in Nigeria. *Journal of Agriculture and Environment for International Development*, *104*(1/2), 25–38.

Eze, E. B., & Neba, N. E. (2006). A conceptual framework for ecologically and socially sustainable land management and agricultural development in Africa. *International Journal of Environment and Sustainable Development*, *5*(3), 275–286. doi:10.1504/IJESD.2006.010897

Famoriyo, S. (1982). Impinging upon land resources management in Nigeria. In Efficient Use of Nigerian Land Resources (pp. 289-300). Academic Press.

Famoriyo, S. (1984). Administration of land allocation in Nigeria. *Land Use Policy*, *1*(3), 217–224. doi:10.1016/0264-8377(84)90065-6

Gefu, J. O., & Gilles, J. L. (1990). Pastoralists, ranchers and the state in Nigeria and North America: A comparative analysis. *Nomadic Peoples*, 34–50.

Gillard, R., Gouldson, A., Paavola, J., & Van Alstine, J. (2016). Transformational responses to climate change: Beyond a systems perspective of social change in mitigation and adaptation. *Wiley Interdisciplinary Reviews: Climate Change*, *7*(2), 251–265. doi:10.1002/wcc.384

Giuliani, G., Chatenoux, B., De Bono, A., Rodila, D., Richard, J. P., Allenbach, K., Dao, H., & Peduzzi, P. (2017). Building an earth observations data cube: Lessons learned from the swiss data cube (SDC) on generating analysis ready data (ARD). *Big Earth Data*, *1*(1-2), 100–117. doi:10.1080/20964471.2017.1398903

Goklany, I. M. (2012). Is climate change the number one threat to humanity? *Wiley Interdisciplinary Reviews: Climate Change*, *3*(6), 489–508. doi:10.1002/wcc.194

Haygarth, P. M., & Ritz, K. (2009). The future of soils and land use in the UK: Soil systems for the provision of land-based ecosystem services. *Land Use Policy*, *26*, S187–S197. doi:10.1016/j.landusepol.2009.09.016

Herrick, J. E., Bestelmeyer, B. T., Archer, S., Tugel, A. J., & Brown, J. R. (2006). An integrated framework for science-based arid land management. *Journal of Arid Environments*, *65*(2), 319–335. doi:10.1016/j.jaridenv.2005.09.003

Huang, Y., Chen, Z. X., Tao, Y. U., Huang, X. Z., & Gu, X. F. (2018). Agricultural remote sensing big data: Management and applications. *Journal of Integrative Agriculture*, *17*(9), 1915–1931. doi:10.1016/S2095-3119(17)61859-8

Ian, H., Cornelius, S., & Carver, S. (2010). *An introduction to GIS*. Pearson Education Limited.

Islam, K., Jashimuddin, M., Nath, B., & Nath, T. K. (2018). Land use classification and change detection by using multi-temporal remotely sensed imagery: The case of Chunati wildlife sanctuary, Bangladesh. *The Egyptian Journal of Remote Sensing and Space Sciences*, *21*(1), 37–47. doi:10.1016/j.ejrs.2016.12.005

Jacobs, S. (2002). Land reform: Still a goal worth pursuing for rural women? *Journal of International Development*, *14*(6), 887–898. doi:10.1002/jid.934

Kasim, O. F., & Agbola, S. B. (2018). Strategies and challenges of land reform in Nigeria. *Journal of Public Administration, 53*(2-1), 541-564.

Kolstad, I., & Søreide, T. (2009). Corruption in natural resource management: Implications for policy makers. *Resources Policy*, *34*(4), 214–226. doi:10.1016/j.resourpol.2009.05.001

Lenderink, G., & Van Meijgaard, E. (2008). Increase in hourly precipitation extremes beyond expectations from temperature changes. *Nature Geoscience*, *1*(8), 511–514. doi:10.1038/ngeo262

Lewis, A., Oliver, S., Lymburner, L., Evans, B., Wyborn, L., Mueller, N., Raevksi, G., Hooke, J., Woodcock, R., Sixsmith, J., Wu, W., Tan, P., Li, F., Killough, B., Minchin, S., Roberts, D., Ayers, D., Bala, B., Dwyer, J., ... Wang, L.-W. (2017). The Australian geoscience data cube-foundations and lessons learned. *Remote Sensing of Environment*, *202*, 276–292. doi:10.1016/j.rse.2017.03.015

Lillesand, T., Kiefer, R. W., & Chipman, J. (2015). *Remote sensing and image interpretation*. John Wiley & Sons.

Mabogunje, A. L. (2009). New land reform in Nigeria. *Science*, *325*(5942), 793.

Mabogunje, A. L. (2010, April). Land Reform in Nigeria: progress, problems & prospects. In *Annual Conference on Land Policy and Administration*. World Bank.

Madu, I. A. (2012). Spatial Impacts of Rural Population Pressure on Agricultural Land Use in Nigeria. *Applied Spatial Analysis and Policy*, *5*(2), 123–135. doi:10.100712061-011-9064-x

Martin, K. S., & Wing, J. (2007). The discourse and discipline of GIS. *Cartographica: The International Journal for Geographic Information and Geovisualization*, *42*(3), 235–248. doi:10.3138/carto.42.3.235

Mashi, S. A., & Shuaibu, H. S. (2018). People and sustainable land management: Assessment of stakeholders' knowledge of the nature of land use/cover change in Abuja, Nigeria. *GeoJournal*, *83*(3), 545–562. doi:10.100710708-017-9782-y

Matemilola, S., Adedeji, O. H., & Enoguanbhor, E. C. (2018). Land Use/Land Cover Change in Petroleum-Producing Regions of Nigeria. In The Political Ecology of Oil and Gas Activities in the Nigerian Aquatic Ecosystem (pp. 257-276). Academic Press.

Mbuyi, K. (2019). Kinshasa: problems of land management, infrastructure, and food supply. In African Cities in Crisis (pp. 148-175). Routledge.

McMichael, A. J. (2013). Globalization, climate change, and human health. *The New England Journal of Medicine*, *368*(14), 1335–1343. doi:10.1056/NEJMra1109341 PMID:23550671

Mmom, P. C., & Arokoyu, S. B. (2010). Mangrove forest depletion, biodiversity loss and traditional resources management practices in the Niger Delta, Nigeria. *Research Journal of Applied Sciences, Engineering and Technology*, *2*(1), 28–34.

Nkwunonwo, U. C. (2015). Digital mapping of soil chemical properties for erosion hazard management in Bayelsa state, Nigeria. *Chemical Science Review and Letters*, *4*(13), 209–217.

Nkwunonwo, U. C., & Okeke, F. I. (2013). GIS-based production of digital soil map for Nigeria. *Ethiopian Journal of Environmental Studies and Management*, *6*(5), 498–506. doi:10.4314/ejesm.v6i5.7

Nkwunonwo, U. C., Okeke, F. I., Chiemelu, E. N., & Ebinne, E. S. (2019). Geospatial Technology Potentials in Reawakening the Consciousness of Soil Distribution in Nigeria's North-Central Region and Mediating the Herdsmen-Farmers Conflicts. *Journal of Geoscience and Environment Protection*, *7*(02), 156–175. doi:10.4236/gep.2019.72011

Nkwunonwo, U. C., Whitworth, M., & Baily, B. (2016). Review article: A review and critical analysis of the efforts towards urban flood risk management in the Lagos region of Nigeria. *Natural Hazards and Earth System Sciences*, *16*(2), 349–369. doi:10.5194/nhess-16-349-2016

Obioha, E. E. (2009). Climate variability, environment change and food security nexus in Nigeria. *Journal of Human Ecology (Delhi, India)*, *26*(2), 107–121. doi:10.1080/09709274.2009.11906172

Ogunjinmi, A. A., Aiyeloja, A. A., Osunsina, I. O. O., & Ogunjinmi, K. O. (2014). Understanding perceptions of protected area management in developing countries: The case of Nigeria National Parks. *Global Environment*, *7*(2), 610–628. doi:10.3197/ge.2014.070213

Olorunfemi, M. A., & Olorunfemi, M. O. (1987). Geophysical investigations for groundwater in Precambrian terrains: A case history from Ikare, southwestern Nigeria. *Journal of African Earth Sciences*, *6*(6), 787–796. doi:10.1016/0899-5362(87)90035-2

Otubu, A. (2018). The land use act and land administration in 21st century Nigeria: Need for reforms. *Journal of Sustainable Development Law and Policy (The)*, *9*(1), 80–108.

Papworth, A., Maslin, M., & Randalls, S. (2015). Is climate change the greatest threat to global health? *The Geographical Journal*, *181*(4), 413–422. doi:10.1111/geoj.12127

Parkinson, B. W., & Gilbert, S. W. (1983). NAVSTAR: Global positioning system—Ten years later. *Proceedings of the IEEE*, *71*(10), 1177–1186. doi:10.1109/PROC.1983.12745

Pavlovskaya, M. (2006). Theorizing with GIS: A tool for critical geographies? *Environment & Planning A*, *38*(11), 2003–2020. doi:10.1068/a37326

Pfister, S., Bayer, P., Koehler, A., & Hellweg, S. (2011). Environmental impacts of water use in global crop production: Hotspots and trade-offs with land use. *Environmental Science & Technology*, *45*(13), 5761–5768. doi:10.1021/es1041755 PMID:21644578

Sabins, F. F. (2007). *Remote sensing: principles and applications*. Waveland Press.

Sala-i-Martin, X., & Subramanian, A. (2013). Addressing the natural resource curse: An illustration from Nigeria. *Journal of African Economies*, *22*(4), 570–615. doi:10.1093/jae/ejs033

Sam, K., Coulon, F., & Prpich, G. (2016). Working towards an integrated land contamination management framework for Nigeria. *The Science of the Total Environment*, *571*, 916–925. doi:10.1016/j.scitotenv.2016.07.075 PMID:27443458

Schuurman, N. (2000). Trouble in the heartland: GIS and its critics in the 1990s. *Progress in Human Geography*, *24*(4), 569–590. doi:10.1191/030913200100189111

Sfeir-Younis, A. (2019). *Land and soil management: Technology, economics, and institutions*. CRC Press. doi:10.1201/9780429041631

Sforna, G. (2019). Climate change and developing countries: From background actors to protagonists of climate negotiations. *International Environmental Agreement: Politics, Law and Economics*, *19*(3), 273–295. doi:10.100710784-019-09435-w

Sharma, S., & Shivakoti, G. (2017). En route to effective management of natural resources for conservation and livelihood advances in central Vietnam. In *Redefining Diversity & Dynamics of Natural Resources Management in Asia* (Vol. 3, pp. 237–248). Elsevier. doi:10.1016/B978-0-12-805452-9.00016-3

Stoll-Kleemann, S., de la Vega-Leinert, A. C., & Schultz, L. (2010). The role of community participation in the effectiveness of UNESCO Biosphere Reserve management: Evidence and reflections from two parallel global surveys. *Environmental Conservation*, *37*(3), 227–238. doi:10.1017/S037689291000038X

Tàbara, J. D., Jäger, J., Mangalagiu, D., & Grasso, M. (2019). Defining transformative climate science to address high-end climate change. *Regional Environmental Change*, *19*(3), 807–818. doi:10.100710113-018-1288-8

Tai, A. P., Martin, M. V., & Heald, C. L. (2014). Threat to future global food security from climate change and ozone air pollution. *Nature Climate Change*, *4*(9), 817–821. doi:10.1038/nclimate2317

Thapa, G. B., & Yila, O. M. (2012). Farmers' land management practices and status of agricultural land in the Jos Plateau, Nigeria. *Land Degradation & Development*, *23*(3), 263–277. doi:10.1002/ldr.1079

Thi, N. B. D., Kumar, G., & Lin, C. Y. (2015). An overview of food waste management in developing countries: Current status and future perspective. *Journal of Environmental Management*, *157*, 220–229. doi:10.1016/j.jenvman.2015.04.022 PMID:25910976

Tripathi, A., Tripathi, D. K., Chauhan, D. K., Kumar, N., & Singh, G. S. (2016). Paradigms of climate change impacts on some major food sources of the world: A review on current knowledge and future prospects. *Agriculture, Ecosystems & Environment*, *216*, 356–373. doi:10.1016/j.agee.2015.09.034

Umar, G. K., Yusuf, D. A., & Mustapha, A. (2019). Urban land use, planning and historical theories: An overview of Kano Metropolis. *World Scientific News*, *118*, 257–264.

Vanlauwe, B., Diels, J., Sanginga, N., & Merckx, R. (2005). Long-term integrated soil fertility management in South-western Nigeria: Crop performance and impact on the soil fertility status. *Plant and Soil*, *273*(1-2), 337–354. doi:10.100711104-005-0194-2

Villmann, T., Merényi, E., & Hammer, B. (2003). Neural maps in remote sensing image analysis. *Neural Networks*, *16*(3-4), 389–403. doi:10.1016/S0893-6080(03)00021-2 PMID:12672434

Vine, H. (1949). Nigerian soils in relation to parent materials. *Commonwealth Bureau of Soil Science and Technical Communication*, *46*, 22–29.

Virkkala, R., Heikkinen, R. K., Kuusela, S., Leikola, N., & Pöyry, J. (2019). Significance of Protected Area Network in Preserving Biodiversity in a Changing Northern European Climate. In *Handbook of Climate Change and Biodiversity* (pp. 377–390). Springer. doi:10.1007/978-3-319-98681-4_23

Wing, M. G., Eklund, A., & Kellogg, L. D. (2005). Consumer-grade global positioning system (GPS) accuracy and reliability. *Journal of Forestry*, *103*(4), 169–173. doi:10.1093/jof/103.4.169

Wright, D. J., Goodchild, M. F., & Proctor, J. D. (1997). Demystifying the persistent ambiguity of GIS as 'tool' versus 'science'. *Annals of the Association of American Geographers*, *87*(2), 346–362. doi:10.1111/0004-5608.872057

Chapter 5
Investigation of Urban Environment Using Tasseled Cap Transformation:
Vegetation and Urban Indices in Indian Sub-Continental City (Kharagpur)

Asraful Alam
https://orcid.org/0000-0001-7683-9206
University of Calcutta, India

Arijit Ghosh
Sidho-Kanho-Birsha University, Purulia, India

Lakshminarayan Satpati
University of Calcutta, India

ABSTRACT

Urban settlements have more complex environments, in unremitting fruition, where most of the world population lives. Most of the cities in developing countries have been developed without a rationale, and the life conditions are repeatedly insufferable. For this research work, NDVI is particularly used to assess the status of vegetation cover. Tasseled cap is another index that creates three band images for this study. Brightness, greenness, wetness are the three bands that represent the area under consideration. The present study aims particularly at comparing high NDVI area and greenness values given by tasseled cap and low NDVI values and high brightness values and status of urban environment. Based on the overlapping of tasseled cap image, an NDVI image is observed in which most of the area of healthy vegetation is located in the north west and south east part of Kharagpur city, which extended from south west to north east and north to south respectively.

DOI: 10.4018/978-1-7998-5027-4.ch005

INTRODUCTION

Urban settlements have more complex environments, in unremitting fruition, where most of the world population lives. Most of the cities in developing countries have been developed without a rationale and the life conditions are repeatedly insufferable. Urban development is an organic process, and which has been analyzing the urban built-up area expansion characteristics is advantageous when determining the urban land use trend and intensive land use degree (*Lin Wang et al. 2018*). The difference between the characteristics of surface cover in urban and non-urbanized areas in terms of 3-d geometry of the built area, heat absorption, the building materials, surface albedo and vegetation cover lead to different air and surface temperatures within a city relative to the surrounding area because of the various band comparison. Therefore, urban expansion can lead to radical changes in the nature of the surface of the urbanized area (*Azad Rasul et al. 2015*). Urban landscapes are characterized by various types of buildings, settlements, roads, gardens, exposed soil, fallow land and water body. In order to understand the heterogeneous landscapes better, the standardized model that describes these component surfaces should be defined. Despite the existing remote sensing instruments are already able to extract useful information in this scenario, so far scarce interest has been focused by the scientific community toward specific applications for developing countries. In this paper we present a project whose goal is the efficient exploiting of existing remote sensing sensors data and Geographical Information System (GIS) facilities in order to provide various statistical numerical instruments to analyze the status of urban monitoring and organization processes in the study area. Updated information on the status and trends of urban ecosystems is needed to develop strategies for reasonable development and to improve the livelihood of cities. Interpretation of the evolution and the management of land features such as NDVI which has been widely used for mapping vegetation and NDBI for interpreting urban (*Y. ZHA, 2003*) were used a technique of combining the NDBI and the NDVI with a specific processes calculation to extract the built-up areas and were retched an accuracy of 92.6%, (*Y. Limin, 2003*). Tasseled Cap Transformation (TCT) components can be used to determine urban environment, vegetation and urban indices. Brightness, wetness and greenness are components of TCT. High reflectance of the various features like built-up areas, man-made features, roads and soil which are the brightness component these topographic features can be observed. The tasselled cap transformation (TCT) is a useful tool for compressing spectral data into a few bands associated with physical scene characteristics with minimal information loss. TCT was originally evolved from the Landsat multi-spectral scanner (MSS) launched in 1972 and is widely adapted to modern sensors (Baig et al., 2015).

Study Area

Kharagpur is the fourth largest city of West Bengal in area and fifth most populated city in West Bengal. Kharagpur city has been selected as a study area which extent up to 22⁰16′12″ to 22⁰21′50″N and 87⁰14′52″ to 87⁰21′35″ E covering an area of about 127 km² in southern part of West Midnapore district of West Bengal (Figure 1). Kharagpur has a tropical savanna climate where summers start in March and are hot and humid, with average temperatures close to 30 °C and total annual rainfall is around 1400mm as like the other part of south Bengal.

Figure 1. Study area

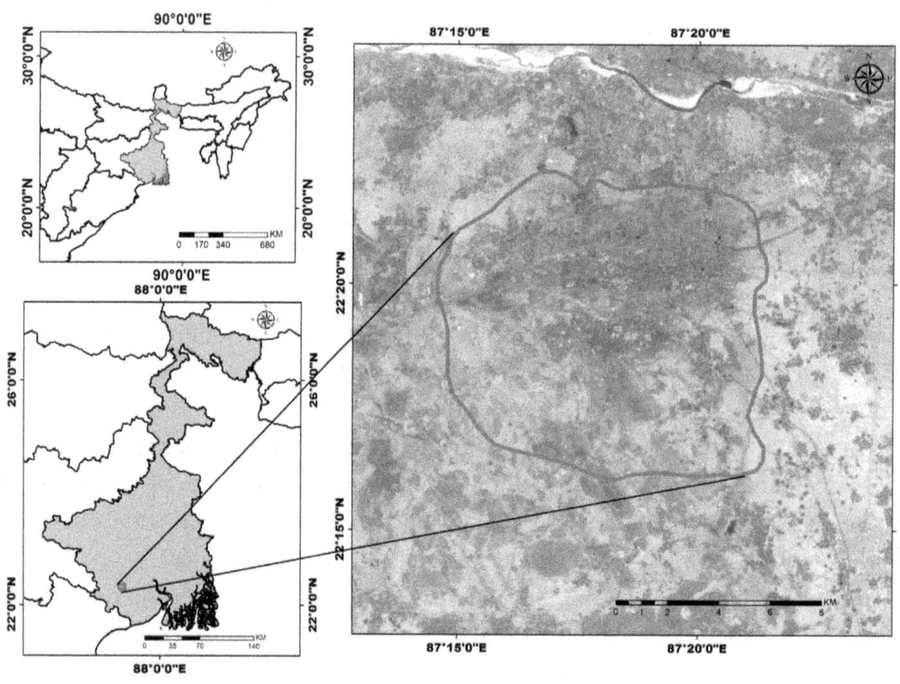

Image Pre-Processing

Landsat-7 ETM+ SLC-off data was gap-filled using segment based method (*USGS, 2004*). Then the Landsat MSS and ETM+ images were geocoded with a root mean square error (RMSE) approximately 6 m. as a single scene could not cover the whole study area, image mosaicking of two adjacent scenes was carried out and color balance was performed in the overlap region between the two scenes. Then the image was subsetted to the boundary of the GSA.

Tasseled Cap Transformation (K-T Transformation)

Kauth and Thomas in 1976 first introduced Tasseled cap transformation for Landsat MSS data (*Kauth & Thomas, 1976*) and was improved and extended to Landsat TM data in the mid-1980s (*Crist & Cicone, 1984a; Crist, 1985; Crist & Kauth, 1986*) and this Tasseled cap transformation has been world widely used in monitoring agriculture, vegetation changes, sea coast change, landscape change etc (*Collins & Woodcock, 1996; Han et al., 2007*); Tasseled cap transformation is a kind of orthogonal transformation, it rotates the original data plane so that the vast majority of data variability is concentrated in the features, i.e., the plane is viewed "head-on" and it presents the most basic structures of the data in the most direct way (*Crist & Cicone, 1984b*). TCT allows for efficient extraction of compound data and the derivation of carrying great weight spectral features, which are easier to grip and interpret than the raw data; for this cause it has become one of the most accepted methods of image transformation for change detection in the remote sensing community (Crist & Kauth, 1986). The original Landsat MSS data space was transformed to a new four-dimensional feature space, that is, the soil brightness index (B), green ess vegetation index (G), yellow stuff index (Y), and non-such (N). In this study, we aimed to extract built-up areas information and B, G and Y were derived.

$$\text{Brightness} = 0.3029M_2 + 0.2786M_3 + 0.4733M_4 + 0.262M_5 + 0.508\,M_6 + 0.1872\,M_7$$
$$\text{Greenness} = -0.2941M_2 - 0.243M_3 - 0.5424M_4 + 0.7276M_5 + 0.0713M_6 - 0.1608M_7$$
$$\text{Yellowness} = 0.1511M_2 + 0.1973M_3 - 0.3283M_4 + 0.3407M_5 - 0.7117M_6 - 0.4559M_7$$

where M 2-7 represent digital numbers (DNs) of band 2 to 7.

For Landsat-7 ETM+ images, Brightness (B), Greenness (G), Wetness (W) and the fourth, fifth, sixth variables were produced and the first three variables were selected. Previous research work has indicated that at-satellite reflectance-based K-T transformation is needed because changing sun illumination geometry affects DNs strongly, and thus affects the derived tasseled cap value. Moreover, at-satellite reflectance-based K-T transformation can differentiate water from land targets better than the DNs and reflectance factor based transformation (*Huang et al., 2005*). DNs should, therefore, be converted to at-satellite reflectance firstly according to formulae 3 and 4 (*Myeong et al., 2006*). After the two steps, the effects owing to different sun angles at different dates were compensated. Correspondingly, the coefficients (*Huang et al., 2005*) here varied from the DNs based transformation.

$$L_\lambda = (DN_\lambda \,.gain_\lambda) + bias_\lambda \quad (1)$$

where L is radiance; λ is the spectral band; gain is he spectral band gain; and bias is the spectral band offset.

$$\rho_\lambda = \frac{\pi . L_\lambda . d^2}{ESUN_\lambda . COS\theta_s} \quad (2)$$

Here, ρ_λ= Planetary TOA reflectance [unitless]; π= Mathematical constant approximately equal to 3.14159 [unitless]; L_λ= Spectral radiance at the sensor's aperture [W/ (m^2 sr μm)]; d= Earth-Sun distance [astronomical units]; $ESUN_\lambda$= Mean exoatmospheric solar irradiance [W/ (m^2 μm)] θs=Solar zenith angle.

Table 1. Coefficients of the Tasseled Cap Transformation applied to Landsat 8

INDEX	BAND 2	BAND 3	BAND 4	BAND 5	BAND 6	BAND 7
Brightness	0.3029	0.2786	0.4733	0.5599	0.508	0.1872
Greeness	-0.2941	-0.243	-0.5424	0.7276	0.0713	-0.1608
Wetness	0.1511	0.1973	0.3283	0.3407	-0.7117	-0.4559

Table 2. Data and Software

Sensor	Path	Row	Date	Source
Landsat 8	139	45	18-01-2018	USGS (GLOVIS)

Result and Discussion

The urban index maps were created using the model aforementioned above (Fig. 2). The legends of these indices can vary from site to site. In general, it ranges from -1 to +1 and in order to extract urban and non-urban areas threshold values should be determined. Threshold values can be obtained from ground data, comparison of a reference image or statistical methods. In all maps, from dark to bright areas the urban details become clearer. Figure 2 shows the colour composites of Brightness, Greenness, and Wetness derived from the 2018 ETM+ images. Built-up areas are highlighted in Red and vegetation is in Green while water is in Blue.

Table 3. Multispectral band of landsat 8

	LANDSAT 8 OLI	Spectral Range	Spatial Resolution(m)
Costal/Aerosol	Band 1	0.433-0.453	30
Blue	Band 2	0.450-0.515	30
Green	Band 3	0.525-0.600	30
Red	Band 4	0.630-0.680	30
NIR	Band 5	0.845-0.885	30
SWIR-1	Band 6	1.560-1.660	30
SWIR-2	Band 7	2.100-2.300	30
Pancromatic	Band 8	0.500-0.680	30
Cirrus	Band 9	1.360-1.390	15
LWIR-1	Band 10	10.30-11.30	30
LWIR-2	Band 11	11.50-12.50	30

Source: USGS Landsat8 Product, 2013

Landsat OLI RGB-654 color composite contains the same 3 bands as the previous, because vegetation reflects higher in the near infrared than the other two bands, the green color dominates. So this combination will make the vegetation look more natural. Figure 2 showed the composite band which unsupervised classification, while Figure 2 showed the composite band that has been classified in a supervised maximum likelihod. The land use / land cover dominated by vegetated area that consists of secondary built-up, settlement, industrial buildings etc. In this area there are still very few areas such as forests, plantations, ponds and others. Interpretation of land use/ land cover (LULC) can be used NDVI, because it can be known the level of density and greenness of vegetation. NDVI values are obtained by the calculation of near infrared with visible light reflected by the plant. Tasselled cap with low red reflectance and higher NIR reflectance was found representative of higher vegetated areas, while the opposite to the top, the flat side of the cap (not perfectly horizontal), was found representative of the bare soil (Ray 1994). The low red reflectance and high NIR reflectance) represented regions of high vegetation and that the flat side of the cap directly opposite the point represented bare soil (Kauth and Thomas 1976). The Tasseled Cap Transformation is a adaptation of the original bands of an image into a new set of bands with defined interpretations that are constructive for vegetation mapping.

Figure 2. Tasseled cap transformation, vegetation and urban indices in Kharagpur city

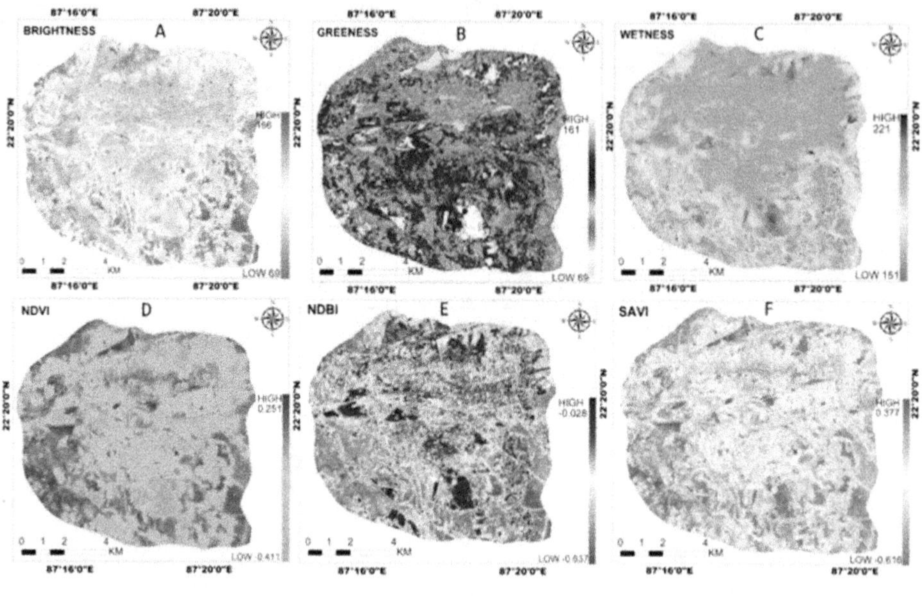

CONCLUSION

This study showed that tasseled cap transformation is an effective method for investigation of urban environment using tasseled cap transformation, vegetation and urban indices in indian sub-continental city (Kharagpur) and the variables derived from tasseled cap transformation. The results also demonstrated that the Kharagpur experienced high-speed impervious surface sprawl over the past years. In this study, the tasseled cap transformation model was created to extract commonly used spectral urban indices and urban environment from Landsat 8 OLI and TIRS data. These indices are obtained from previous Landsat mission images namely TM and ETM+; however, in this study these indices were adapted to Landsat 8 data. The obtained results show that these indices can be obtained efficiently using Landsat 8 data. City planners and decision makers can use these effective indices to observe the changes in urban areas. For further study, it will be investigated which urban index gives the best results to present the urban areas.

ACKNOWLEDGMENT

I am sincerely indebted to Indian Council of Social Science Research (Ministry of Human Resource Development, New Delhi, 110067, India), for the Post Doctoral Fellowship (PDF).

REFERENCES

Baig, M. H. A., Zhang, L., Shuai, T., & Tong, Q. (2014). Derivation of a tasselled cap transformation based on Landsat 8 at-satellite reflectance. *Remote Sensing Letters*, *5*(5), 423–431. doi:10.1080/2150704X.2014.915434

Bouhennache, R., & Bouden, T. (2014). Change Detection In Urban Land Cover Using Landsat Images Satellites, A Case Study In Algiers Town. *IEEExplore*, *14*, 4799–562.

Collins, J. B., & Woodcock, C. E. (1996). An assessment of several linear change detection techniques for mapping forest mortality using multitemporal landsat TM data. *Remote Sensing of Environment*, *56*(1), 66–77. doi:10.1016/0034-4257(95)00233-2

Crist, E. P. (1985). A TM Tasseled Cap equivalent transformation for reflectance factor data. *Remote Sensing of Environment*, *17*(3), 301–306. doi:10.1016/0034-4257(85)90102-6

Crist, E. P., & Cicone, R. (1984a). Application of the tasseled cap concept to simulated Thematic Mapper data. *Photogrammetric Engineering and Remote Sensing*, *50*(3), 343–352.

Crist, E. P., & Cicone, R. C., 1984b. A physically-based transformation of thematic mapper data---The TM Tasseled Cap. *Geoscience and Remote Sensing, IEEE Transactions on*, *22*(3), 256-263.

Crist, E. P., & Kauth, R. J. (1986). The tasseled cap de-mystified. *Photogrammetric Engineering and Remote Sensing*, *52*(1), 81–86.

Fung, T. (1990). An assessment of TM imagery for land-cover change detection. *IEEE Transactions on Geoscience and Remote Sensing*, *28*(4), 681–684. doi:10.1109/TGRS.1990.572980

Han, T., Wulder, M., White, J., Coops, N., Alvarez, M., & Butson, C. (2007). An Efficient Protocol to Process Landsat Images for Change Detection With Tasselled Cap Transformation. *Geoscience and Remote Sensing Letters, IEEE*, *4*(1), 147–151. doi:10.1109/LGRS.2006.887066

Huang, C., Wylie, B., Zhang, L., Homer, C., & Zylstra, G. (2005). *tasseled.pdf (application/pdf Object)*. Retrieved September 11, 2009, from http://landcover.usgs.gov/pdf/tasseled.pdf

Jensen, J. R. (2005). *Introductory Digital Image Processing A Remote Sensing Perspective* (3rd ed.). Prentice Hall.

Kauth, R., & Thomas, G. 1976. The tasselled cap-a graphic description of the spectral-temporal development of agricultural crops as seen by LANDSAT. In *Proceedings of the Symposium on Machine Processing of Remotely Sensed Data*. Purdue University.

Limin, Y., & George, X. (2003, September). Urban Land-Cover Change Detection through Sub-Pixel Imperviousness Mapping Using Remotely Sensed Data. *Photogrammetric Engineering and Remote Sensing*, *69*(9), 1003–1010. doi:10.14358/PERS.69.9.1003

Lu, D., Mausel, P., Brondízio, E., & Moran, E. (2004). Change detection techniques. *International Journal of Remote Sensing*, *25*(12), 2365–2401. doi:10.1080/0143116031000139863

Rasula, A. (2015). *Heiko Balztera, Claire Smitha, 2015, Spatial variation of the daytime Surface Urban Cool Island during the dry season in Erbil, Iraqi Kurdistan, from Landsat 8*. Urban Climate. doi:10.1016/j.uclim.2015.09.001

Ray, T. W. (1994). *A FAQ on Vegetation in Remote Sensing*. Accessed March 23, 2014. http://www. yale.edu/ceo/Documentation/rsvegfaq.html

Wang, L., Zhu, J., Xu, Y., & Wang, Z. (2018). Urban Built-Up Area Boundary Extraction and Spatial-Temporal Characteristics Based on Land Surface Temperature Retrieval. *Remote Sensing*, *10*(3), 473. doi:10.3390/rs10030473

Zha, Y., Gao, J., & Ni, S. (2003). Use of normalized difference built-up index in automatically mapping urban areas from TM imagery. *International Journal of Remote Sensing*, *24*(3), 583–594. doi:10.1080/01431160304987

Chapter 6
Efficacy of Advanced Remote Sensing (Hyperspectral and LIDAR) in Enhancing Forest Resources Management

Laxmikant Shrama
https://orcid.org/0000-0003-2911-2893
Central University of Rajasthan, India

Rajit Gupta
Central University of Rajasthan, India

Rajani Kant Verma
https://orcid.org/0000-0002-3169-5832
Central University of Rajasthan, India

ABSTRACT

Sustainable management of natural forest resources is a vital requirement in the changing climatic conditions on Earth. Two advances techniques, hyperspectral remote sensing (HRS) and LIDAR (light detection and ranging) remote sensing (LRS), provide more enhanced and accurate measurements than that of conventional optical remote sensing (ORS). Hyperspectral sensor like AVIRIS, which has hundreds of narrow bands, have advantages over a broadband multispectral sensor. In addition, the fusion of HRS and LRS can play an essential role in assessing biophysical and biochemical variables of forest species. In this chapter, the authors reviewed the extant literature and tried to understand the position of HRS, LRS, and their integration with the machine and deep learning algorithms for the effective monitoring and management of natural forest resources. Further, scopes and challenges are also discussed to enhance the effectiveness of these techniques in natural forest resources management.

DOI: 10.4018/978-1-7998-5027-4.ch006

1. INTRODUCTION

Sustainable management of natural forest resources is the inescapable demand in the present changing climatic scenarios. Recent developments in the remote sensing & Geographical Information system (GIS) offers improved methods to investigate and develop research approaches to manage natural forest resources. Moreover, the data and techniques offered by the remote sensing & GIS have gained much popularity in the monitoring and management of natural forest resources. Optical remote sensing (ORS) comes under conventional remote sensing which is widely used and has several application in forestry. Some application which is quickly evaluated with good accuracy using ORS includes assessment of area under forest cover, change detection, and forest health. However, species-level classification of forests, estimation of biophysical parameters, generation of tree inventory data is quite a non-satisfactorily with ORS (Wulder, 1998). Further, another issue in case of ORS is acquiring consistent cloud-free datasets over large areas (Avitabile et al., 2012). The accuracy of forest structural properties estimated in case of optical ORS typically decreases with increasing biomass and LAI (Foody et al., 2001; Hudak et al., 2002; Duncanson et al., 2010). According to Lefsky et al., 1998; Duncanson et al., 2010 the measurement of vertical vegetation structure using ORS data is constrained on the assumption that there is a predictable relationship between the two-dimensional structural properties of forests that can be sensed by ORS, and the three-dimensional structural properties of a forest that required for the estimation of forest volume and above-ground biomass (AGB). In general, ORS provide a promising opportunity to derive essential outputs which are very useful in forest monitoring and management; however, there is still some extents which cannot be addressed with ORS. To overcome these lacks, advanced remote sensing techniques such as LIDAR (Light Detection and Ranging) and hyperspectral offers a remarkable prospect in enhancing both the monitoring and management of natural forest resources.

Advancements in sensors technology have made it possible to collect the hundreds of spectral bands in remote sensing imagery. This technology is known as Hyperspectral remote sensing (HRS) or imaging spectroscopy. HRS is an advanced remote sensing technique, which comprises of hundreds of narrow and contiguous spectral bands covering the range from the visible to the infrared region of the electromagnetic spectrum (Chang, 2007). These spectral bands due to their narrow width and continuity provide ample scope to identify targets and extraction of information with high accuracy. Each spectral signature in hyperspectral data is based on the chemical and physical properties of specific objects. Therefore, the spectral signature is unique for every vegetation and because of this, HRS has the dynamic ability of monitoring and management of forest natural resources. HRS data provides accurate species-level classification for large areas based on a quite small amount of training datasets (Hycza et al., 2018). Hyperspectral data can be applied to evaluate the biophysical variables like LAI, FPAR and biochemical changes in vegetation at a leaf level and mapping as well as to monitor the relative changes in forests due to the impact of climate-induced disturbances. Hyperspectral data was able to identify low stressed forests which usually represent normal and healthy vegetation and high stressed forests which represent some adverse site factors or climate-induced disturbances. Besides, HRS gives an essential idea on disturbances in forests at a leaf level. Overall, this technique provides an opportunity for mapping and monitoring of natural resources over the sizeable geographical region at a very high spectral resolution. According to recent studies, the outcomes generated using hyperspectral data could be used to derive essential relations and has direct applications in natural resource management and conservation. Thenkabail et al., 2011 presented a wide-ranging work on HRS and its applications in a book, namely "Hyperspectral remote sensing of vegetation". Another advanced technique is Lidar remote sensing (LRS) which has the capability to penetrate forest canopy. LRS represents one of the best sources of investigation for the detailed vertical structure and measuring forest structural parameters such as stem volume, basal area, canopy cover or tree height (Dubayah and Drake, 2000; Tonolli et al., 2011; Urbazaev et al., 2018). Lidar is a laser altimeter that calculates the distance from the instrument to the target by assessing the time passed between the pulse emission and the reflected signal. The reflected signal may identify multiple returns originating from vegetation and calculates vegetation height. Waveform LiDAR is able to record and digitize the full-returned signal holds much potential for the detailed vertical structure of vegetation (Wu et al., 2009). Further, Lidar data provides accurate estimates of aboveground biomass (AGB) without signal saturation (Urbazaev et al., 2018). According to Man et al., 2014, there is no clear relation between the forest biophysical and biochemical parameters derived from hyperspectral data and the structural attributes from LiDAR data. Several studies show that the combined use of HRS and LRS has been beneficial to get improved results in forestry applications. Moreover, the integrated use of hyperspectral and LIDAR could provide structural and spectral information of targets at the same time. According to Lucas et al. (2008), the combined use of LIDAR and hyperspectral data offers improved estimates of forest biomass. Besides, the use of machine learning and deep learning algorithms with HRS and LRS offers new ways to proceed for the sustainable management of natural forest resources. This chapter discusses the efficacy of advance remote sensing such as HRS and LIDAR in natural forests resource management. These two techniques are quite useful in evaluating and quantifying the detailed monitoring and analysis of natural forest resources. Besides, the fusion of HRS and LRS, integration with advanced techniques such as machine learning and deep learning provide outputs which are useful in forest resource management in an effective manner. On reviving the extant literature, it is demonstrated that both HRS and LRS has a tremendous opportunity in the field of spatial science for natural forest resources management. Scope of these technologies lies in their multiple data integration along with using modelling and machine learnings algorithms which could provide the best use of these technologies in the sustainable management of natural resources.

2. HYPERSPECTRAL REMOTE SENSING (HRS)

HRS has become recognised as an analytical technique for Earth observations and applications since the 1980s (Goetz et al., 1985; Tong et al., 2014). Hyperspectral data has many, narrow and contiguous spectral bands (**Fig. 1**) throughout the visible, short wave infrared and thermal infrared portions of the electromagnetic spectrum (Govender et al., 2008). HRS has a great ability to enhance natural forest monitoring and management. HRS data has two main characteristics, one is narrow spectral bandwidth, and another is continuous arrangements of bands. Therefore, the information or the spectral signature of the target is collected in a continuous mode, which gives a better examination of the target than that of the spectral signature obtained through conventional remote sensing. According to Jong et al., 2003 a high spectral resolution data in the range of 400 - 2500 nm of the electromagnetic spectrum would improve the evaluation of vegetation, forest, soil, hydrology, mineral and other natural resources due to the absorption in the reflectance curve (**Fig. 2**). These reflectance curves are called spectral signatures or fingerprints or spectra, which is used to recognised objects in hyperspectral imagery. The source of hyperspectral data can be ground-based, airborne, or space-borne sensors such as imaging spectrometer, Airborne Visible/Infrared Imaging Spectrometer (AVIRIS) and EO-1 Hyperion respectively. This technology gains much attention and curiosity among researchers worldwide. Satellite-based sensors such as Hyperion, EnMap and airborne hyperspectral sensors such as CASI, HyMap, AVIRIS, AVIRIS-NG plays an essential role in forestry (**Table 1**).

2.1 Hyperspectral Sensors

Hyperspectral sensors, also called imaging spectrometers, are a systems technology that collected images of a scene in tens to hundreds of narrow and contiguous spectral bands. There are different types of hyperspectral sensors which differs in their spectral, spatial, radiometric and temporal resolution. According to the sensors mounted position, these can be ground-based, airborne and space-borne. In the present, there is very less number of hyperspectral sensors as listed in

Figure 1. A three-dimensional cube of hyperspectral data(AVIRIS-NG)
Source: SAC (Ahmedabad)

Figure 2. Broad application of HRS in forestry

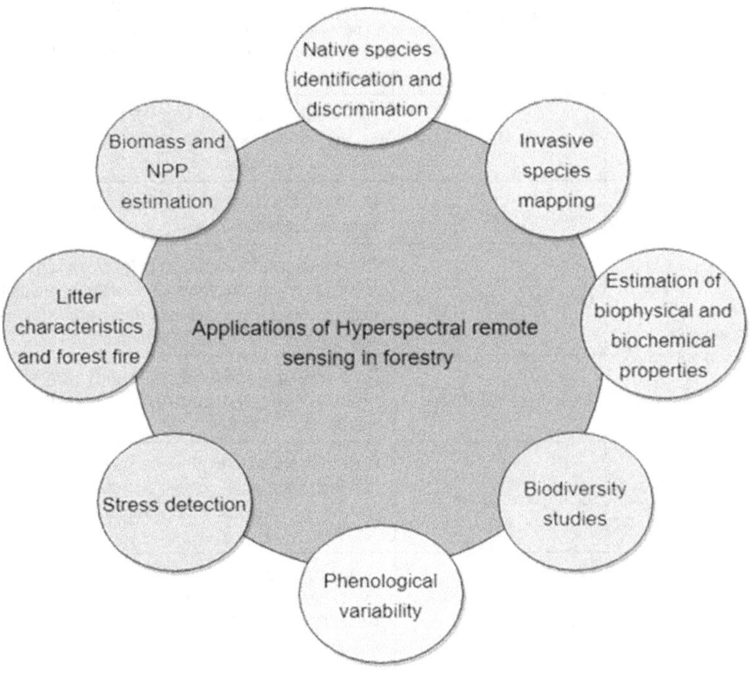

Table 1. List of hyperspectral sensors, studies name and their significant finding/ contribution in forestry

Hyperspectral sensor	Study name	Finding/contribution
Airborne		
Airborne Visible InfraRed Imaging Spectrometer (AVIRIS)	Martin and Aber (1997)	This study aimed to derive relationships between AVIRIS data and field observed canopy biochemical composition. Further, these relationships were used to develop spatial estimations of canopy chemistry as drivers for ecosystem models
	Martin et al., 1998	This study used the relationship between the chemical composition of foliage and species identification to demonstrate that high spectral resolution data can be used to classify forest species
	Serrano et al., 2002	This study determines the canopy Nitrogen and lignin concentrations in chaparral vegetation using AVIRIS data. They proposed two VI, Normalized Difference Nitrogen Index (NDNI) and Normalized Difference Lignin Index (NDLI) to estimate the N and Lignin in native shrubs
	Smith et al., 2002	They estimated the potential of AVIRIS data to determine foliar N concentration which in turn useful for the estimation of forest productivity across a regional level
	Lee et al., 2004	This study provides an assessment of different methods for the estimation of water-related biophysical and biochemical properties in vegetation using AVIRIS and MODIS.
	Carlson et al., 2007	They mapped species diversity in lowland tropical forests ecosystem. They stated that the hyperspectral data are a crucial component for species diversity mapping due to its excellent spectral resolution, and it allows great flexibility in wavelength choice which is not offered from multispectral data
	Huesca et al., 2016	They studied the potential of AVIRIS data to assess canopy structure and to discriminate structural canopy types in a heterogeneous forest
CASI-1500Compact Airborne Spectrographic Imager (CASI)	Sampson et al., 2003	They used CASI to estimate Chlorophyll content in a managed tolerant hardwood
	Gong et al., 1995	They attempted to assess the potential of CASI to determine coniferous forests LAI using three techniques namely, univariate regression, multivariate regression and vegetative indices based LAI estimation
	Matsuki et al., 2015	The proposed methodology to classify trees of mixed forests using hyperspectral data and LiDAR-derived canopy height models
	Zhang et al., 2008	This study aimed to explore the methods to retrieve the total leaf chlorophyll content from CASI data for both open spruce and closed forests and tested for open forest canopies.
HyMap (Hyperspectral Mapper)	Schlerf and Atzberger, 2006	They estimated different canopy structural variables using forest reflectance model Invertible Forest Reflectance Model (INFORM) by inverting hyperspectral HyMap data and with neural network-based approach

continues on following page

Taable 1. Continued

Hyperspectral sensor	Study name	Finding/contribution
Airborne		
Airborne Imaging Spectroradiometer for Applications (AISA)	Narumalani et al., 2009	They used AISA data for detecting and mapping the four invasive species distribution by spectral angle mapping technique
Spaceborne		
Hyperion EO-1 (launched November 21, 2000; decommissioned March 2017)	Gong et al., 2003	They evaluated twelve VIs, which were constructed using bands across the spectral range of 0.4–2.5m. They identified some important bands with potential for improving LAI estimation at the large regional.
	Castro-Esau et al., 2004	They attempted to discriminate lianas and trees by observing differences between hyperspectral signatures at the leaf level using pattern recognition techniques
	Vyas et al., 2011	They attempted to compare the performance of three different classifiers (SAM, SVM, ANN) over highly diverse tropical forest species. They suggested that hyperspectral data provides encouraging outcomes for the discrimination of tropical forest species
	Vyas et al., 2013	Derived canopy level estimation of chlorophyll and leaf area index (LAI) for two tropical species
CHRIS (PROBA)	Barnsley et al., 2000	They presented key features of the CHRIS sensor and derived the estimates on variables like leaf biochemistry, LAI, FAPAR and Net Primary Productivity (NPP) relating to vegetation canopies

Table 1. Some hyperspectral sensors were decommissioned, and some has been planned to launch in the upcoming future. Transon et al., 2017 did a review on twenty years of researches and applications of spaceborne HRS by identifying the significant limitations of current and future hyperspectral space-borne sensors for systematic Earth observations. They identified eight space-borne hyperspectral sensors. PRISMA (Precursore IperSpettrale della Missione Applicativa), HYSI (Hyperspectral short-wave Infrared radiometer), EnMAP (Environmental Mapping and Analysis Program), HISUI (Hyperspectral Imager Suite), SHALOM (Spaceborne Hyperspectral Applicative Land and Ocean Mission and Hyspiri (Hyperspectral Infrared Imager) are the hyperspectral sensors which are planned to be launched in the near future by different countries. Two sensors, one is Hyperion, and another was TG-1 (TianGong-1) are already decommissioned.

2.2 Application of HRS in Forest Management

HRS emerged as a useful technique for forestry to estimate different forest parameters that are quite incapable of estimating or can be estimated with some missing information using conventional remote sensing. Hyperspectral data provides an advanced method to explore forest applications (**Fig. 1**). Many researchers retrieved forest biochemical parameters using data from various hyperspectral sensors (**Table 1**).

The retrieval of biophysical and biochemical parameters of vegetation using broadband vegetative Indices (VIs) can be unstable due to the underlying soil types, leaf and canopy properties, and atmospheric effects. For instance, NDVI asymptotically saturates after a specific biomass density and for a specific range of LAI (Mutanga and Skidmore, 2004). Hence, it reduces the accuracy of outputs significantly (Gao et al. 2000; Thenkabail et al. 2000; Adam et al., 2010). A wide range of hyperspectral remote sensing applications has been identified in forestry. **Fig. 2** shows some typical applications of HRS in forestry. **Fig. 3** shows the spectral signature of different targets which could be beneficial in target identification and classification. Accurate forest species identification and classification has been required for the effective monitoring of species diversity and sustainable management of forest resources. HRS emerged as an advanced technology that provides a practical approach for species identification and classification. Several studies perform LAI estimation (Gong et al., 2003; Lee et al., 2004; Schlerf et al., 2005; Bulcock and Jewitt, 2010; Darvishzadeh et al., 2011; Vyas et al., 2013; Atzberger et al., 2015; Wang et al., 2016), FPAR (Peng et al., 2018; Tan et al., 2018; Qin et al., 2018), forest species identification (Martin et al. 1998; Goodwin et al. 2005; Dalponte et al., 2008; Lucas et al. 2008; van Aardt and Norris- Rogers, 2008; Peerbhay et al., 2013; Ballanti et al., 2016; Hycza et al., 2018), growth rates (Caughlin et al., 2016), vegetation AGB and parameters (Clark et al., 2011; Laurin et al., 2014; Brovkina et al., 2017; Schlerf and Atzberger, 2006; Maire et al., 2008; Casas et al., 2014; Huesca et al., 2016; Halme et al., 2019), invasion of alien species (Ustin et al., 2002, Underwood et al., 2003; Glenn et al., 2005; Lawrence et al., 2006; Andrew and Ustin, 2008, 2009; Peerbhay et al., 2015, 2016; Skowronek et al., 2017) using HRS data and its integration with other remote sensing datasets.

Shen et al., 2010 concluded that high spatial resolution hyperspectral imagery has an excellent prospect to investigate and demonstrate the mapping potential of the forests ecosystem at a species level. They aimed to verify the mapping performances of eight different classification methods and to understand the effects of noise treatments and leaf growth periods on the classification performance. Dalponte et al., 2008 proposed a novel method for the classification of complex forest areas by the combined use of hyperspectral and LIDAR data. Jones et al., 2010 assessed the utility of hyperspectral AISA imagery and small footprint, multi return LiDAR data for mapping 11 tree species. Also assessed the impact of integrating AISA data at the pixel-level with LiDAR computed concurrently, small footprint, and multi-return forest height and volume data. They stated that the pixel-level fusion of hyperspectral and LiDAR data could significantly enhance and improve current forest inventory approaches, providing detailed information pertinent to a wide

Figure 3. Spectral signature of different objects

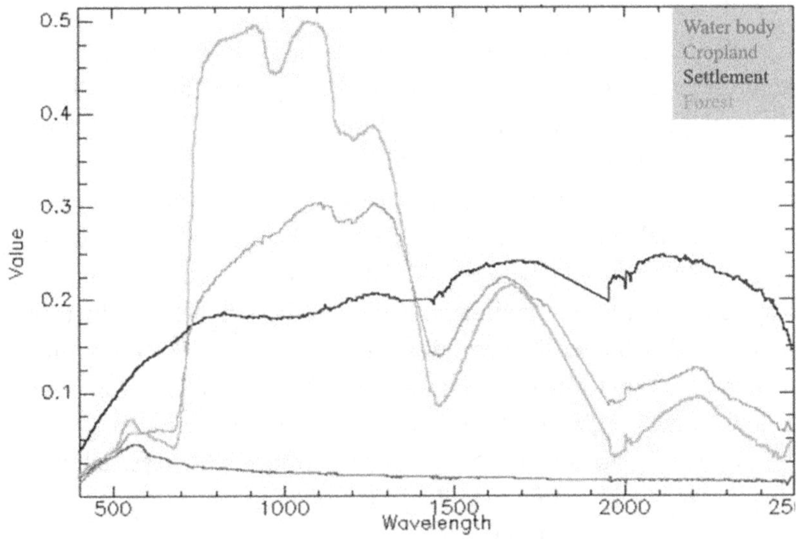

range of forest management tasks, and the maximum utilisation of information derived from conventional aerial photograph interpretation. According to Asner et al. (2008), and integration of HRS and LiDAR data provides enormous flexibility and analytical potential to understand and examine current as well as possible future invasion of species within the terrestrial ecosystems.

2.3. Challenges and Scope of HRS in Forestry

Although HRS provides reasonable estimates of many forest variables, it has the number of complex challenges in forestry, which needs to be further explored. Optimal hyperspectral bands, which best characterise different vegetation classes, should be incorporated into multispectral systems through sensor modifications or spectral filters. Using this approach researchers will no longer be restricted to existing sensor specifications, but can tailor remote sensing model inputs to a particular application. Less data availability, need for large storage space and processing methodologies makes HRS is still an emerging field of advanced remote sensing. The ultimate goal is to move from impractical large data volumes and expensive sensor development towards practical operational low data volumes, high temporal resolution and cheaper sensor development (Govender et al., 2008).

3. LIDAR REMOTE SENSING (LRS)

LiDAR is also an advanced remote sensing technology which can monitor and measure forest natural resources. LRS provides an accurate and effective means of measuring and monitoring vegetation structure. Lidar systems directly measure the three-dimensional structure of forest canopies, including tree and canopy height, density, and indirectly aboveground biomass, as well as topographic structures (Nilsson, 1996; Nelson et al., 1997; Lefsky et al., 2002; Khalefa et al., 2013). According to Dubayah and Drake (2000) the three important characteristics of LiDAR systems for forestry applications are (a) the manner in which the return signal is recorded; whether they record the range to the first or last return or fully digitise the return signal (2) footprint size; whether the small (a few centimetres) or large (tens of meters); and (3) sampling rate and scanning pattern.

LRS data can be generated from spaceborne, airborne or terrestrial instruments. According to Dubayah and Drake, 2000, one of the major strength of LRS lies in the direct measurement of vertical forest structure. They added that the direct measurement of canopy heights, sub-canopy topography, and the vertical distribution of intercepted surfaces beneficial in attaining the data for better forest characterisation and management. Likewise HRS data, LIDAR data has limited availability due to its very high cost and requirement of large storing volume. Further, a few spaceborne LIDAR, for instance, ICESat 1 and 2 data is available from LIDAR spaceborne sensors. ICESat 1 data is based on waveform LiDAR, capable of recording and digitising the full returned pulse at high vertical resolution (Wu et al., 2009) whereas ICESat-2, launched in September 2018, use a new photon-counting approach adopted for the first time on a spaceborne platform (Chen et al., 2019). ICESat-2 provides ALT08 product for measuring land vegetation and individual tree components such as basal area, canopy height and canopy cover (**Fig. 4c**). However, airborne and terrestrial laser altimetry datasets have high point density which is quite useful in various applications with reasonable accuracy. In forest resources management, LIDAR shows a tremendous utility as demonstrated in different studies. Good use of LIDAR in forest application is the quantification of forest inventory variables. Inventory data generated from LIDAR remote sensing could be useful in estimating forest biomass and productivity. Further, integration of LiDAR data with hyperspectral could provide structural and spectral information of targets.

3.1 Applications in Forest Management

Lidar remote sensing is a breakthrough technology for forestry applications (Dubayah and Drake). In recent years LIDAR has become recognised as a valuable remote sensing tool for forest inventory and

Figure 4. a) 2014 USFS Tahoe National Forest Lidar airborne data (source: https://opentopography.org/) b) Spaceborne ICESat-2 (source: https://nsidc.org/data/icesat-2; c) Surface elevation and canopy height in meters obtained from ICESat-2 data (https://www.openaltimetry.org/)

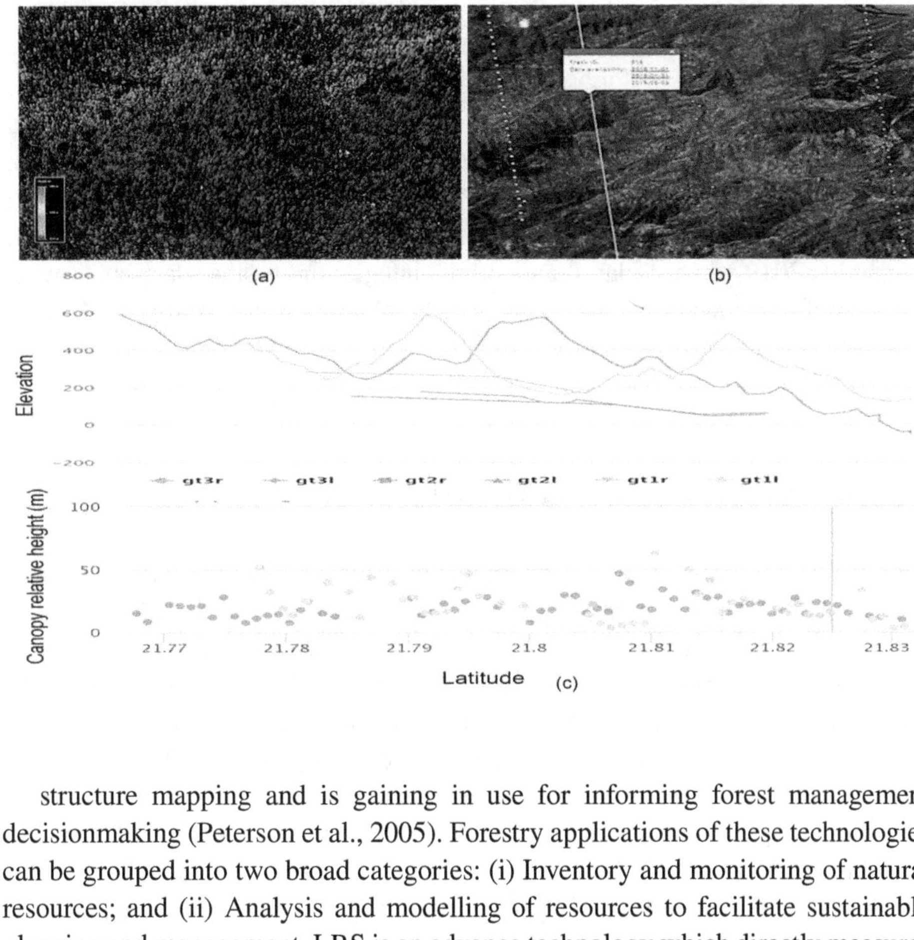

structure mapping and is gaining in use for informing forest management decisionmaking (Peterson et al., 2005). Forestry applications of these technologies can be grouped into two broad categories: (i) Inventory and monitoring of natural resources; and (ii) Analysis and modelling of resources to facilitate sustainable planning and management. LRS is an advance technology which directly measures both horizontal and vertical forest structure and offers many forestry applications (Dubayah and Drake, 2000; Lim et al., 2003). According to Lim et al. (2003), the procedure with which the Lidar system retrieved the 3-D canopy's structure in the vertical and horizontal direction; LRS systems for forestry uses can be as either 'discrete return' (Fig. 3b) or 'full-waveform' systems. LRS data can be generated from Spaceborne, Airborne and terrestrial instruments, which offers data from plot to regional scale. In spaceborne, currently, there is only availability of ICESat – based on the full waveform and ICESat-2 – based on photon counting Lidar technology LRS data. In airborne and terrestrial, the availability of LRS data is largely limited due to a very high cost. Fig. 4 shows the canopy height retrieval and surface elevation from recently launched ICESat- 2.

In recent years, LRS has become recognized as an important remote sensing tool for forest structure measurements and generating forest inventory data such as canopy heights, stand volume, basal area, DBH, canopy cover and assessing above-ground biomass (Peterson et al., 2005). Many recent studies such as Lefsky et al., 2005; Rosette et al., 2008; Nelson et al., 2009; Chen 2010; Duncanson et al., 2010; Mcinerney et al., 2010; Simard et al., 2011; Tonolli et al., 2011; Ahmed et al., 2013; Hayashi et al., 2013; Peterson and Nelson, 2014; Fayad et al., 2016; Neuenschwander and Magruder, 2016; Pourrahmati, 2016; Dhanda et al., 2017; Pourrahmati et al., 2018; Urbazaev et al., 2018; Zhang et al., 2019; Chen et al., 2019; Tang et al., 2019 where Lidar along with its integration in other datasets provide a quite useful estimation of canopy height, basal area, timber volume and AGB. Therefore, it is a good opportunity to use the LRS datasets in different perspectives and generate forest inventory data.

4. INTEGRATED USE OF HRS AND LRS

Data fusion is a promising method for producing estimates with improved quality and dimensionality. Hyperspectral data contains spectral information about forests while LiDAR data delivers the structural properties of forests. On fusing these two datasets, we get an improved estimate of forest characteristics. LiDAR and hyperspectral data fusion cab be of three levels, comprising pixel level, feature level, and decision level (Man et al., 2014). The way by which data is fused is depended on the variables to predict (Tusa et al., 2020). The integration of data from hyperspectral sensors and lidar shows improved results in García-Sopo et al., 2015; Chu et al., 2016; Huang and Zhu, 2013; Zhang, 2014; Liu et al., 2017; Liao et al., 2018. According to Zhang (2014), the integration of hyperspectral data with all lidar-derived features and based on an ensemble of three machine learning classifiers attained the best outcome with overall accuracy and Kappa coefficient of 86% and 0.82 respectively. Liu et al., 2017 classified the urban vegetation using a combination of airborne hyperspectral and LiDAR data. According to them, the integration of structural and spectral information delivered an enhanced classification accuracy as compared when used distinctly. Further, the variables derived from LiDAR data added more to accurate prediction than hyperspectral variables. The overall accuracy of 51.1%, 61%, and 70% were attained using hyperspectral, LiDAR and the combined data respectively. Liao et al., 2018 used deep learning to fuse hyperspectral and LiDAR data for vegetation classification. They developed a two-stage deep learning fusion method for hyperspectral and LiDAR data fusion. The results from complex, closed forest canopies revealed that two-stage deep fusion method is effective in fusing hyperspectral and LiDAR data for mapping. According to Zhang and Qiu (2012), the integration of Lidar data with hyperspectral data are not only proficient of identifying individual trees and estimating their attributes, but also detecting the species types with the proposed algorithm. Anderson et al. (2008) used the Minimum Noise Fraction to transform AVIRIS bands and combined with Lidar data. They stated an 8–10% increase in accuracy of forest biomass prediction for northeastern temperate forests La et al., 2015 examined the result of combination between hyperspectral and LiDAR data on individual tree extraction based on segmentation technique. The study added that individual tree detection with fusion data is marginally improved than single data. They found the fusion data can accurately identify 70% and 92% trees in high and low forest density areas.

5. CONCLUSION

This chapter is attempted to discuses the role of Hyperspectral, LIDAR and their fusion in enhancing the natural forest resources management. Many studies on hyperspectral, LIDAR and their fusion methods revealed that accurate and enhanced results were obtained on fusing this two techniques. These two techniques could be useful in identification, classification, and extraction of individual tree species features. Moreover, the application of these could be further extended to forest fire management, forest changes detection, biophysical and biochemical variables estimation.

LRS is useful in deriving vegetation properties such as height, DBH and canopy cover which proved a useful way to estimate the forest aboveground biomass. Biomass estimation from lidar includes different methods such as regression, simulation, and machine learning techniques. Different classification techniques such as Support vector machine, machine learning, deep learning, random forest, decision tree, artificial neural network are used in HRS to derive accurate classification outcome from hyperspectral data. Since LiDAR delineates the vertical structure of forests and hyperspectral data can provide detailed spectral information on biophysical parameters of forests; therefore, the fusion of LiDAR and hyperspectral data has received increased attention in the recent time. Besides, the fusion of lidar and hyperspectral with an ensemble of machine learning approaches could prove an enhanced estimate of forest characteristics. However, there is a need to make efforts to continuously incres the possibituies in enhancing the application of this method and techniques. Besides, there are some lacks such as data availability, high-cost non-global coverage, and storage problems need further attention to make these techniques are widely used for the sustainable management of the natural forest.

Figure 5. Broad application of LRS in forestry

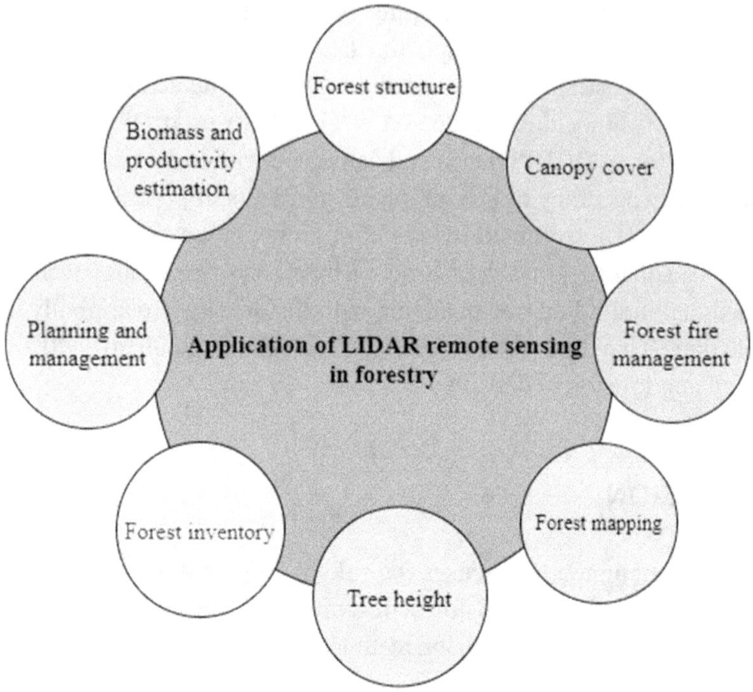

REFERENCES

Adam, E., Mutanga, O., & Rugege, D. (2010). Multispectral and hyperspectral remote sensing for identification and mapping of wetland vegetation: A review. *Wetlands Ecology and Management*, *18*(3), 281–296. doi:10.100711273-009-9169-z

Ahmed, R., Siqueira, P., & Hensley, S. (2013). A study of forest biomass estimates from lidar in the northern temperate forests of New England. *Remote Sensing of Environment*, *130*, 121–135. doi:10.1016/j.rse.2012.11.015

Anderson, J. E., Plourde, L. C., Martin, M. E., Braswell, B. H., Smith, M. L., Dubayah, R. O., Hofton, M., & Blair, J. (2008). Integrating waveform lidar with hyperspectral imagery for inventory of a northern temperate forest. *Remote Sensing of Environment: Remote Sensing Data Assimilation Special Issue*, *112*(4), 1856–1870. doi:10.1016/j.rse.2007.09.009

Andrew, M. E., & Ustin, S. L. (2008). The role of environmental context in mapping invasive plants with hyperspectral image data. *Remote Sensing of Environment*, *112*(12), 4301–4317. doi:10.1016/j.rse.2008.07.016

Andrew, M. E., & Ustin, S. L. (2009). Habitat suitability modelling of an invasive plant with advanced remote sensing data. *Diversity & Distributions*, *15*(4), 627–640. doi:10.1111/j.1472-4642.2009.00568.x

Asner, G. P., Knapp, D. E., Kennedy-Bowdoin, T., Jones, M. O., Martin, R. E., Boardman, J., & Hughes, R. F. (2008). Invasive species detection in Hawaiian rainforests using airborne imaging spectroscopy and LiDAR. *Remote Sensing of Environment*, *112*(5), 1942–1955. doi:10.1016/j.rse.2007.11.016

Atzberger, C., Darvishzadeh, R., Immitzer, M., Schlerf, M., Skidmore, A., & Maire, G. (2015). Comparative analysis of different retrieval methods for mapping grassland leaf area index using airborne imaging spectroscopy. *Int. J. Appl. L. Earth Obs.*, *43*, 19–31. doi:10.1016/j.jag.2015.01.009

Avitabile, V., Baccini, A., Friedl, M. A., & Schmullius, C. (2012). Capabilities and limitations of Landsat and land cover data for aboveground woody biomass estimation of Uganda. *Remote Sensing of Environment*, *117*, 366–380. doi:10.1016/j.rse.2011.10.012

Ballanti, L., Blesius, L., Hines, E., & Kruse, B. (2016). Tree Species Classification Using Hyperspectral Imagery: A Comparison of Two Classifiers. *Remote Sensing*, *8*(6), 445. doi:10.3390/rs8060445

Barnsley, M. J., Lewis, P., O'Dwyer, S., Disney, M. I., Hobson, P., Cutter, M., & Lobb, D. (2000). On the potential of CHRIS/PROBA for estimating vegetation canopy properties from space. *Remote Sensing Reviews*, *19*(1-4), 171–189. doi:10.1080/02757250009532417

Brovkina, O., Novotny, J., Cienciala, E., Zemek, F., & Russ, R. (2017). Mapping forest aboveground biomass using airborne hyperspectral and LiDAR data in the mountainous conditions of Central Europe. *Ecological Engineering*, *100*, 219–230. doi:10.1016/j.ecoleng.2016.12.004

Bulcock, H. H., & Jewitt, G. P. W. (2010). Spatial mapping of leaf area index using hyperspectral remote sensing for hydrological applications with a particular focus on canopy interception. *Hydrology and Earth System Sciences*, *14*(2), 383–392. doi:10.5194/hess-14-383-2010

Carlson, K. M., Asner, G. P., Hughes, R. F., Ostertag, R., & Martin, R. E. (2007). Hyperspectral remote sensing of canopy biodiversity in Hawaiian lowland rainforests. *Ecosystems (New York, N.Y.)*, *10*(4), 536–549. doi:10.100710021-007-9041-z

Casas, A., Riaño, D., Ustin, S. L., Dennison, P., & Salas, J. (2014). Estimation of water-related biochemical and biophysical vegetation properties using multitemporal airborne hyperspectral data and its comparison to MODIS spectral response. *Remote Sensing of Environment, 148*, 28–41. doi:10.1016/j.rse.2014.03.011

Castro-Esau, K. L., Sanchez-Azofeifa, G. A., & Caelli, T. (2004). Discrimination of lianas and trees with leaf level hyperspectral data. *Remote Sensing of Environment, 90*(3), 353–372. doi:10.1016/j.rse.2004.01.013

Caughlin, T. T., Graves, S. J., Asner, G. P., van Breugel, M., Hall, J. S., Martin, R. E., Ashton, M. S., & Bohlman, S. A. (2016). A hyperspectral image can predict tropical tree growth rates in single-species stands. *Ecological Applications, 26*(8), 2367–2373. doi:10.1002/eap.1436

Chang, C. I. (2007). *Hyperspectral Data Exploitation: Theory and Applications.* John Wiley & Sons. doi:10.1002/0470124628

Chen, B., Pang, Y., Li, Z., North, P., Rosette, J., Sun, G., Suárez, J., Bye, I., & Lu, H. (2019). Potential of Forest Parameter Estimation Using Metrics from Photon Counting LiDAR Data in Howland Research Forest. *Remote Sensing, 11*(7), 856. doi:10.3390/rs11070856

Chen, Q. (2010). Retrieving vegetation height of forests and woodlands over mountainous areas in the Pacific Coast region using satellite laser altimetry. *Remote Sensing of Environment, 114*(7), 1610–1627. doi:10.1016/j.rse.2010.02.016

Chu, H.-J., Wang, C.-K., Kong, S.-J., & Chen, K.-C. (2016). Integration of full-waveform LiDAR and hyperspectral data to enhance tea and areca classification. *Journal GIScience & Remote Sensing, 53*(4), 542–559. doi:10.1080/15481603.2016.1177249

Clark, M. L., Roberts, D. A., Ewel, J. J., & Clark, D. B. (2011). Estimation of tropical rain forest aboveground biomass with small-footprint lidar and hyperspectral sensors. *Remote Sensing of Environment, xxx*(11), xxx–xxx. doi:10.1016/j.rse.2010.08.029

Dalponte, M., Bruzzone, L., & Gianelle, D. (2008). Fusion of Hyperspectral and LIDAR Remote Sensing Data for Classification of Complex Forest Areas. *IEEE Transactions on Geoscience and Remote Sensing, 46*(5), 1416–1426. doi:10.1109/TGRS.2008.916480

Darvishzadeh, R., Atzberger, C., Skidmore, A., & Schlerf, M. (2011). Mapping grassland leaf area index with airborne hyperspectral imagery: A comparison study of statistical approaches and inversion of radiative transfer models. *ISPRS Journal of Photogrammetry and Remote Sensing*, *66*(6), 894–906. doi:10.1016/j.isprsjprs.2011.09.013

De Jong, S. M., Pebesma, E. J., & Lacaze, B. (2003). Above-ground biomass assessment of Mediterranean forests using airborne imaging spectrometry: The DAIS Peyne experiment. *International Journal of Remote Sensing*, *24*(7), 1505–1520. doi:10.1080/01431160210145560

Dhanda, P., Nandy, S., Kushwaha, S.P.S., Ghosh, S., Murthy, Y.V.N. K., & Dadhwal, V.K. (2017). Optimizing spaceborne LiDAR and very high resolution optical sensor parameters for biomass estimation at ICESat/GLAS footprint level using regression algorithms. *Prog Phys Geog*., 1-21. Doi:10.1177/0309133317693443

Dubayah, R. O., & Drake, J. B. (2000). Lidar remote sensing for forestry. *Journal of Forestry*, *98*, 44–46.

Duncanson, L. I., Niemann, K. O., & Wulder, M. A. (2010). Estimating forest canopy height and terrain relief from GLAS waveform metrics. *Remote Sensing of Environment*, *114*(1), 138–154. doi:10.1016/j.rse.2009.08.018

Fayad, I., Baghdadi, N., Bailly, J.-S., Barbier, N., Gond, V., Hérault, B., Hajj, M. E., Fabre, F., & Perrin, J. (2016). Regional Scale Rain-Forest Height Mapping Using Regression-Kriging of Spaceborne and Airborne LiDAR Data: Application on French Guiana. *Remote Sensing*, *8*(3), 240. doi:10.3390/rs8030240

Foody, G. M., Cutler, M. E., McMorrow, J., Pelz, D., Tangki, H., Boyd, D. S., & Douglas, I. (2001). Mapping the biomass of Bornean tropical rain forest from remotely sensed data. *Global Ecology and Biogeography*, *10*(4), 379–387. doi:10.1046/j.1466-822X.2001.00248.x

Gao, X., Huete, A., Ni, W., & Miura, T. (2000). Optical-biophysical relationships of vegetation spectra without background contamination. *Remote Sensing of Environment*, *74*(3), 609–620. doi:10.1016/S0034-4257(00)00150-4

García-Sopo, M. Á., Cuartero, A., Rodríguez, P. G., & Plaza, A. 2015. Hyperspectral And Lidar Data Integration And Classification. *Proceedings of the International Geoscience and Remote Sensing Symposium*, 57–60. 10.1109/IGARSS.2015.7325696

Glenn, N. F., Mundt, J. T., Weber, K. T., Prather, T. S., Lass, L. W., & Pettingill, J. (2005). Hyperspectral data processing for repeat detection of small infestations of Leafy spurge. *Remote Sensing of Environment, 95*(3), 399–412. doi:10.1016/j.rse.2005.01.003

Goetz, A. F. H., Vane, G., Solomon, J. E., & Rock, B. N. (1985). Imaging spectrometry for earth remote-sensing. *Science, 228*(4704), 1147–1153. doi:10.11 26cience.228.4704.1147

Gong, P., Pu, R., Biging, G. S., & Larrieu, M. R. (2003). Estimation of Forest Leaf Area Index Using Vegetation Indices Derived from Hyperion Hyperspectral Data. *IEEE Transactions on Geoscience and Remote Sensing, 41*(6), 1–8. doi:10.1109/TGRS.2003.812910

Gong, P., Pu, R., & Miller, J. R. (1995). Coniferous Forest Leaf Area Index Estimation along the 0regon Transect Uslng Compact Airborne Spectrogtaphic lmager Data. *Photogrammetric Engineering and Remote Sensing, 61*(9).

Goodwin, N., Turner, R., & Merton, R. (2005). Classifying eucalyptus forests with high spatial and spectral resolution imagery. An investigation of individual species and vegetation communities. *Australian Journal of Botany, 53*(4), 337–345. doi:10.1071/BT04085

Govender, M., Chetty, K., Naiken, V., & Bulcock, H. (2008). A comparison of satellite hyperspectral and muyltispectral remote sensing imagery for improved classification and mapping of vegetation. *Water S.A., 34*(2), 147–154. doi:10.4314/wsa.v34i2.183634

Halme, E., Pellikka, P., & Mõttus, M. (2019). Utility of hyperspectral compared to multispectral remote sensing data in estimating forest biomass and structure variables in Finnish boreal forest. *Int J Appl Earth Obs Geoinformation, 83*, 101942. doi:10.1016/j.jag.2019.101942

Hayashi, M., Saigusa, N., Oguma, H., & Yamagata, Y. (2013). Forest canopy height estimation using ICESat/GLAS data and error factor analysis in Hokkaido, Japan. *ISPRS Journal of Photogrammetry and Remote Sensing, 81*, 12–18. doi:10.1016/j.isprsjprs.2013.04.004

Huang, R., & Zhu, J. (2013). Integration of Hyperspectral Image and Lidar Data through Tri-training for Land Cover Semi-supervised Classification. In F. Bian, Y. Xie, X. Cui, & Y. Zeng (Eds.), *Geo-Informatics in Resource Management and Sustainable Ecosystem. Communications in Computer and Information Science* (Vol. 398). Springer. doi:10.1007/978-3-642-45025-9_57

Hudak, A. T., Lefsky, M. A., Cohen, W. B., & Berterretche, M. (2002). Integration of LiDARand Landsat ETM+ data for estimating and mapping forest canopy height. *Remote Sensing of Environment*, *82*(2-3), 397–416. doi:10.1016/S0034-4257(02)00056-1

Huesca, M., García, M., Roth, K. L., Casas, A., & Ustin, S. L. (2016). Canopy structural attributes derived from AVIRIS imaging spectroscopy data in a mixed broadleaf/conifer forest. *Remote Sensing of Environment*, *182*, 208–226. doi:10.1016/j.rse.2016.04.020

Hycza, T., Stereńczak, K., & Bałazy, R. (2018). Potential use of hyperspectral data to classify forest tree species. *New Zealand Journal of Forestry Science*, *48*(18), 1–13. doi:10.118640490-018-0123-9

Jones, G., Coops, N. C., & Sharma, T. (2010). Assessing the utility of airborne hyperspectral and LiDAR data for species distribution mapping in the coastal Pacific Northwest, Canada. *Remote Sensing of Environment*, *114*(12), 2841–2852. doi:10.1016/j.rse.2010.07.002

Khalefa, E. I., Smit, P. J., Nickless, A., Archibald, S., Comber, A., & Balzter, H. (2013). Retrieval of Savanna Vegetation Canopy Height from ICESat-GLAS Spaceborne LiDAR With Terrain Correction. *IEEE Geoscience and Remote Sensing Letters*, *10*(6), 1439–1443. doi:10.1109/LGRS.2013.2259793

La, H. P., Eo, Y. D., Chang, A., & Kim, C. (2015). Extraction of Individual Tree Crown using Hyperspectral Image and LiDAR Data. *KSCE Journal of Civil Engineering*, *19*(4), 1078–1087. doi:10.100712205-013-1178-z

Laurin, G. V., Chen, Q., & Lindsell, J. A. (2014). Above ground biomass estimation in an African tropical forest with lidar and hyperspectral data. *ISPRS Journal of Photogrammetry and Remote Sensing*, *89*, 49–58. doi:10.1016/j.isprsjprs.2014.01.001

Lawrence, R. L., Wood, S. D., & Sheley, R. L. (2006). Mapping invasive plants using hyperspectral imagery and breiman cutler classifications (randomforest). *Remote Sensing of Environment*, *100*(3), 356–362. doi:10.1016/j.rse.2005.10.014

Lee, K.-S., Cohen, W. B., Kennedy, R. E., Maiersperger, T. K., & Gower, S. T. (2004). Hyperspectral versus multispectral data for estimating leaf area index in four different biomes. *Remote Sensing of Environment*, *91*(3-4), 508–520. doi:10.1016/j.rse.2004.04.010

Lefsky, M. A., Cohen, W. B., Acker, S. A., Spies, T. A., Parker, G. G., & Harding, D. (1998). LiDAR remote sensing of forest canopy structure and related biophysical parameters at the H.J. Andrews Experimental Forest, Oregon, USA. *Geoscience and Remote Sensing Symposium Proceedings, IGARSS*, 1252–1254.

Lefsky, M. A., Cohen, W. B., Harding, D. J., Parker, G. G., Acker, S. A., & Gower, S. T. (2002). Lidar remote sensing of above-ground biomass in three biomes. *Global Ecology and Biogeography*, *11*(5), 393–399. doi:10.1046/j.1466-822x.2002.00303.x

Lefsky, M.A., Harding, D.J., Keller, M., 2005. Estimates of forest canopy height and aboveground biomass using ICESat. *Geophysical Research Letters, 32*(L22S02), 1-4. doi:10.1029/2005GL023971

Liao, W., Gao, L., Zhang, B., & Chanusso, J. (2018). Deep Learning for Fusion of APEX Hyperspectral and Full-Waveform LiDAR Remote Sensing Data for Tree Species Mapping. *IEEE Access : Practical Innovations, Open Solutions*, *6*, 68716–68729.

Lim, K., Treitz, P., Wulder, M., St-Onge, B., & Flood, M. (2003). LiDAR remote sensing of forest structure. *Progress in Physical Geography*, *27*(1), 88–106. doi:10.1191/0309133303pp360ra

Liu, L., Coops, N. C., Aven, N. W., & Pang, Y. (2017). Mapping urban tree species using integrated airborne hyperspectral and LiDAR remote sensing data. *Remote Sensing of Environment*, *200*, 170–182. doi:10.1016/j.rse.2017.08.010

Lucas, R. M., Lee, A. C., & Bunting, P. J. (2008). Retrieving forest biomass through integration of CASI and LIDAR data. *International Journal of Remote Sensing*, *29*(5), 1553–1577. doi:10.1080/01431160701736497

Maire, G., François, C., Soudani, K., Berveiller, D., Pontailler, J.-Y., Bréda, N., Genet, H., Davi, H., & Dufrêne, E. (2008). Calibration and validation of hyperspectral indices for the estimation of broadleaved forest leaf chlorophyll content, leaf mass per area, leaf area index and leaf canopy biomass. *Remote Sensing of Environment*, *112*(10), 3846–3864. doi:10.1016/j.rse.2008.06.005

Man, Q., Dong, P., Guo, H., Liu, G., & Shi, R. (2014). Light detection and ranging and hyperspectral data for estimation of forest biomass: A review. *Journal of Applied Remote Sensing*, *8*(1), 081598. doi:10.1117/1.JRS.8.081598

Martin, M. E., & Aber, J. D. (1997). High spectral resolution remote sensing of forest canopy lignin, nitrogen, and ecosystem processes. *Ecological Applications*, *7*(2), 431–443. doi:10.1890/1051-0761(1997)007[0431:HSRRSO]2.0.CO;2

Martin, M. E., Newman, S. D., Aber, J. D., & Congalton, J. C. (1998). Determining forest species composition using high spectral resolution remote sensing data. *Remote Sensing of Environment, 65*(3), 249–254. doi:10.1016/S0034-4257(98)00035-2

Matsuki, T., Yokoya, N., & Iwasaki, A. (2015). Hyperspectral tree species classification of Japanese complex mixed forest with the aid of LiDAR data. *IEEE Journal of Selected Topics in Applied Earth Observations and Remote Sensing, 8*(5), 2177–2187. doi:10.1109/JSTARS.2015.2417859

Mcinerney, D. O., Suarez-Minguez, J., Valbuena, R., & Nieuwenhuis, M. (2010). Forest canopy height retrieval using LiDAR data, medium-resolution satellite imagery and kNN estimation in Aberfoyle, Scotland. *Forestry, 83*(2), 1–12. doi:10.1093/forestry/cpq001

Mutanga, O., & Skidmore, A. K. (2004). Narrow band vegetation indices overcome the saturation problem in biomass estimation. *International Journal of Remote Sensing, 25*(19), 3999–4014. doi:10.1080/01431160310001654923

Narumalani, S., Mishra, D. R., Wilson, R., Reece, P., & Kohler, A. (2009). Detecting and mapping four invasive species along the floodplain of North Platte River, Nebraska. *Weed Technology, 23*(1), 99–107. doi:10.1614/WT-08-007.1

Nelson, R., Oderwald, R., & Gregoire, T. G. (1997). Separating the ground and airborne laser sampling phases to estimate tropical forest basal area, volume, and biomass. *Remote Sensing of Environment, 60*(3), 311–326. doi:10.1016/S0034-4257(96)00213-1

Nelson, R., Ranson, K. J., Sun, G., Kimes, D. S., Kharuk, V., & Montesano, P. (2009). Estimating Siberian timber volume using MODIS and ICESat/GLAS. *Remote Sensing of Environment, 113*(3), 691–701. doi:10.1016/j.rse.2008.11.010

Neuenschwander, A. L., & Magruder, L. A. (2016). The Potential Impact of Vertical Sampling Uncertainty on ICESat-2/ATLAS Terrain and Canopy Height Retrievals for Multiple Ecosystems. *Remote Sensing, 8*(12), 1039. doi:10.3390/rs8121039

Nilsson, M. (1996). Estimation of tree heights and stand volume using an airborne LiDAR system. *Remote Sensing of Environment, 56*(1), 1–7. doi:10.1016/0034-4257(95)00224-3

Peerbhay, K., Mutanga, O., & Ismail, R. (2016). The identification and remote detection of alien invasive plants in commercial forests: An Overview. *South African Journal of Geomatics, 5*(1).

Peerbhay, K. Y., Mutanga, O., & Ismail, R. (2013). Commercial tree species discrimination using airborne AISA Eagle hyperspectral imagery and partial least squares discriminant analysis (PLS-DA) in KwaZulu–Natal, South Africa. *SPRS J Photogramm Remote Sens.*, *79*, 19–28. doi:10.1016/j.isprsjprs.2013.01.013

Peerbhay, K. Y., Mutanga, O., & Ismail, R. (2015). Random forests unsupervised classification: The detection and mapping of solanum mauritianum infestations in plantation forestry using hyperspectral data. *IEEE Journal of Selected Topics in Applied Earth Observations and Remote Sensing*, *8*(6), 3107–3122. doi:10.1109/JSTARS.2015.2396577

Peng, D., Zhang, H., Yu, L., Wu, M., Wang, F., Huang, W., Liu, L., Sun, R., Li, C., Wang, D., & Xu, F. (2018). Assessing spectral indices to estimate the fraction of photosynthetically active radiation absorbed by the vegetation canopy. *International Journal of Remote Sensing*, *39*(22), 1–19. doi:10.1080/01431161.2018.1479795

Peterson, B., Dubayah, R., Hyde, P., Hofton, M., Blair, J. B., & Fites-Kaufman, J. 2005. Use of LIDAR for Forest Inventory and Forest Management Application. *Proceedings of the Seventh Annual Forest Inventory and Analysis Symposium, 193*-202.

Peterson, B., & Nelson, K. J. (2014). Mapping Forest Height in Alaska Using GLAS, Landsat Composites, and Airborne LiDAR. *Remote Sensing*, *6*(12), 12409–12426. doi:10.3390/rs61212409

Pourrahmati, M. R. (2016). *Investigating the possibility of forest height/volume estimation using lidar, radar and optical images: case study: Nowshahr Forests in Mazindaran*. Electronics. Université Montpellier.

Pourrahmati, M. R., Baghdadi, N., Darvishsefat, A. A., Namiranian, M., Gond, V., Bailly, J.-S., & Zargham, N. (2018). Mapping Lorey's height over Hyrcanian forests of Iran using synergy of ICESat/ GLAS and optical images. *European Journal of Remote Sensing*, *51*(1), 100–115. doi:10.1080/22797254.2017.1405717

Qin, H., Wang, C., Zhao, K., & Xi, X. (2018). Estimation of the fraction of absorbed photosynthetically active radiation (fPAR) in maize canopies using LiDAR data and hyperspectral imagery. *PLoS One*, *13*(5), e0197510. doi:10.1371/journal.pone.0197510

Rosette, J. A. B., North, P. R. J., & Suárez, J. C. (2008). Vegetation height estimates for a mixed temperate forest using satellite laser altimetry. *International Journal of Remote Sensing*, *29*(5), 1475–1493.

Sampson, P. H., Zarco-Tejada, P. J., Mohammed, G. H., Miller, J. R., & Noland, T. L. (2003). Hyperspectral remote sensing of forest condition: Estimating chlorophyll content in tolerant hardwoods. *For. Sci.*, *49*, 381–391. doi:10.1093/forestscience/49.3.381

Schlerf, M., & Atzberger, C. (2006). Inversion of a forest reflectance model to estimate structural canopy variables from hyperspectral remote sensing data. *Remote Sensing of Environment*, *100*(3), 281–294. doi:10.1016/j.rse.2005.10.006

Schlerf, M., Atzberger, C., & Hill, J. (2005). Remote sensing of forest biophysical variables using HyMap imaging spectrometer data. *Remote Sensing of Environment*, *95*(2), 177–194. doi:10.1016/j.rse.2004.12.016

Serrano, L., Peñuelas, J., & Ustin, S. L. (2002). Remote sensing of nitrogen and lignin in Mediterranean vegetation from AVIRIS data: Decomposing biochemical from structural signals. *Remote Sensing of Environment*, *81*(2-3), 355–364. doi:10.1016/S0034-4257(02)00011-1

Shen, G., Sakai, K., & Hoshino, Y. (2010). High Spatial Resolution Hyperspectral Mapping for Forest Ecosystem at Tree Species Level. *Agric. Inf. Res.*, *19*(3), 71–78. doi:10.3173/air.19.71

Simard, M., Pinto, N., Fisher, J. B., & Baccini, A. (2011). Mapping forest canopy height globally with spaceborne lidar. *Journal of Geophysical Research*, *116*(G4), G04021. Advance online publication. doi:10.1029/2011JG001708

Skowronek, S., Asner, G. P., & Feilhauer, H. (2017). Performance of one-class classifiers for invasive species mapping using airborne imaging spectroscopy. *Ecological Informatics*, *37*, 66–76. doi:10.1016/j.ecoinf.2016.11.005

Smith, M.-L., Ollinger, S. V., Martinm, M. E., Aber, J. D., Hallett, R. A., & Goodale, C. L. (2002). Direct estimation of aboveground forest productivity through hyperspectral remote sensing of canopy nitrogen. *Ecological Applications*, *12*(5), 1286–1302. doi:10.1890/1051-0761(2002)012[1286:DEOAFP]2.0.CO;2

Tan, C., Wang, D., Zhou, J., Du, Y., Luo, M., Zhang, Y., & Guo, W. (2018). Remotely Assessing Fraction of Photosynthetically Active Radiation (FPAR) for Wheat Canopies Based on Hyperspectral Vegetation Indexes. *Front. Plant Sci.*, *9*(776), 1–9. doi:10.3389/fpls.2018.00776

Tang, H., Armston, J., Hancock, S., Marselis, S., Goetz, S., & Dubayah, R. (2019). Characterizing global forest canopy cover distribution using spaceborne lidar. *Remote Sensing of Environment*, *231*, 111262. doi:10.1016/j.rse.2019.111262

Thenkabail, P. S., & Lyon, J. G. (2011). *Hyperspectral Remote Sensing of Vegetation*. CRC Press.

Thenkabail, P. S., Smith, R. B., & de Pauw, E. (2000). Hyperspectral vegetation indices and their relationships with agricultural crop characteristics. *Remote Sensing of Environment*, *71*(2), 158–182. doi:10.1016/S0034-4257(99)00067-X

Tong, Q., Xue, Y., & Zhang, L. (2014). Progress in Hyperspectral Remote Sensing Science and Technology in China Over the Past Three Decades. *IEEE Journal of Selected Topics in Applied Earth Observations and Remote Sensing*, *1*(1), 70–90. doi:10.1109/JSTARS.2013.2267204

Tonolli, S., Dalponte, M., Neteler, M., Rodeghiero, M., Vescovo, L., & Gianelle, D. (2011). Fusion of airborne LiDAR and satellite multispectral data for the estimation of timber volume in the Southern Alps. *Remote Sensing of Environment*, *115*(10), 2486–2498. doi:10.1016/j.rse.2011.05.009

Tusa, E., Laybros, A., Monnet, J.-M., & Mura, M. D. (2020). *Fusion of hyperspectral imaging and LiDAR for forest monitoring* (Vol. 32). Data Handling in Science and Technology. doi:10.1016/B978-0-444-63977-6.00013-4

Underwood, E., Ustin, S., & Dipietro, D. (2003). Mapping nonnative plants using hyperspectral imagery. *Remote Sensing of Environment*, *86*(2), 150–161. doi:10.1016/S0034-4257(03)00096-8

Urbazaev, M., Thiel, C., Cremer, F., Dubayah, R., Migliavacca, M., Reichstein, M., & Schmullius, C. (2018). Estimation of forest aboveground biomass and uncertainties by integration of field measurements, airborne LiDAR, and SAR and optical satellite data in Mexico. *Carbon Balance and Management*, *13*(5), 1–20. doi:10.118613021-018-0093-5

Ustin, S. L., DiPietro, D., Olmstead, K., Underwood, E., & Scheer, G. J. (2002). Hyperspectral remote sensing for invasive species detection and mapping. *IEEE Geoscience and Remote Sensing Symposium*, *3*, 1658 – 1660 10.1109/IGARSS.2002.1026212

Van Aardt, J., & Norris-Rogers, M. (2008). Spectral-age interactions in managed, even-aged eucalyptus plantations: Application of discriminant analysis and classification and regression trees approaches to hyperspectral data. *International Journal of Remote Sensing*, *29*(6), 1841–1845. doi:10.1080/01431160701874546

Vyas, D., Christian, B., & Krishnayya, N. S. R. (2013). Canopy level estimations of chlorophyll and LAI for two tropical species (teak and bamboo) from Hyperion (EO1) data. *International Journal of Remote Sensing*, *34*(5), 1676–1690. doi:10.1080/01431161.2012.725484

Vyas, D., Krishnayya, N. S. R., Manjunath, K. R., Ray, S. S., & Panigrahy, S. (2011). Evaluation of classifiers for processing Hyperion (EO-1) data of tropical vegetation. *International Journal of Applied Earth Observation and Geoinformation*, *13*(2), 228–235. doi:10.1016/j.jag.2010.11.007

Wang, Z., Wang, T., Darvishzadeh, R., Skidmore, A., Jones, S., Suarez, L., Woodgate, W., Heiden, U., Heurich, M., & Hearne, J. (2016). Vegetation Indices for Mapping Canopy Foliar Nitrogen in a Mixed Temperate Forest. *Remote Sensing*, *8*(6), 491. doi:10.3390/rs8060491

Wu, J., van Aardt, J. A. N., Asner, G. P., Mathieu, R., Kennedy-Bowdoin, T., Knapp, D., Wessels, K., Erasmus, B. F. N., & Smit, I. (2009). Connecting the dots between laser waveform and herbaceous biomass for assessment of land degradation using small-footprint waveform LIDAR data. *Proc. IEEE Int. Geosci. Remote Sens. Symp.*, 2, 334–337. 10.1109/IGARSS.2009.5418078

Wulder, M. (1998). Optical remote-sensing techniques for the assessment of forest inventory and biophysical parameters. *Progress in Physical Geography*, *22*(4), 449–476. doi:10.1177/030913339802200402

Zhang, C. (2014). Combining Hyperspectral and Lidar Data for Vegetation Mapping in the Florida Everglades. *Photogrammetric Engineering and Remote Sensing*, *80*(8), 9–19. doi:10.14358/PERS.80.8.733

Zhang, C., & Qiu, F. (2012). Mapping Individual Tree Species in an Urban Forest Using Airborne Lidar Data and Hyperspectral Imagery. *Photogrammetric Engineering and Remote Sensing*, *78*(10), 1079–1087.

Zhang, L., Shao, Z., Liu, J., & Cheng, Q. (2019). Deep Learning Based Retrieval of Forest Aboveground Biomass from Combined LiDAR and Landsat 8 Data. *Remote Sensing*, *11*(12), 1459. doi:10.3390/rs11121459

Zhang, Y. Q., Chen, J. M., Miller, J. R., & Noland, T. L. (2008). Leaf chlorophyll content retrieval from airborne hyperspectral remote sensing imagery. *Remote Sensing of Environment*, *112*(7), 3234–3247. doi:10.1016/j.rse.2008.04.005

Chapter 7
Synergy of Satellite-Derived Drought Indices for Agricultural Drought Quantification and Yield Prediction

Dipti Ladli
Central University of Jharkhand, India

Kanhaiya Lal
Central University of Jharkhand, India

Kiran Jalem
National Institute of Rural Development and Panchayati Raj, Hyderabad, India

Avinash Kumar Ranjan
National Institute of Technology, Rourkela, India

ABSTRACT

The present study was conducted over Jharkhand state (India) for assessing the drought condition and corresponding yield of paddy (district-level) during Kharif 2018. Vegetation drought indices, namely Vegetation Condition Index (VCI), Temperature Condition Index (TCI), Vegetation Health Index (VHI), and vegetation indices (VI) anomaly, were derived from different VI (i.e., NDVI, EVI) to assess the paddy health condition during drought year (2018) and non-drought year (2017). Later, the correlation between the DES-based yield data and derived drought indices (for the year 2017) were made to develop the district-level paddy yield model for

DOI: 10.4018/978-1-7998-5027-4.ch007

the drought year 2018. The key results of the study shown that VCI derived from EVI data was found to be more reasonable to depict the drought condition, wherein ~21% area was under severe drought condition, 43% area under moderate drought condition, and 36% area under no drought condition. In addition, the yield prediction model derived from VCI (EVI-based) was found to be promising for predicting the paddy yield for Kharif 2018 with fair R2 of 0.53.

1. INTRODUCTION

Agriculture and allied sectors have a significant contribution to the development of countries economy, wherein, it provides nutritious need to the billions of the world's populations. During last few decades, the agriculture sector has been dramatically affected by the climate change events e.g. unseasonal rainfall, drought, hailstorms, strong wind surges, etc. (Patel et al., 2012; Dutta et al., 2015). These undesirable weather event poses loss in agriculture production; and subsequently, it leads to many socio-economic issues, such as, escalation in global commodity values, remortgage or sale of productive croplands to endorse the real estate industry, migration of peoples from rural area to municipal centre, increases the rashness and social offence, etc. (Vaani and Porchelvan., 2018).

Agricultural drought is one of the serious and undesirable events, which poses huge loss in agricultural productivity (especially during Kharif season) (Dalezios et al., 2014). Basically, drought is a phenomenon of a long period of dry conditions where a region observes a deficit in its water supply, whether surface or underground water. Droughts are recurring phenomena in India owing to moisture stress; and comes under the chronically drought-prone areas, and they receive less than ~750 mm rainfall, while, 35% of the region comes under the drought-prone area, and they receive ~750-1125 mm rainfall. Altogether, ~68% of the country falls under the drought-prone region and experiences severe drought conditions (PACS Programme 2001-2008). The agricultural drought badly affects the economy of many agrarian dominant countries (i.e. India, China, Bangladesh, Vietnam), where more than 65% of people are dependent on the agriculture and allied sector. In India, about 16% of the agricultural area comes under the drought-prone region, wherein, ~50 million people are annually affected by the drought events (DAC 2009). However, continuous decline in the share of agriculture and allied sectors in the Gross Value Added (GVA) is observed from 18.6% (during 2013-14) to 17.4% (during 2016-2017) (DAC annual Report 2017-18).

Giving importance to the agriculture sector, it is very crucial to have detailed and timely information on agricultural drought so that the impact of the drought can be mitigated by making the appropriate decision. Albeit, it is quite hard to acquire detailed, timely, and accurate information on drought by traditional survey-based methods. Typically, survey-based methods are cumbersome, required large man-power, costly, and large time-taking. In this line, remote sensing (RS) technology and Geographical Information System (GIS) have proven its effectiveness to monitor the agriculture system by its specific capabilities of spectral, temporal, radiometric, and spatial characteristics (Murthy et al., 2007; Ranjan et al., 2016a, 2016b). The RS-based methods provide a cost-effective, robust, reliable, easy-access, and time-saving aspect to monitor, mapping, and policy-making opportunities (Ranjan and Parida, 2019).

Basically, RS-based methods are used for crop acreage mapping and yield estimation, cropping pattern analysis, crop water management, agricultural drought assessment and monitoring, horticultural enlargement, precision agriculture farming, soil types mapping, watershed management and monitoring, assessment of climatic events on agriculture, and so on (Navalgund and Ray, 2000; Panigrahy and Ray, 2006; Navalgund et al., 2007; Parida and Ranjan, 2019a). In this regard, wide-range of study has been conducted over the globe using remote sensing technique to monitor the agriculture system (Ji et al., 2003; Jain et al., 2015; Sholihah et al., 2016). Nandeesha and Ramu (2015) conducted a study using 250m resolution MODIS TERRA images over central dry agro-climatic zone to assess the agricultural drought situation. Further, Sahu and Patel (2016) have utilized MODIS Normalized Difference Vegetation Index (NDVI) product from 2002-2015 in Bundelkhand region (India) for monitoring agricultural drought. Vaani and Porchelvan (2018) have also used the long-term NDVI of Global Inventory Modeling and Mapping Studies (GIMMS) for the period of 20 years (1984-2003) to derive Vegetation Condition Index (VCI), and thereby, to recognize the vegetation vigour and the fortnightly deviation of VCI during major crop growing period (June to September). Das et al., (2017) utilized the MODIS-based NDVI product for the years 2006 (Kharif season) and 2010 (both Kharif and Ravi season) over Puruliya district, West Bengal to assess the agriculture drought.

In the same way, numerous studies have been successfully deployed the various vegetation indices for agricultural drought assessment (Muthumanickam et al., 2011; Dutta et al., 2015; Amalo et al., 2017; Wu et al., 2017; Lim et al., 2019; Zuo et al., 2019; Parida and Ranjan, 2019b). However, mostly, coarser spatial resolution satellite data (i.e. NOAA-AVHRR, MODIS) have been used for drought assessment over the larger spatial coverage area due to larger area coverage (Boken et al., 2004; Hua et al., 2019). In addition, coarser spatial resolution datasets are typically available in high temporal resolution (i.e. 8-days, 16-days, composite). Such a specific ability of satellite data offers a dynamic way for quantitative assessment of agricultural drought. Thereby, the overarching objective of the present case study is: (i) to quantify the district-level agricultural drought (Kharif) over Jharkhand state (India) using various satellite-derived drought index (i.e. VCI, Temperature Condition Index (TCI), Vegetation Health Index (VHI), NDVI and Enhanced Vegetation Index (EVI) anomaly) for the year 2018 as compared to 2017; (ii) to predict the paddy yield at district-level over Jharkhand state for the year 2018 using drought indices and DES-based APY data.

2. STUDY AREA

The Jharkhand state of India is selected for this case study, as Jharkhand has experienced unprecedented drought condition during Kharif 2018 due to insufficient rainfall. District-wise rainfall trend along with corresponding NDVI trend (of a crop) is comparatively plotted for the year 2017 (non-drought year) and 2018 (a drought year) as shown in **Fig. 1** and **2**. Notably, we can see that during the year 2018 rainfall is less over the entire district (as compared to 2017), consequently, less NDVI is also noted during 2018 the entire district. These things indicate that there was agricultural drought condition during 2018, and thereby we have selected Jharkhand state as a study area.

Figure 1. District-wise rainfall over JH state for the year 2017 (normal year) & 2018 (a drought year)

Source: CHIRPS precipitation (Climate Hazards Group Infrared Precipitation with Station data) (ftp://ftp.chg.ucsb.edu/pub/org/chg/products/CHIRPS-2.0/)

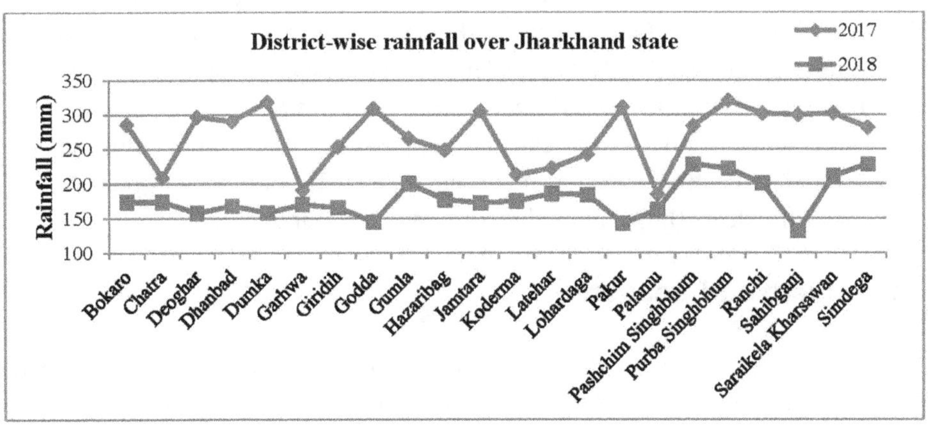

Figure 2. District-wise NDVI over JH state derived from MODIS data for the year 2017 (normal year) & 2018 (drought year)
Source: MNCFC (Mahalanobis National Crop Forecast Centre) (http://14.139.50.199:8080/ncfc/download.html).

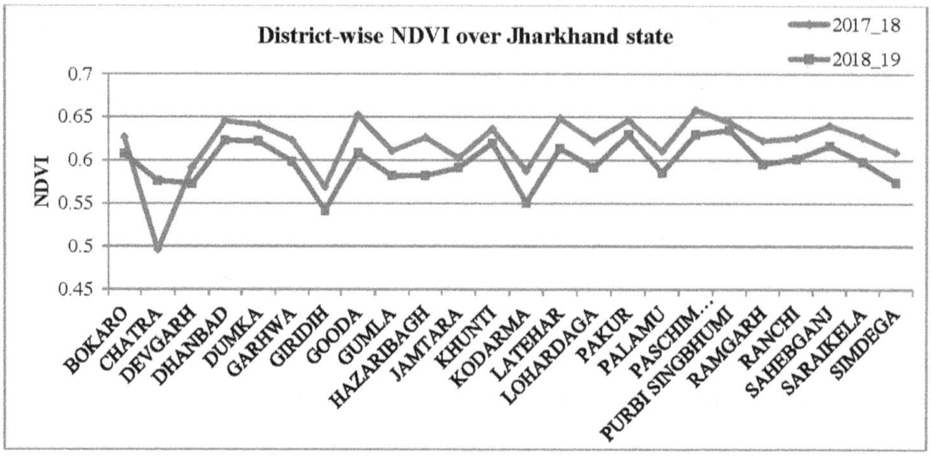

2.1. Geography of the Study Area

The name Jharkhand comes from two words “Jhar” and “Khand” tends to “Bushland” or the land of forest (https://en.wikipedia.org/wiki/Jharkhand). The geographical coordinates of Jharkhand State (India) ranges between 21° 55’ to 25° 35’ North (Latitude) and 83° 20’ to 88° 02’ East (Longitude) (**Fig. 3**). The administrative border of Jharkhand states shares with Bihar (in North), Uttar Pradesh (in Northwest), Chhattisgarh (in West), Odisha (in South), and West Bengal (in East). Most of the portion of Jharkhand state lies under the Chota Nagpur Plateau. In addition, many rivers and tributaries (i.e., Damodar, Ganga, North Koel, South Koel, Sankh, Brahmani, Gumani, and Subarnarekha rivers) pass/originates through this plateau. The hilly and undulating plateau is predominantly characterized by tropical forests and tribal settlements (https://en.wikipedia.org/wiki/Jharkhand).

2.2. Climate and Season

The climate condition of Jharkhand states diverges from humid subtropical (over north) to tropical wet and dry (over south-east). In Jharkhand state, there are mainly three-season i.e. summer, rainy, and winter; moreover, autumn and spring season are also experienced. The summer season goes on from mid-April to mid-June and highest average temperature during this season goes ~38 °C and low temperatures around 25°C (JAPCC 2014). The southwest monsoon (from mid-June to October)

Figure 3. Location map of the JH state (India) with district boundary

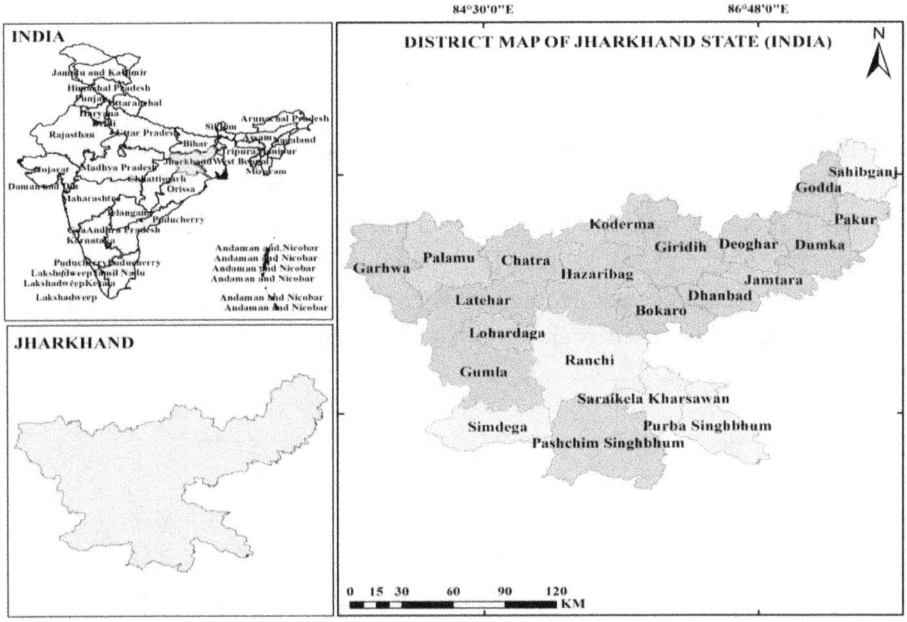

carries annual rainfall nearly all portion of the state (ranges from ~ 1,000 mm to 1,500 mm) (JAPCC 2014). Almost maximum annual precipitation occurs in the month of July and August, and hence paddy is prime agricultural crop sown over the entire state over large extent.

2.3. Agriculture Status in Jharkhand

In Jharkhand, agriculture and allied sectors are the backbone for the ~80% of the rural population (JSADP, 2012). Agriculture and related deeds are their employment and primary source of income. Roughly 91% agriculture area of Jharkhand state is under food crops (~84% under cereals and ~7% under pulses). In addition, cash crops found~3% followed by oilseeds, sugarcane, and tobacco. Whereas, ~6% of the cropped area comes under the vegetables and fruits (JSDMP 2011). The Jharkhand state's agricultural economy is characterized and dependent on nature, low investment, paddy monoculture as the dominant crop, inadequate irrigation facilities, limited and marginal holdings

3. MATERIALS AND METHODOLOGY

3.1. Data Used

In this study, basically, MODIS-based products i.e. MODIS NDVI & EVI data (MOD13Q1), MODIS LST & Emissivity (E) data (MOD11A2), MODIS Land Cover Product (MCD12Q1) were used to accomplished the respective objectives. The MODIS products were downloaded from the United State Geological Survey (USGS) earth explorer[1]. Moreover, DES-based district-wise area, production, and yield (APY) data of paddy were collected for the year 2017 and 2018. The details of the data used, their characteristics, and purpose are briefly described in corresponding sections.

MODIS NDVI and EVI Data (MOD13Q1)

In this study, MODIS based MOD13Q1 product was preferred, as it provides NDVI and EVI data in 16 days composite with 250m of spatial resolution. These datasets were acquired from USGS earth explorer for Kharif 2017 and 2018. MODIS NDVI and EVI data offers to assess the vegetation status during a particular season. The EVI reduces the canopy-soil discrepancies and enhance the sensitivity over dense vegetation condition. These time-series NDVI & EVI datasets were used to derive the various drought indices.

MODIS LST and Emissivity(E) Data (MOD11A2)

The MOD11A2 (v6) product offers 8-days composite (average) Land Surface Temperature and Emissivity (LST&E) data with 1 km of spatial resolution. The datasets were downloaded from USGS earth explorer corresponding to Kharif 2017 and 2018. The day and night-time surface temperature bands are associated with quality control assessments, observation times, view zenith angles, and clear-sky coverage's along with bands 31 and 32 (emissivity). The MOD11A2 datasets were used to derive the temperature condition index for agricultural drought assessment.

MODIS Land Cover Product (MCD12Q1)

The MODIS MCD12Q1 product provides land use land cover data. The MODIS land cover product is a global land cover classification data, which layer is annually produced in 500m of spatial resolution. The MODIS-based land cover product is usually pre-classified as per the International Geosphere-Biosphere Programme (IGBP) classification scheme. In this study, the MODIS land cover product was used to mask out the non-agricultural land from the derived agriculture drought map. So that the agricultural drought over the study area can be depicted precisely.

DES-Based Area, Production, and Yield (APY) Data

District-wise APY data was acquired from the Directorate of Economic Survey (DES)[2]. The DES-based yield data was used to correlate with various drought indices, and to drive a linear regression equation/model for yield prediction. The DES-based yield data were plotted against the corresponding drought indices of 2017 (district-level) and yield prediction model was derived to predict the yield for Kharif 2018.

Table 1. **D***ata used in the present study*

Data	Spatial Resolution	Temporal Resolution	Time (Period)	Use	Source
MCD12Q1	500 m	Yearly composite	2017	LU/LC Map	USGS
MOD13Q1	250 m	16 Days composite	2017, 2018	VCI, VHI, NDVI Anomaly Map	USGS
MOD11A2	1 km	8 Days composite	2017, 2018	TCI, VHI Map	USGS
APY data	District-level	Yearly	2017	Yield Model	DES

3.2. Methods

The methods section includes the satellite data pre-processing and processing, derivation of various drought indices, drought assessment, yield model development, and district wise yield prediction of paddy for the year 2018. The self-explanatory flowchart of the methodology is provided in **Fig. 4**. The brief of some key steps are explained under corresponding subsections.

Satellite Data Pre-Processing

Typically, MODIS products are having a sinusoidal projection, so, these datasets were reprojected to geographical latitude and longitude projection and the datum is assigned as WGS 1984. Further, all images were transformed into the typical format by multiplying with the scale factor (i.e. 0.0001). Few miscellaneous pre-processing i.e. AOI subsetting, resampling, were also done. Nevertheless, maximum value

Figure 4. Flow chart of the methodology adopted in the present study

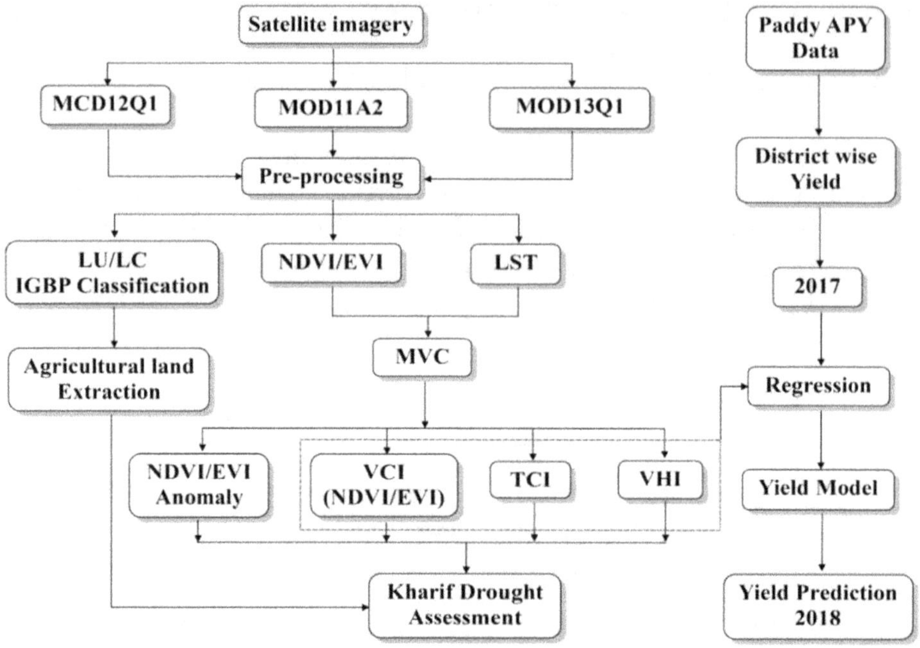

composite (MVC) was done to get a single image file (for entire Kharif season) from time-series satellite data (Holben et al., 1986).

Vegetation Condition Index (VCI)

In this study, the Vegetation Condition Index (VCI) was used for the evaluation of Kharif drought condition over the study area. VCI provides information on how the current vegetation status is compared to the maximum and minimum historical status. It helps to isolate the short-term weather signals from the long-term environmental signal in the NDVI/EVI (Kogan, 1997a; 1997b). The VCI can be calculated by following **Eq. 1** and **2** (using NDVI and EVI, respectively); its scale typically ranges from 0-100 and is expressed in percentage (%). Extremely dry month and poor vegetation may be around 0 percent, and healthy vegetation may be around 50 percent. Although, VCI values between 50 and 100 percent indicate conditions that are ideal or above average.

$$\mathbf{VCI}_{NDVIi} = (NDVIi - NDVI_{MINi})/(NDVI_{MAXi} - NDVI_{MINi})*100 \quad (1)$$

$$\mathbf{VCI}_{\mathbf{EVIi}} = (EVIi - EVI_{MINi})/(EVI_{MAXi} - EVI_{MINi})*100 \quad (2)$$

Where, $NDVI_i/EVI_i$ = NDVI/EVI at the time i; $NDVI_{MINi}/EVI_{MINi}$ = historic minimum NDVI/EVI; $NDVI_{MAXi}/EVI_{MAXi}$ = historic maximum NDVI/EVI.

Temperature Condition Index (TCI)

TCI was suggested by Kogan (1995), and can be calculated by using **Eq.3.** TCI is used to assess the vegetation response to temperature, where maximum temperature indicates more drought. TCI is based on LST normalization for each pixel by using maximum and minimum temperature in a given time series. It. The TCI scale ranges from 0 to 100 and is expressed in percentage (%). The 50 percent TCI value suggests mild or no drought. The TCI value of 100% reflects the optimal / above average situation and TCI values below 50% reflect different degrees of severe drought.

$$\mathbf{TCI}_{\mathbf{LSTi}} = (LST_i - LST_{MINi})/(LST_{MAXi} - LST_{MINi})*100 \quad (3)$$

Where, LST_i= LST at the time i; LST_{MINi}= historic minimum LST; LST_{MAXi}= historic maximum LST.

Vegetation Health Index (VHI)

VHI projects the overall vegetation/crops health condition in a given time and space (Kogan, 2000). VHI is being used for various purpose, such as, drought detection, drought severity assessment, early drought monitoring (Seiler et al., 1998; Du et al., 2018), crop yield estimation during the growing season (Unganai and Kogan, 1998), biomass and vegetation density estimation (Gitelson et al., 1998), evaluation of irrigated areas (Boken et al., 2004), and estimation wetness condition (Unganai and Kogan, 1998). VHI is usually calculated using VCI and TCI (**Eq. 4),** and its value ranges from 0-100 (expressed in percentage). The higher value of VHI indicates more heat stress while the lower value of VHI reflects less heat stress on the crop.

$$\mathbf{VHI} = 0.5(VCI) + 0.5(TCI) \quad (4)$$

NDVI and EVI Anomaly

NDVI/EVI can be used as an index to assess crop condition through analysis of NDVI/EVI anomaly (Murali et al., 2008). Vegetative drought index was determined using the values of NDVI/EVI. Maximum NDVI/EVI and long term mean maximum

NDVI/EVI in the long term was computed in order to derive NDVI/EVI anomaly. These can be calculated using **Eq. 5** and **6**.

NDVI anomaly = [(NDVI Max - Mean NDVI Max)/ (Mean NDVI Max)]*100 (5)

EVI anomaly= [(EVI Max - Mean EVI Max)/ (Mean EVI Max)]*100 (6)

Where, NDVI max =Maximum NDVI of the year; Mean NDVI max=long term mean maximum NDVI of the range of the year. EVI max =Maximum EVI of the year; Mean EVI max=long term mean maximum EVI of the range of the year.

Linear Regression-Based Yield Model

In linear regression analysis, one can predict the value of an individual variable from the second variable. The predicted variable is referred to as Y, and the variable which are basing the predictions on is called the predictor variable (referred to X). In this study, we have plotted drought indices (i.e. VCI, TCI, VHI, etc.) against the DES-based yield of 2017 for linear regression analysis. Consequently, derive equation is followed as a yield model for 2018 as written in **Eq. 7**.

$$\mathbf{Y} = a + bX \quad (7)$$

Where, X=drought indices; Y=predicted yield for 2018; b=coefficient of X, a=intercept.

4. RESULT AND DISCUSSION

4.1. Land Use Land Cover (LULC) and cropland masking

The LULC map of Jharkhand was prepared from the MODIS land cover product (MCD12Q1) which was pre-classified based on the International Geosphere-Biosphere Programme (IGBP) classification scheme. The various LU/LC features overs the study area are found as Forest, Savannas, Grasslands, Croplands, Urban and Built-up lands, Natural vegetations, Barren land, and Waterbodies as represented in **Fig. 5**. Notably, cropland covers a larger geographical area of the state as compared to other LU/LC classes.

Figure 5. LULC map of JH state derived from MCD12Q1 based on the IGBP classification

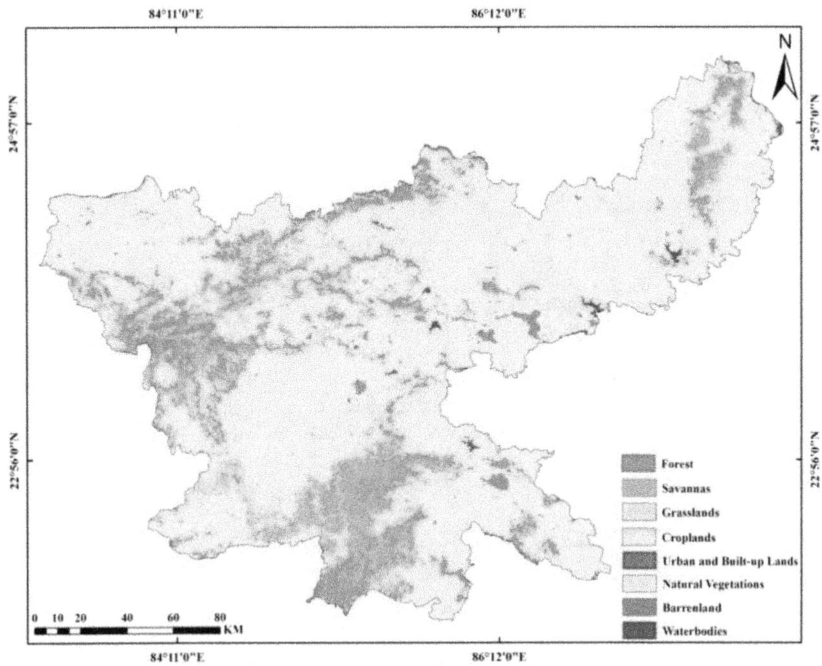

Further, agricultural land over the state was extracted from the derived LU/LC map as shown in **(Fig.6)**. As this study is concentrated on agricultural drought we masked out the other features from the land cover map except for agriculture land. This agricultural map was further used to extract the agricultural land from further derived maps i.e. VCI, TCI, VHI, etc.

4.2. Drought Assessment Using Vegetation Condition Index (VCI)

The district-wise VCI maps of the state are prepared based on both NDVI and EVI datasets as presented in **Fig. 7** and **8.** The maps were broadly classified into three classes i.e. severe drought, moderate drought, and no drought. Wherein, VCI value between 0-30, 30-50, and 50-100 were considered as a severe drought, moderate drought, and no drought, respectively. The result of VCI derived from NDVI data for the normal year 2017 (**Fig. 7**(**A**)) shows that the area with severe, moderate, and no drought was 2188, 8980, and 7645 sq.km, respectively. While area coverage of severe, moderate, and no drought during 2018 (drought year in JH state) was

Figure 6. Agriculture area masked from MODIS LULC map

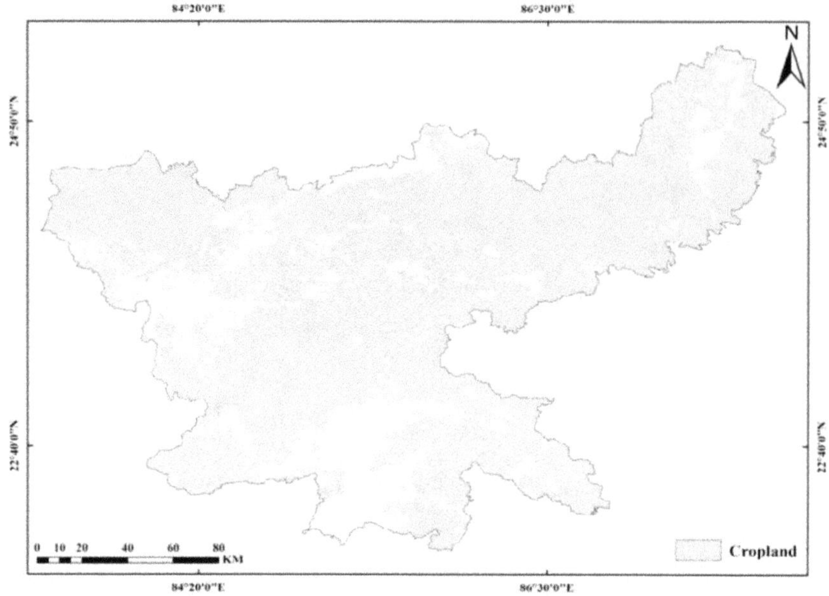

observed as 3562, 9010, and 6242 sq.km, respectively. Thereby, it can be said that most of the area (~48% area of the state) during 2018 was covered with moderate drought condition. Wherein severe drought regions were detected over 19% area. Although, 33% of the area was found under no drought condition. Interestingly, 48% of the area in 2017 has been also detected under moderate drought condition. These ambiguities may be driven due to the coarser spatial resolution of the MODIS satellite data and cloud impacts on data.

Furthermore, EVI-based VCI map was generated for the year 2017 (normal year) and 2018 (drought year) as presented in **Fig. 8(A)**, & (**B**), respectively. The result shows that ~2624 and 3982 sq.km area was covered by severe drought condition during the year 2017 and 2018, respectively. Moreover, ~7562 and 8063 sq.km area was detected under moderate drought condition during the year 2017 and 2018, respectively. The area under no drought condition in 2017 and 2018 was found as ~8624 and 6766 sq.km, respectively. Notably, the spatial coverage of severe drought & moderate drought region has shown increment and no drought region shows a decrement in the drought year 2018 as compared to the normal year 2017. The VCI map derived from EVI is perceived as more consistent than NDVI based VCI map. Notably, NDVI could show better result than EVI data, as EVI is more responsive to the crop canopy structure and leaf area index of the crops; wherein, NDVI is more responsive to the chlorophyll content present in the foliage.

Figure 7. Vegetation condition index (VCI) map of JH state derived from MODIS NDVI data

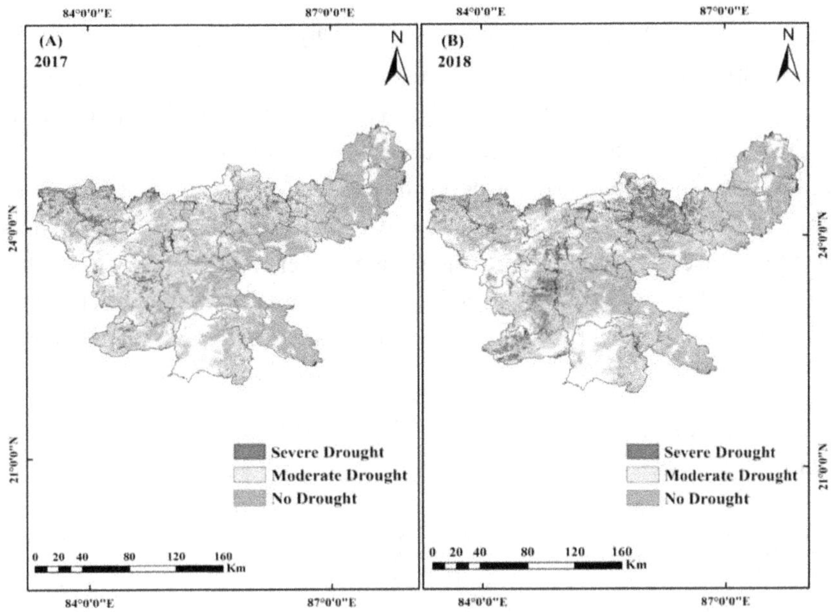

Figure 8. Vegetation condition index (VCI) map of JH state derived from MODIS EVI data

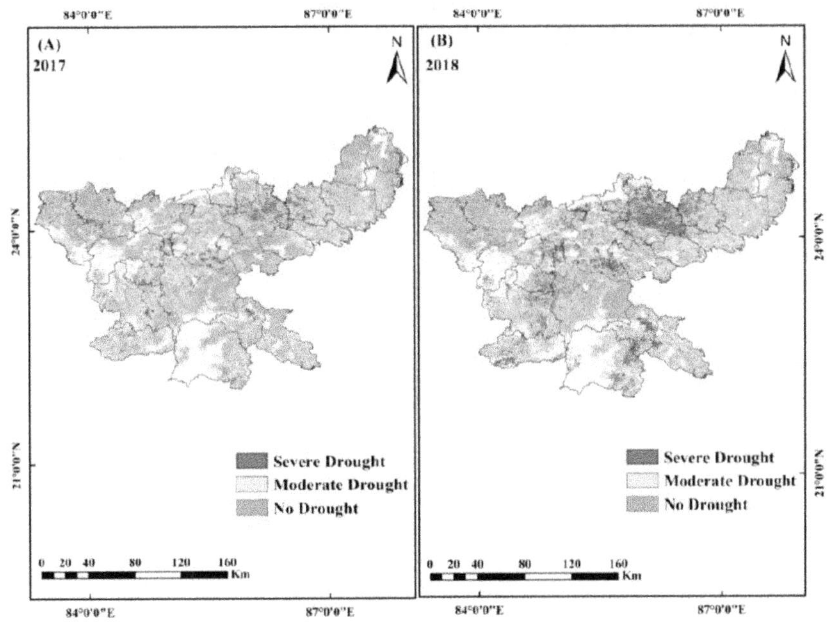

4.3. Drought Assessment using Temperature Condition Index (TCI)

The TCI maps (**Fig.9** (**A**) & (**B**)) shows that, in 2017, the area covered by severe drought (0-30), moderate drought (30- 50) & no drought (50 -100) were approx. 4352, 7944 and 6549 sq.km respectively. Whereas, in 2018, the area covered by severe drought, moderate drought & no drought was approx. 5594, 8231 and 5020 sq.km, respectively. Maximum agricultural land was observed under the moderate drought condition in the year 2018, also it was analyzed that about 1242 and 287 sq.km of the total coverage area shows increment under severe and moderate drought condition as well as 217 sq.km of total converge area shows decrement under no drought condition in 2018 (drought area) as compared to 2017 (normal year).

Figure 9. Temperature condition index (TCI) map of JH state derived from MODIS LST data

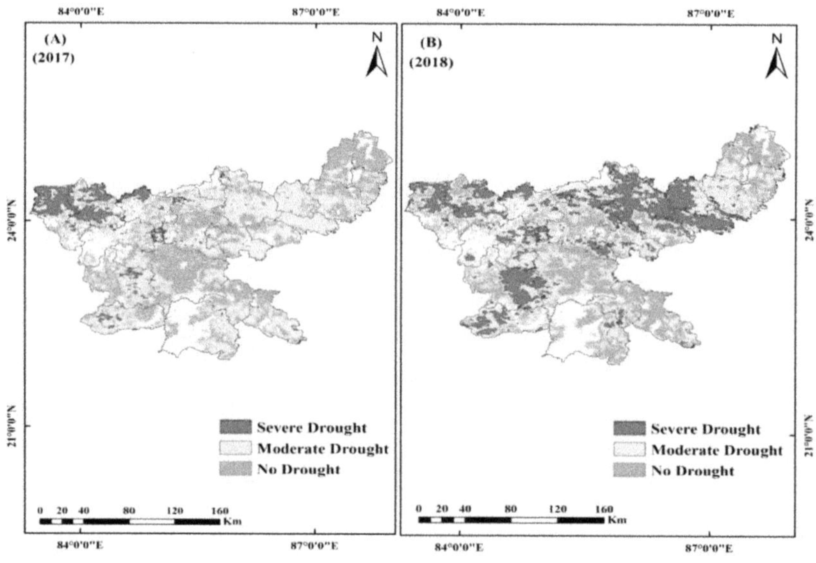

4.4. Drought Assessment using Vegetation Health Index (VHI)

The VHI maps derived from NDVI & LST data (**Fig. 10** (**A**) & (**B**)) shown that in 2018 most of the agricultural land was under the moderate drought condition. Also, it is noted that the total coverage area of severe and moderate drought got increased and no drought gets decreased in 2018 as compared to 2017. VHI result in 2018 (a drought year) indicates that ~ 26% (4751 sq.km) & 45% (8117 sq.km) area of Jharkhand state were under the severe and moderate drought condition and about 29% (5199 sq.km) area was under no drought condition.

Figure 10. Vegetation health index (VHI) map of JH state derived from MODIS NDVI and LST data

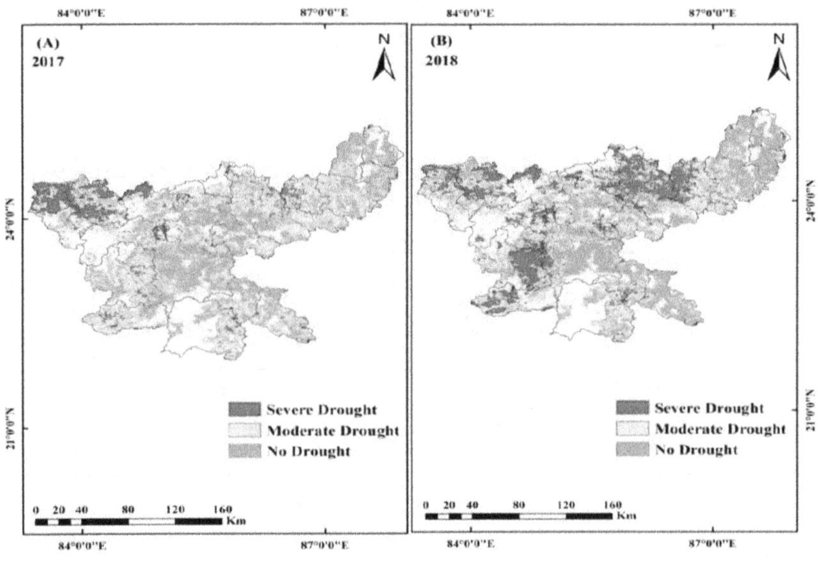

Further, VHI maps derived from EVI & LST data (**Fig. 11** (**A**) & (**B**)) shows that in 2018 (a drought year) approx. 26% (4763 sq.km) & 50% (8995 sq.km) area of Jharkhand state was under severe and moderate drought condition and about 24% (4314 sq.km) area was under no drought condition. The statistics for drought assessment from various indices/approaches have been well tabulated in **Table 2.**

Figure 11. Vegetation health index (VHI) map of JH state derived from MODIS EVI and LST data

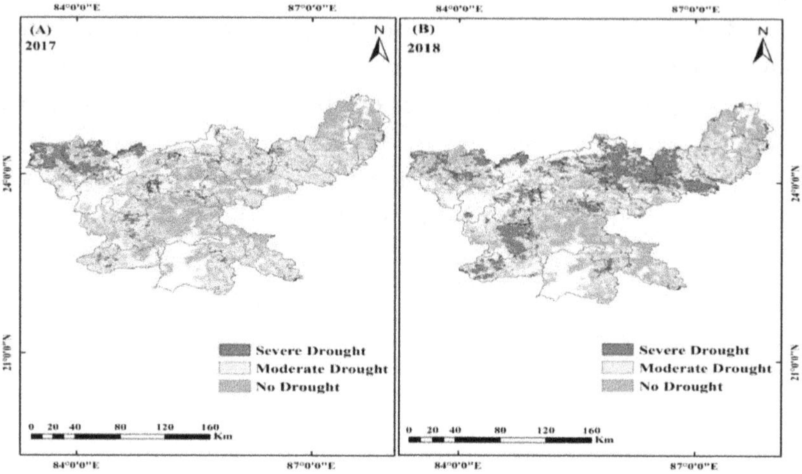

Table 2. Areal coverage of agricultural drought severity classes based on various drought indices for the normal year (2017) and drought year (2018)

CLASS	VCI (NDVI) sq.km		VCI (EVI) sq.km		TCI (LST) sq.km		VHI (NDVI& LST) sq.km		VHI (EVI& LST) sq.km	
	2017	2018	2017	2018	2017	2018	2017	2018	2017	2018
Severe Drought (0-30%)	2188	3562	2624	3982	4352	5594	2679	4751	3199	4763
Moderate Drought (30-50%)	8980	9010	7562	8063	7944	8231	7906	8117	8421	8995
No-Drought (50-100%)	7645	6242	8624	6766	6549	5020	7482	5199	6452	4314
TOTAL	**18813**	**18814**	**18810**	**18811**	**18845**	**18845**	**18067**	**18068**	**18072**	**18072**

4.5. Drought Assessment using NDVI and EVI Anomaly

4.5.1. NDVI Anomaly

The map (Fig. 12(A) & (B)) shows the spatial patterns of agricultural drought events, and the level of drought severity ranges from low to severe in each year i.e. 2017 and 2018. However, the extent of severe drought condition was detected over small areas. The majority of the study area was stricken by moderate agricultural drought. It is noted that in 2017 agricultural area covered by severe drought & moderate drought was ~16% (2888 sq.km) and 44% (8310 sq.km) & no drought was ~40% (7615 sq.km); whereas, in 2018 the area covered by severe drought (-120 to -60) & moderate drought (-60 to -30) was ~24% (4562 sq.km) and 48% (8979 sq.km) & no drought was ~28% (5273 sq.km). Therefore, it is clearly exposed that the area covered by severe & moderate drought was minimum in the 2017 and maximum in 2018 during Kharif season. And the area covered by no drought was maximum in 2017 & minimum in 2018.

Figure 12. NDVI anomaly map of JH state derived from MODIS NDVI data

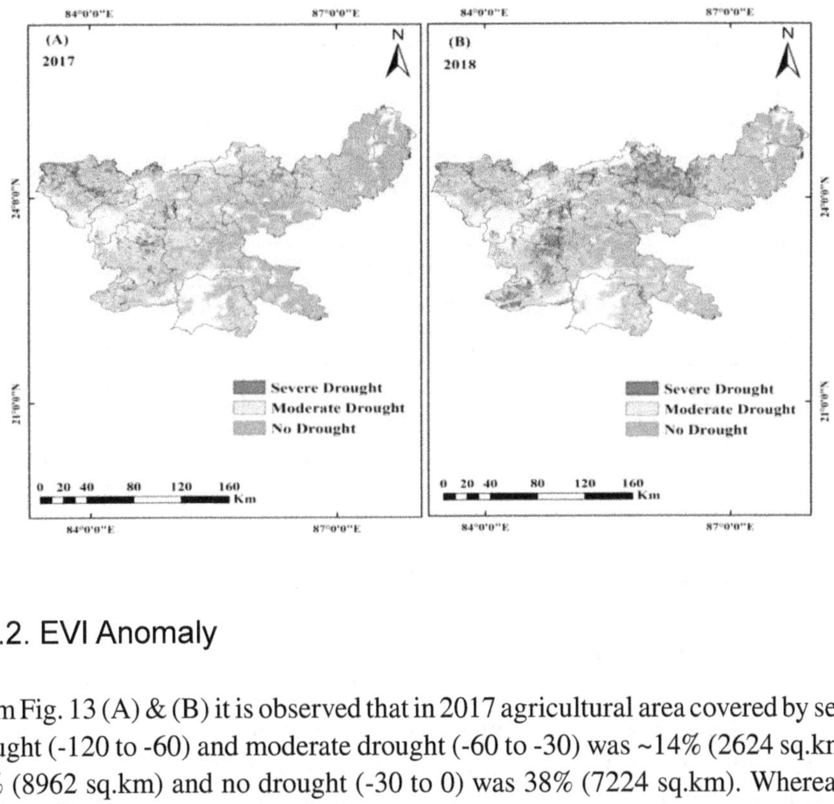

4.5.2. EVI Anomaly

From Fig. 13 (A) & (B) it is observed that in 2017 agricultural area covered by severe drought (-120 to -60) and moderate drought (-60 to -30) was ~14% (2624 sq.km) & 48% (8962 sq.km) and no drought (-30 to 0) was 38% (7224 sq.km). Whereas, in 2018 the agricultural area covered by severe & moderate drought was ~23% (4282 sq.km) & 50% (9466 sq.km), and no drought is about 27% (5063 sq.km), respectively.

Based on NDVI & EVI anomaly the total area coverage of severe drought (-120 to -60), moderate drought (-60 to -30), and no drought (-30 to 0) for the year 2017 (normal year in JH state) & 2018 (drought year in JH state) is presented in **Table 3.**

Figure 13. EVI anomaly map of JH state derived from MODIS EVI data

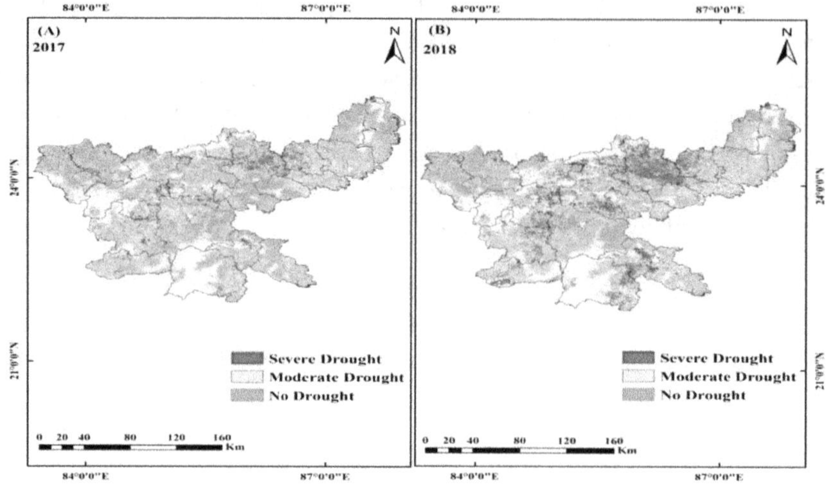

Table 3. Spatial coverage of agricultural drought severity classes, based on NDVI & EVI anomaly, for JH state for the normal year (2017) and drought year (2018)

Class	NDVI anomaly (sq.km)		EVI anomaly (sq.km)	
	2017	2018	2017	2018
Severe Drought (-120 to -60)	2888	4562	2624	4282
Moderate Drought (-60 to -30)	8310	8979	8962	9466
No-Drought (-30 to 0)	7615	5273	7224	5063
TOTAL	**18813**	**18814**	**18810**	**18811**

4.6. Linear Regression-Based Yield Modelling

The results from the correlation can be used to make a predictive yield model based on the average VCI (NDVI), VCI (EVI), TCI (LST), VHI (NDVI & LST) & VHI (EVI & LST) scores. Scatterplot was prepared from mean zonal statistics of each drought indices vs. DES-based yield (for the year 2017) for all the districts of Jharkhand state. A yield model was derived by fitting a linear regression curve to the data, where the coefficient of determination (R^2) was observed for each correlation (**Table 4**). The yield prediction model which have the utmost correlation coefficient was considered for yield prediction for the drought year 2018. A better correlation was observed with VCI (EVI) vs. DES yield (**Fig. 14**), wherein, rest has shown statistically weak correlation. The coefficient of determination was not excellent but statistically, it was significant for yield prediction. Consequently, we have used VCI (derived from EVI) based yield model to predict the paddy yield for the year 2018.

Figure 14. Linear correlation between district wise VCI (derived from MODIS EVI data) and DES-based paddy yield (2017)

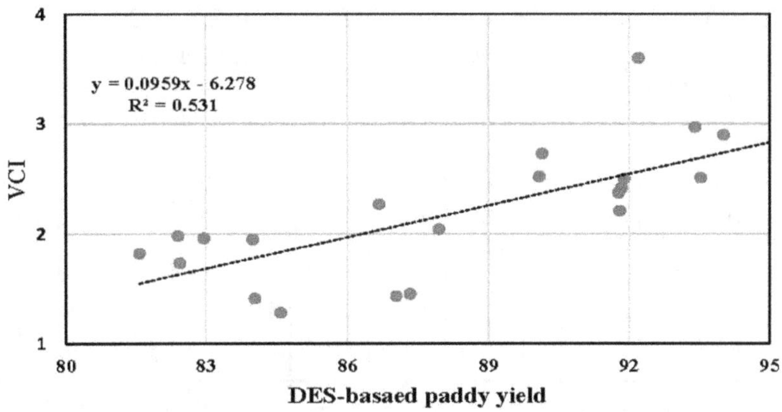

Table 4. Coefficient of determination (R^2) for each correlation computed during the development of a yield model using different drought indices & crop yield data of the year 2017

Sl. No.	Drought indices	R^2(Coefficient of determination)
1	VCI (NDVI)	0.491
2	VCI (EVI)	0.531
3	TCI (LST)	0.425
4	VHI (NDVI & LST)	0.444
5	VHI (EVI & LST)	0.485

4.7. Paddy Yield Prediction for the Year 2018

The paddy yield for the year 2018 was estimated based on the linear regression analysis between the district wise drought indices and DES-based paddy (2017). As, the EVI-based model has shown the utmost correlation with R^2 value of 0.53, it was considered for yield prediction. The R^2 value for this yield model was not so strong, but statistically, it was significant. The linear regression-based yield models have been well shown in the previous table. Sahibganj district was found with highest paddy yield of 2.8 ton/hec and Deoghar district was found with least paddy yield of 0.93 ton/hec during the drought year (2018). Total paddy production of Jharkhand state was predicted as 38.46 ton/hec during the drought year 2018. Further, district-wise paddy yield for the entire Jharkhand state during the drought year 2018 was estimated as shown in **Fig. 15.**

Figure 15. District-wise predicted paddy yield for drought year 2018, wherein, red bar is showing least yield and green bar is showing highest yield amongst the all district of the JH state

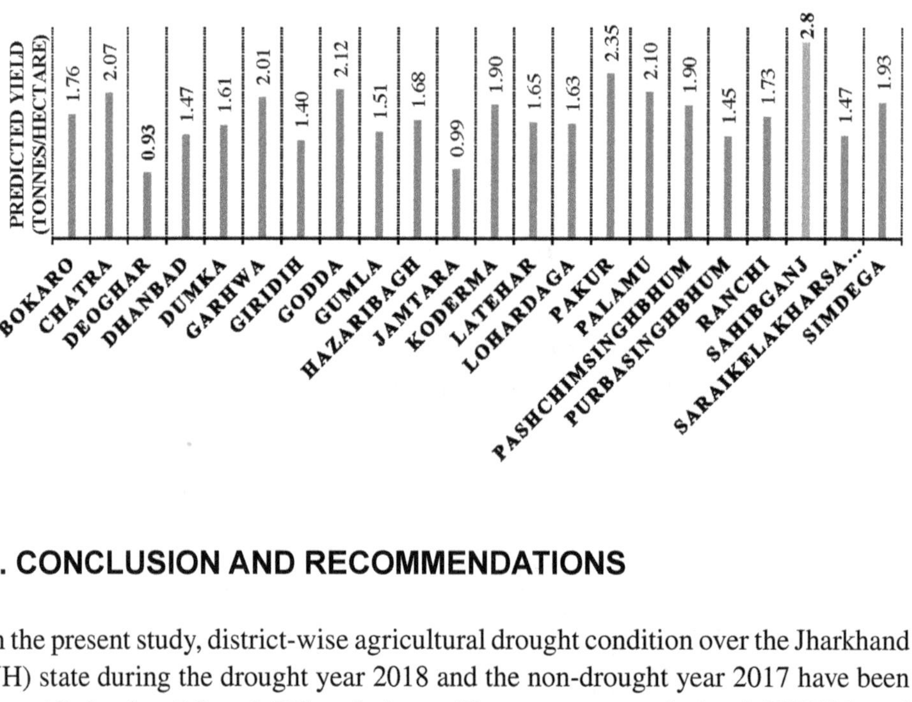

5. CONCLUSION AND RECOMMENDATIONS

In the present study, district-wise agricultural drought condition over the Jharkhand (JH) state during the drought year 2018 and the non-drought year 2017 have been quantified using RS and GIS techniques. Here, coarser-resolution MODIS-based vegetation indices were used to assess the district-wise drought-affected area. Different indices (i.e. NDVI, EVI) were employed in order to understand the spatio-temporal distribution of agricultural drought by deriving various drought indices (i.e. VCI, TCI, VHI, etc.).

By quantifying the drought maps of the Jharkhand, it was noticed that ~ 21% area (based on VCI derived from EVI) was badly affected by severe drought condition. Wherein, ~43% of the area was detected under moderate drought condition. The districts namely, Giridih, Deoghar, Godda, Lohardaga, Dumka etc. were observed under severe drought condition. Moreover, Hazaribagh, Dhanbad, Palamu, Latehar etc., were found under moderate drought condition and rests were under no/very less drought condition.

Additionally, paddy yield for the corresponding district during the drought year 2018 was predicted. The role of the satellite-derived drought indices (i.e. VCI (NDVI), VCI (EVI), TCI, VHI (NDVI&LST), VHI (EVI&LST), NDVI Anomaly & EVI) was exemplified with crop yield and a strong linear relationship found between them. From the study, it is observed that VCI derived from EVI is an important parameter which shows the significant correlation with crop yield for most of the districts of Jharkhand state. Hence, it was used as an operational predictor for the estimation of paddy yield over the state.

From the yield model, it was estimated that the approximate paddy yield loss over the entire Jharkhand state in 2018 (a drought year) was 10.12 ton/hec as compared to 2017 (normal year). However, Sahibganj district was found with highest paddy yield of 2.8 ton/hec, wherein, Deoghar district was found with very less paddy yield of 0.93 ton/hec during the drought year. Consequently, total paddy production of Jharkhand state was estimated as 38.46 ton/hec during the drought year 2018.

Hence, this study demonstrates that the utilization of remote sensing data for drought monitoring is reliable, cost-effective, and time-saving technique. The RS-based methods can be preferred for such study at the regional-national level; because the conventional methods are too costly required a large database, time-consuming and required large manpower. Nevertheless, space technology is widely being used to provide vital information within a short time period in a cost-effective manner.

Nonetheless, few drawbacks of the study are noted which may be taken into consideration for more accurate and reliable drought assessment. Finer or medium spatial resolution satellite data (i.e. Sentinel 2, LISS III, IV, Landsat series) could provide more details drought assessment as compared to the coarser resolution satellite data (i.e. MODIS, AVHRR). But these datasets are highly prone to cloud contamination, as paddy crops are grown under the rainfed condition in the study area. Further, only broadband vegetation indices were used in this study, wherein, some narrowband satellite data could provide some dynamic vegetation indices (e.g. Normalized Multi-band Drought Index (NMDI), Chlorophyll Absorption Ratio Index (CARI), etc.) which are highly responsive to the crop health conditions. Though, Soil Moisture Active Passive (SMAP) satellite data with a higher spatial resolution of soil moisture products can be deployed for more accurate and reliable agricultural drought assessment in sub-region (village level/field level). Eventually, the relation between VCI and yield was not so good for deriving yield model but statistically, it was significant. For such inconsistency, high-resolution data and proper ground-truthing can improve accuracy.

ACKNOWLEDGMENT

Author's thanks to United State Geological Survey (USGS) and Directorate of Economics & Statistics, Ministry of Agriculture and Farmers Welfare, Govt. of India, New Delhi, for providing the MODIS product and paddy APY data, respectively.

Conflict of Interest: Authors declare no potential conflict of interest.

REFERENCES

Amalo, L. F., Hidayat, R., & Haris. (2017). Comparison between remote-sensing based drought indices in East Java. *Earth and Environmental Science*, *54*(1), 012009. Advance online publication. doi:10.1088/1755-1315/54/1/012009

Boken, V. K., Hoogenboom, G., Kogan, F. N., Hook, J. E., Thomas, D. L., & Harrison, K. A. (2004). Potential of using NOAA-AVHRR data for estimating irrigated area to help solve an inter-state water dispute. *International Journal of Remote Sensing*, *25*(12), 2277–2286. doi:10.1080/01431160310001618077

DAC. (2009). *Manual for Drought Management*. Department of Agriculture & Cooperation, Ministry of Agriculture, Government of India.

Dalezios, N. R., Blanta, A., Spyropoulos, N. V., & Tarquis, A. M. (2014). Risk identification of agricultural drought for sustainable Agroecosystems. *Natural Hazards and Earth System Sciences*, *14*(9), 2435–2448. doi:10.5194/nhess-14-2435-2014

Das, P., Palchaudhuri, M., Biswas, S., & Deka, C. (2017). *Analysis of Agricultural Drought Severity Using MODIS Satellite Images*. Academic Press.

Du, T. L. T., Bui, D. D., Nguyen, M. D., & Lee, H. (2018). Satellite-Based, Multi-Indices for Evaluation of Agricultural Droughts in a Highly Dynamic Tropical Catchment, Central Vietnam. *Water (Basel)*, *10*(5), 659. doi:10.3390/w10050659

Dutta, D., Kundu, A., Patel, N. R., Saha, S. K., & Siddique, A. R. (2015). Assessment of agricultural drought in Rajasthan (India) using remote sensing derived Vegetation Condition Index (VCI) and Standardized Precipitation Index (SPI). *The Egyptian Journal of Remote Sensing and Space Sciences*, *18*(1), 53–63. doi:10.1016/j.ejrs.2015.03.006

Gitelson, A. A., Kogan, F., Zakarin, E., Spivak, L., & Lebed, L. (1998). Using AVHRR data for quantitative estimation of vegetation conditions: Calibration and validation. *Advances in Space Research*, *22*(5), 673–676. doi:10.1016/S0273-1177(97)01129-0

Holben, B. (1986). Characteristics of maximum-value composite images from temporal AVHRR data. *International Journal of Remote Sensing*, *7*(11), 1417–1434. doi:10.1080/01431168608948945

Hua, L., Wang, H., Sui, H., Wardlow, B., Hayes, M. J., & Wang, J. (2019). Mapping the Spatial-Temporal Dynamics of Vegetation Response Lag to Drought in a Semi-Arid Region. *Remote Sensing*, *11*(16), 1873. doi:10.3390/rs11161873

Jain, V. J., Pandey, R. P., Jain, M. K., & Byun, H. R. (2015). Comparison of drought indices for appraisal of drought characteristics in the Ken River Basin. *Weather and Climate Extremes*, *8*, 1–11. doi:10.1016/j.wace.2015.05.002

JAPCC (Jharkhand- Action Plan on Climate Change). (2014). Government of Jharkhand.

Ji, L., & Peters, A. J. (2003). Assessing vegetation response to drought in the northern Great Plains using vegetation and drought indices. *Remote Sensing of Environment*, *87*(1), 85–98. doi:10.1016/S0034-4257(03)00174-3

JSADP (Jharkhand State Agriculture Development Plan). (2012). *Report by Nabard Consultancy Services*. Jharkhand Regional Office Ranchi.

JSDMP (A draft on Jharkhand State Disaster Management Plan). (2011). Disaster Management Department MDI Building, Dhurwa, Ranchi, Jharkhand.

Kogan, F. N. (1995). Application of vegetation index and brightness temperature for drought detection. *Advances in Space Research*, *15*(11), 91–100. doi:10.1016/0273-1177(95)00079-T

Kogan, F. N. (1997a). *Contribution of Remote Sensing to Drought Early Warning*. https://drought.unl.edu/monitor/EWS/ch7_Kogan.pdf

Kogan, F. N. (1997b). Global drought watch from space. *Bulletin of the American Meteorological Society*, *78*(4), 621–636. doi:10.1175/1520-0477(1997)078<0621:GDWFS>2.0.CO;2

Kogan, F. N. (2000). Satellite-observed sensitivity of world land ecosystems to El Nino/La Nina. *Remote Sensing of Environment*, *74*(3), 445–462. doi:10.1016/S0034-4257(00)00137-1

Lim, C.H., Kim, S.H., Chun, J.A., Kafatos, M.C., & Lee, W.K. (2019). *Assessment of Agricultural Drought Considering the Hydrological Cycle and Crop Phenology in the Korean Peninsula*. . doi:10.3390/w11051105

Murali, K., Ravikumar, G., & Krishnaveni, M. (2008). Remote sensing based agricultural drought assessment in Palar basin of Tamil Nadu State, India. *Photonirvachak (Dehra Dun)*, *37*, 9–20.

Murthy, C. S., Sai, M. V. R. S., Kumari, V. B., & Roy, P. S. (2007). Agricultural drought assessment at disaggregated level using AWiFS/WiFS data of Indian Remote Sensing satellites. *Geocarto International*, *22*(2), 127–140. doi:10.1080/10106040701205039

Muthumanickam, D., Kannan, P., Kumaraperumal, R., Natarajan, S., Sivasamy, R., & Poongodi, C. (2011). Drought assessment and monitoring through remote sensing and GIS in western tracts of Tamil Nadu, India. *International Journal of Remote Sensing*, *32*(18), 5157–5176. doi:10.1080/01431161.2010.494642

Nandeesha & Ramu, D. (2015). Agricultural drought assessment in central dry zone of agro climatic zones of Karnataka using Gis and Remote Sensing. *International Journal of Information Technology and Computer Science, 3*(2), 575-581. http://eprints.uni-mysore.ac.in/id/eprint/4842

Navalgund, R. R., Jayaraman, V., & Roy, P. S. (2007). Remote sensing applications: An overview. *Current Science*, *93*(12), 1747–1766.

Navalgund, R. R., & Ray, S. S. (2000). Geomatics in natural resources management. Proceedings of Geomatics-2000, NR1-NR14.

Panigrahy, S., & Ray, S. S. (2006). Remote Sensing. In K. L. Chadha & M. S. Swaminathan (Eds.), *Environment and Agriculture* (pp. 361–375). Malhotra Publishing House.

Parida, B. R., & Oinam, B. (2015). Unprecedented drought in North East India compared to Western India. *Current Science*, *109*(11), 2121–2126. doi:10.18520/cs/v109/i11/2121-2126

Parida, B. R., & Ranjan, A. K. (2019a). Wheat Acreage Mapping and Yield Prediction Using Landsat-8 OLI Satellite Data: A Case Study in Sahibganj Province, Jharkhand (India). *Remote Sensing in Earth Systems Sciences*, *2*(2-3), 96–107. doi:10.100741976-019-00015-9

Parida, B. R., & Ranjan, A. K. (2019b). Up-scaling paddy yield at satellite-footprint scale using satellite data in conjunction with CCE data in Sahibganj district, Jharkhand. *The International Archives of the Photogrammetry, Remote Sensing and Spatial Information Sciences*, *17*(W6), 235–238. doi:10.5194/isprs-archives-XLII-3-W6-235-2019

Patel, N. R., Parida, B. R., Venus, V., Saha, S. K., & Dadhwal, V. K. (2012). Analysis of Agricultural Drought Using Vegetation Temperature Condition Index (VTCI) from Terra/MODIS Satellite Data. *Environmental Monitoring and Assessment, 184*(12), 7153–7163. doi:10.100710661-011-2487-7 PMID:22200944

Ranjan, A.K., Anand, A., Vallisree, S., & Singh, R.K. (2016b). LU/LC Change Detection and Forest Degradation Analysis in Dalma Wildlife Sanctuary Using Geospatial Technology: A Case Study in Jamshedpur-India. *AIMS Geosciences, 2*(4), 273-285. doi:10.3934/geosci.2016.4.273

Ranjan, A. K., & Parida, B. R. (2019). *Paddy Acreage Mapping and Yield Prediction Using Sentinel-Based Optical and SAR Data in Sahibganj District*. Spatial Information Research. doi:10.100741324-019-00246-4

Ranjan, A. K., Vallisree, S., & Singh, R. K. (2016a). Role of Geographic Information System and Remote Sensing in Monitoring and Management of Urban and Watershed Environment: Overview. *J of Remote Sens & GIS, 7*, 1–14.

Sahu, R., & Patel, N. R. (2016). Agricultural drought assessment using satellite derived vegetation condition index and rainfall anomaly index over Bundelkhand region. Academic Press.

Seiler, R. A., Kogan, F., & Sullivan, J. (1998). AVHRR-Based vegetation and temperature condition indices for drought detection in Argentina. *Advances in Space Research, 21*(3), 481–484. doi:10.1016/S0273-1177(97)00884-3

Sholihah, R. I., Trisasongko, B. H., Shiddiq, D., Iman, L. O. S., Kusdaryanto, S., Manijo, & Panuju, D. R. (2016). Identification of agricultural drought extent based on vegetation health indices of Landsat data: Case of Subang and Karawang, Indonesia. *Procedia Environmental Sciences, 33*, 14–20. doi:10.1016/j.proenv.2016.03.051

Unganai, L. S., & Kogan, F. N. (1998). Drought monitoring and corn yield estimation in Southern Africa from AVHRR data. *Remote Sensing of Environment, 63*(3), 219–232. doi:10.1016/S0034-4257(97)00132-6

Vanni, N., Porchelvan, P. (2018). *Monitoring of agricultural drought using fortnightly variation of vegetation condition index (VCI) for the state of Tamil Nadu, India*. doi:10.5194/isprs-archives-XLII-4-W9-159-2018

Wu, H., Qian, H., Chen, J., & Huo, C. (2017). Assessment of Agricultural Drought Vulnerability in the Guanzhong Plain, China. *Water Resources Management, 31*(5), 1557–1574. doi:10.100711269-017-1594-9

Zuo, D., Cai, S., Xu, Z., Peng, D., Kan, G., Sun, B. P., & Yang, H. (2019). Assessment of meteorological and agricultural droughts using in-situ observations and remote sensing data. *Agricultural Water Management*, *222*, 125–138. doi:10.1016/j.agwat.2019.05.046

ENDNOTES

1 United States Geological Survey earth explorer (https://earthexplorer.usgs.gov/).

2 Directorate of Economic Survey (DES) (http://aps.dac.gov.in/APY/Index.htm).

Chapter 8
Comparative Flood Inundation Mapping Utilizing Multi-Temporal Optical and SAR Satellite Data Over North Bihar Region:
A Case Study of 2019 Flooding Event Over North Bihar

Gaurav Tripathi
https://orcid.org/0000-0002-1114-7660
Department of Geoinformatics, Central University of Jharkhand, Brambe, India

Arvind Chandra Pandey
Department of Geoinformatics, Central University of Jharkhand, Brambe, India

Bikash Ranjan Parida
Department of Geoinformatics, Central University of Jharkhand, Brambe, India

Achala Shakya
https://orcid.org/0000-0003-2880-5743
Department of Computer Engineering, National Institute of Technology, Kurukshetra, India

DOI: 10.4018/978-1-7998-5027-4.ch008

ABSTRACT

Floods are investigated to be the utmost frequent and destructive phenomena among all other types of natural calamities worldwide. Thus, flood events need to be mapped to understand their impact on the affected region. The present case study is intended to examine and analyze the flood events occurred in July-August 2019 over the Northern Bihar region situated in Kosi and Gandak river basins. Furthermore, a comparative study was carried out to map the satellite based near real time flood inundation using multi-temporal Sentinel–1A (SAR) and MODIS NRT Flood data (optical and 3-day composite). Optical (MODIS) and Sentinel-1 SAR data were acquired to compare their flood inundation extent and the result shows overestimation in MODIS flood data due to varying spatial resolutions.

1. INTRODUCTION

Floods are the utmost destructive, pervasive, recurring meteorological, and hydrological hazard occurred due to natural and anthropogenic reasons. Notably, in the South Asian region, floods are the most vulnerable natural calamity with regard to human and economic losses because of the very dense population in the region. It was indicated (EM-DAT, 2016) that by the year 2016, Asia has experienced 46.7% natural disaster followed by America, Africa, Europe, and Oceania which experienced 24.3%, 16.9%, 8.2%, and 3.9% respectively. Among all natural disasters, flash floods, rain, and landslides were the most significant fatalities caused by higher run-off and higher rainfall in a concise period (Modrick and Georgakakos 2015). So, various mitigation practices were taken into consideration to restrict flooding. Improper planning in the roads construction, barrages, bunds across the river as well as the public efficacies inside the floodplain has led to a severe effect over the normal flow of the river or drainage system. Consequently, hasty siltation in the riverbed tends to the overflow of flooding water (Kumar et al. 2014). Inundations are frequent and a serious concern to human society that consequently leads to power quantification and flood occurrence. Over the decades, flood events were rising (Freer et al. 2013), so the advancement of flood mapping and applying flood inundation models using remotely sensed satellite data has become an essential application in monitoring and assessing the impact of floods. The inadequate resources in heavily populated countries led to the effective supervision in forecasting of floods (Wu et al. 2014). Over large areas, remote sensing (RS) data works as a substitute to provide useful information on progression of floods because the collection of information through in situ observation is impractical. Remote sensing data provides a classical solution for assessing a spatio-temporal flood evolution. The flood mapping based on RS data has shown the significant breakthroughs in research over the last few decades. RS-based flood maps not only provide up-to-date and spatial information (Schumann et al. 2016) but also utilized hydrodynamic model calibration to enhance forecast prediction (Schumann and Domeneghetti 2016).

RS data maps flood over a wide range of geographical areas and is highly effective where *in-situ* data is narrow. In last few years, remotely sensed data of flood events was made available qualitatively to the stakeholders for improving flood monitoring and mapping. The high spatio-temporal resolution of the remote sensing data advanced the evolution of near-real-time mapping of flood approaches (Martinis et al. 2015; Twele et al. 2016). The remote sensing data contributes crucial information required for calibrating and validating the hydrodynamic models namely, Hydrologic Engineering Center-River Analysis System (HEC-RAS 2-D (Two dimensional), MIKE3 3-D (Three dimensional), TUFLOW (3-D), and so forth (Teng et al. 2017a; Afshari et al. 2018a; McGrath et al. 2018a; Speckhann et al. 2018a; Wan et al. 2019; Shakya and Kumar 2019). The usage of remote sensing data has improved the understanding of flood dynamics, prediction accuracy as well as forecasting (Grimaldi et al. 2016; Mure-Ravaud et al. 2016). In recent literature, optical and synthetic aperture radar (SAR) data has been widely used to monitor surface water dynamics. The passive sensors of optical satellite data have limitations due to the appearance of cloud content in the data for monitoring and mapping of flood events (Gan et al. 2012). SAR systems overcome the inability of optical sensors by enabling data collection with the presence of cloud cover in the land cover. (Alsdorf et al. 2007a; Schlaffer et al. 2015a). SAR data can distinguish the Land/Water interface to depict the flood extent and its management. Therefore, studies operated SAR data for temporal mapping of flood extent. For example, in 2008 RADARSAT–1 data was used for mapping spatio-temporal flood dynamics for Kosi river basin (Singh et al. 2011). The only shortcoming of the SAR sensor is the temporal resolution that results in the unavailability of satellite images during the flood events.

1.1 Related Work

Several studies were introduced to map flood dynamics over the study region and demonstrated the flood occurrences over Northern Bihar districts from 2000 to 2008 using multiple remote sensing satellite data to highlight flood risk zones (Singh et al. 2011; Pandey et al. 2014). Resourcesat-1 (LISS-3 and AWIFS) and Landsat satellite images were used for mapping of waterlogged zones at temporal basis to differentiate between permanent and seasonal waterlogging areas for flood risk zone identification (Pandey et al. 2014). The image segmentation method was applied over Radarsat-1 (SAR) satellite images to extract flood pixels in the Kosi River basin (Singh et al.

2011). Optimum threshold technique was employed on Radarsat-2 images of the Darbhanga district to map flood water across the district (Manjusree et al. 2012). Other structured technique like Analytical Hierarchy Process (AHP) was coupled with Geographic Information System (GIS) to map previous flooding events during 2003 and 2006, where flood risk zones were also identified inside the Kosi river basin (Sinha et al. 2008). Self-organizing Kohonen's maps were applied by coupling with Artificial Neural Network to delineate flood inundated zones (Wahab and Muhamad Ludin 2018; Costache 2019). Although, Moderate Resolution Imaging Spectro-radiometer (MODIS) data having coarser resolution satellite products but was often used to prepare near real-time flood maps over districts of North Bihar because of its higher temporal resolution (Pandey et al. 2010). Annual flooding was also investigated based on temporal MODIS datasets by assessing time-wise changes in flood inundation extent in countries like Vietnam and Cambodia (Sakamoto et al. 2007). Open Water Likelihood (OWL) algorithm was introduced to quantify the stagnant water bodies using MODIS data (Van Dijk et al. 2011). Literature survey suggested various hydrological (1-dimensional, 2-dimensional, and 3-dimensional) flood models as a frequent operative tool to estimate different flooding parameters in pre, post and during flood events and for the management purposes (Afshari et al. 2018b; Lian et al. 2019; Wan et al. 2019).

The historical references and web resources indicated various significant flood events in North Bihar region, during 1987, 1995, 1998, 2000, 2001, 2003, 2004, 2008, 2010, 2013, 2016, 2017, and 2018 (Sinha et al. 2008; Pandey et al. 2010; Singh et al. 2011; Tripathi et al. 2019). Recently, during July-August 2019, more torrential rainfall in the upper catchment caused disastrous flooding conditions for lower areas, which inundated several North Bihar districts in which Patna, Darbhanga and Sitamarhi were the most flood affected districts during August-September 2019. Flooding frequency was increased in the last decade, which extensively affected human lives, stopped traffic on several roads, public property, and forced administrators to re-evaluate the strategies made for river/ basin management. The information over near-real time flood inundation would act as an accurate spatial data during crisis management as well as strategic planning to reduce flooding effect and flood induced economic losses (Matgen et al. 2007).

Nowadays, Earth Observation (EO) datasets is used as an adequate choice for near real-time inundation monitoring, mapping (Brivio et al. 2002) and these flood maps are used in planning, prevention and mitigation exercises. Sensor's capability to overpass massive cloud cover, severe rain, insufficient sunlight and presence of haze, etc. during and post-flood events, forced decision-makers to choose SAR data over optical data. SAR data is independent of cloud cover and provides an essential input in real-time flood damage assessment (Matgen et al. 2007). The most crucial advantage of using SAR images is its capability to distinguish land and water precisely. For SAR sensors flat earth surface acts as a specular reflector, this causes to store dark pixels as water areas and rest as non-water (Brivio et al. 2002). Backscatter intensity, polarimetry, and interferometric coherence were considered as essential inputs in the case of SAR data (Tsyganskaya et al. 2018). A detailed review on small- and large-scale flood inundation mapping, flood dynamics behavior and radar signal behavior over waterbody using SAR data was discussed (Matgen et al. 2007 and Tsyganskaya et al. 2018).

1.2 Objectives

The present study concentrated on evaluating the advantage of Sentinel-1A data over MODIS flood data for mapping of floodwater propagation during the July-August 2019 flood event over the North Bihar districts. The flood occurrence which started in the mid-July 2019 due to massive rain upstream of the areas and these flooded areas stayed for more than a month having few water pools yet present till last of August. Therefore, the specific aim of the present case study indicated to map and compare the spatio-temporal flood extent using Sentinel-1A and MODIS flood data with the help of flood inundation maps. Flood extent were compared using multi-temporal Sentinel-1A and MODIS flood data over Darbhanga district of North Bihar state, which provided information on flood prognosis. The whole flooded area can be traced and allow some fascinating insight into the post-flood mapping process. Results were produced from a high resolution multi-temporal maps of the flood occurrences.

2. STUDY AREA

The North Bihar, India was chosen for the study, which consists of 22 districts having geographical extension between 82.3þE to 88.9þE longitude and 24.8þN to 29.3þN latitude. The region falls under upper ganga basin which consists of various rivers namely, Kosi river (basin is of 19,809 sq.km) and Burhi Gandak river (basin is of 87,165 sq.km) inside Northern Bihar (Figure 1). There are various regional tributaries/dhars present namely, Gandak, Bagmati, Adhwara Group, Kamla Balan, Bhutahi Balan river, Dhemama dhar, etc. which majorly drain the main river channel during monsoon season.

From the foothill of the Himalayas, Kosi river originates at an altitude of 7000 m and comprises three major dhars, namely, the Sun Kosi, Arun, and Tamur. During the monsoon period, each dhars pours sufficient amount of water to the main river channel. A major portion, estimated around 80% of the Kosi river catchment lies inside Nepal and Tibet and the remaining 20% lies inside India (North Bihar state).

Baghmati and Kamla are the main rivers of the Kosi and Gandak river catchment that also forms a sub basin in the Kosi and Gandak basin. Several seasonal channels/ tributaries/ dhars, namely, Sugarwe, Balan, Kamla, Thomani, flows through the districts and caused severe inundation. The North Bihar districts, such as, Saharsa, Darbhanga, Khagaria, Bhagalpur, Muzzafarpur, Araria, Kishanganj are located along the Kosi and Burhi Gandak river. A comparative analysis was performed and analyzed over Darbhanga district which is one of the most flood-affected district among all 22 districts of North Bihar, typically situated between Kosi and Burhi Gandak river. River in the monsoon season makes higher cumulative runoff due to a substantial amount of rainfall and subsequently, inundated lower catchment areas (Sinha et al. 2008), which are prone to flooding as well as waterlogging (Pandey et al. 2010).

Thereby, every coming year flooding occurred drastically and created inundation over large stretch of North Bihar. The Kosi river is known as "Sorrow of Bihar" due to the recurring occurrence of floods over the Kosi river basin. The main reason for Bihar floods is due to the breach of bank alongside the river Kosi owing to excessive rainfall throughout the monsoon period. Plenty of curative measures were executed by the Bihar (India) and Nepal government to limit regular flooding and to guard the marshy areas of Northern Bihar and Nepal (Macchione et al. 2019). The curative measures, such as dams, water channels, storage ponds, etc. have been built to reduce and control the massive flow of water during flooding events. Despite all these flood preventive actions, the Kosi river's overflow happens around every coming year and creates critical harm to the infrastructure and human lives.

3. DATA USED AND METHODOLOGY

Satellite data used in this research incorporates various space-based satellite images including (i) Three Sentinel-1A SAR C-band dual-polarized (HH & VH), utilized for flood inundation mapping, (ii) MODIS NRT flood products of July-August 2019 were also used to compare with the result of S1-A data, (iii) One Sentinel-2A (S-2A) image of path 184 and row 034 were considered to validate the flood areal extent over the district (Table 1). Various optical and SAR satellite images were processed and analyzed using SNAP (ver.7) and ERDAS IMAGINE (ver. 2015) software, while the geo-spatial analysis over spatial data was done using ArcGIS (ver.10.5). Vector data as ESRI shapefiles of the nation, state, district, river, and basin were used to separate the boundary over the entire study area.

Figure 1. ESRI shapefiles of the district and river is overlaid to show location map of the study area

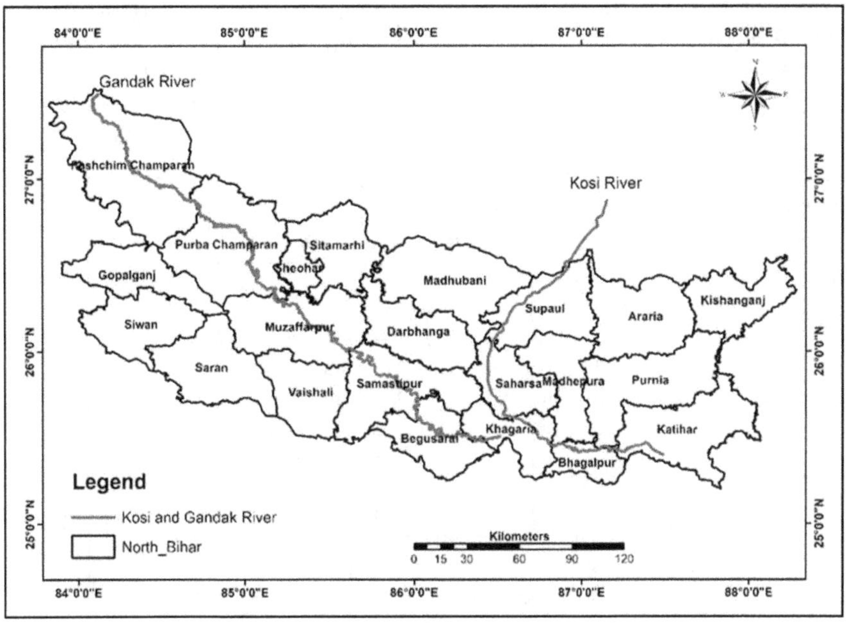

Table 1. Primary and secondary data used in the present study

Name of dataset	Spatial Resolution	Resolution Temporal	Acquisition date	Source	Purpose
Sentinel–2A	10m	10 day	17th Jul, 2019	USGS	LULC
Sentinel–1A	10m	12 days	19th July, 31st July and 12th August	ASF	Flood Inundation
MODIS NRT Flood data	250m	24 hrs	July-August, 2019	NRT Flood Product	Flood Inundation

3.1 Methodology

The methods used in this study are presented in Figure 2. Specifically, Sentinel image classification for flood mapping, with initial pre-processing was carried out to mitigate the SAR-typical speckle noise signatures from the images. During the pre-processing step, radiometric and geometric distortions were rectified due to the characteristics of the imaging system and performed radiometric corrections to improve visualization and interpretation for flood mapping. The pre-processing steps were implemented using ESA's Sentinel Application Platform (SNAP). The

Figure 2. Methodology adopted in this study

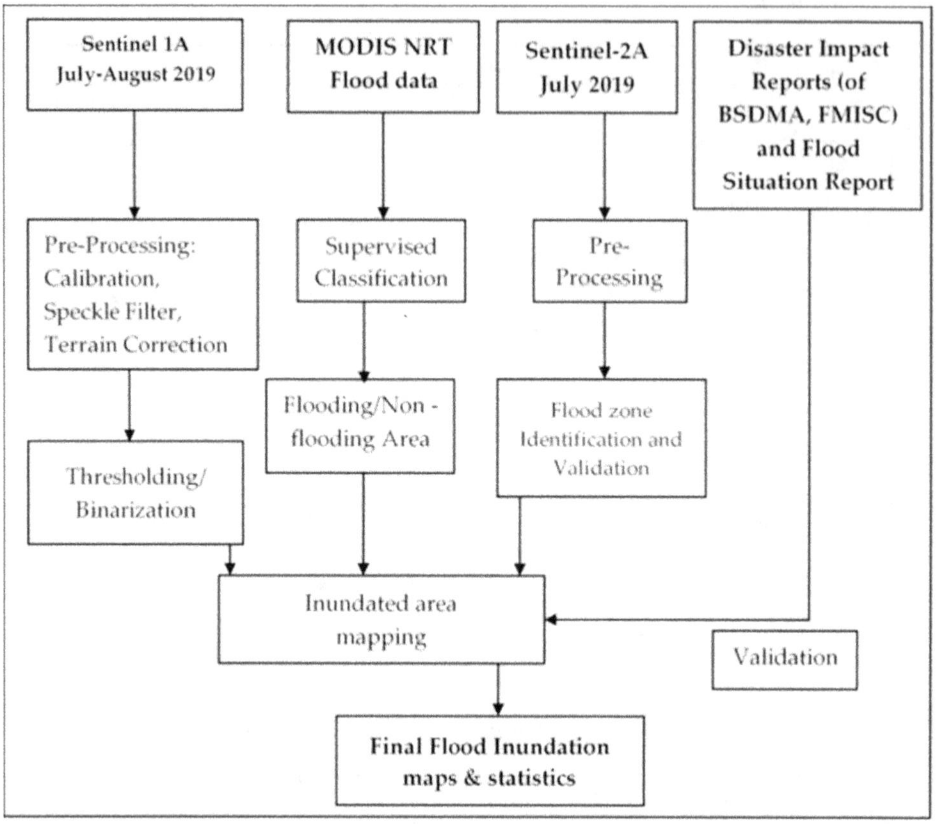

open-access SNAP toolbox is capable of reading, pre-processing, and visualizing Sentinel-1 SAR images.

Sentinel–1A SAR data processing: The pre-processing of Sentinel–1A SAR data includes few processing steps namely, importing of the data, calibration (radiometric), speckle filtering, terrain correction, linear to backscattering coefficient decibel scaling (dB) transformation, binarization process for flood pixels extraction, and the thresholding approach as described in (Matgen et al. 2007; Manjusree et al. 2012; Ajadi et al. 2016). The threshold value of –22.8, -23.7, and -23.6 dB was utilized for extracting the flood pixels.

In the course of Sentinel-1 pre-processing, Level-1 data was firstly imported in the SNAP Desktop tool. Secondly, Sentinel-1 images were corrected radiometrically by employing annotated image calibration constants to reach physically essential radar backscattered pixel values. Next, the speckle filtering to SAR data was applied to lessen the granular noise component present in the data. Then, the multi-looking operation was carried out for reduction in the speckles and thus improved the image interpretation. After that, the correction of geometric distortions present in the SAR data was performed by the coordinate transformation into a convenient reference frame and the ratio band of VH/VV was generated dividing the VH by the VV band. In the final step, a radiometric transformation from a linear scale to a dB scale was performed by utilizing the subsequent expression:

$$\sigma_{dB}^{0} = 10.\log_{10}^{0}\sigma_{1}^{0} \tag{1}$$

where, σ0 (dB) = backscattering image in dB, σ0 = Sigma naught image.

Binarization analysis was performed on the preprocessed images and thresholding technique was applied to derive flood inundated areas by taking the dB values in the constant range. Suitable threshold, value provides more accurate value which can be further used in other operations e.g., feature extraction, image segmentation, target recognition, etc. 1-D Ostu thresholding method was used to separate image (I (x, y)) which contains brighter object from the darker background image as follows (Tripathi et al. 2019; Equation 2).

$$I(x, y) = \{1\ I(x, y) > T;\ 0\ I(x, y) < T\} \tag{2}$$

where, T represents the value for threshold

Literature suggests that the backscattering binarization investigation was the fittest procedure concerning flood pixels extraction. Prominently, over the Northern Bihar region, image binarization (Sentinel–1A SAR data) techniques was applied for extracting the flood pixels. Additionally, rainfall-runoff modeling, hydrological models, and machine learning approaches were utilized to distinguish flood situations.

The MODIS-based NRT product (raster data) was utilized for creating flood maps that involves a total of four classes such as, 0 (no data), 2 (water but not flood), and 3 (water detected and likely flood). The maps for flood inundation were produced primarily from class 3 and the class-wise statistics of this class estimated the presence of quantified regions under flood.

Over the entire Northern Bihar region, image classification (MODIS NRT data) and binarization (Sentinel-1A SAR data) techniques were utilized for flood pixels extraction. The other hydrological modeling, and machine learning algorithms are also available to demarcate flooding (Teng et al. 2017b; Uddin et al. 2019). The field survey strategy can be exercised for flood mapping, and are notably cumbersome.

Figure 3. Spatio temporal flood inundation extent during an extreme flooding event from 19th July to 12th August, 2019 as derived from MODIS NRT Flood product (major flooding dates were considered)

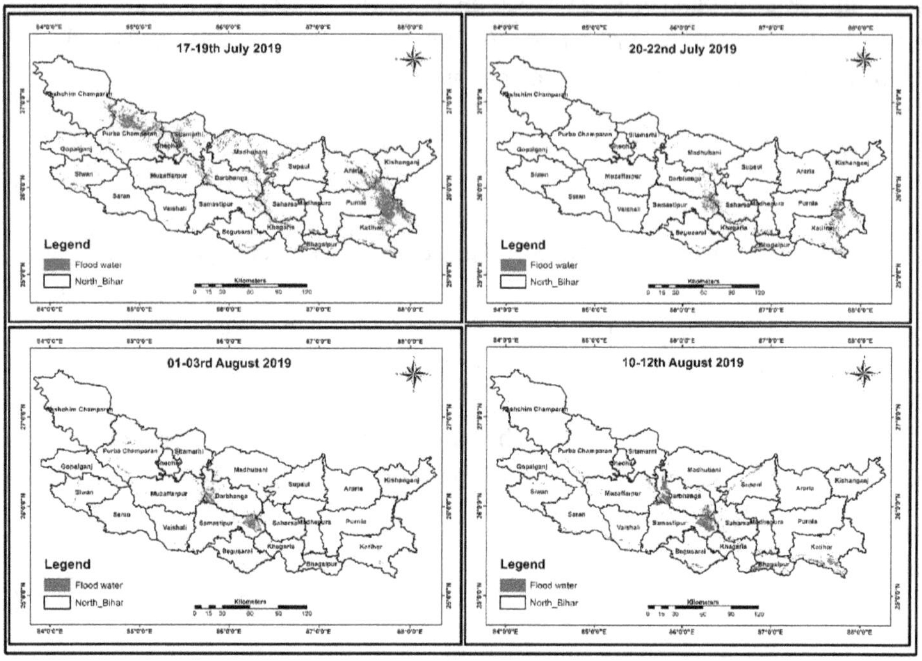

4. RESULTS

4.1 MODIS NRT Based Flood Inundation Assessment Over North Bihar

During the flood crises in July-August 2019, the regressive as well as progressive flood extent based on MODIS-NRT data is displayed in Figure 3. The flood maps during 17–19th August, revealed the flood-affected inundated areas of 574.22 sq.km over Northern Bihar. As an outcome, it engulfed the eastern and up to some extent the northern districts of Bihar, such as Katihar, Purnea, Araria, and Purbi -Champaran. During 20–22nd August, flood-affected inundated regions were lessened and estimated as 484.41 sq.km area. The flood-affected inundated extent of 416.86 sq.km decreased in 23–25th July, which further reduced in 26-28th July and estimated as 287.11 sq.km. The inundation extent finally reduced during 29–31st July and flooded area was estimated as 240.62 sq.km. During 01-03rd August, flood inundation extent expanded rapidly and predicted as 481.89 sq.km. Finally, during

Table 2. 3-day composite MODIS NRT Flood based Flood Inundation from 17th July to 12th August over North Bihar (Bold indicates major inundation)

Flooding date (3–day composite)	Flood Extent of MODIS NRT Flood data (3–day composite in sq.km)
17–19th July	**574.22**
20–22nd July	**484.41**
23–25th July	**416.86**
26th– 28th July	287.11
29th – 31st July	240.62
1st– 3rd August	**481.89**
4th– 6th August	**433.89**
07th–09th August	305.66
10th–12th August	**452.45**

04-06th August, extent was reduced and estimated as 433.89 sq.km along with 305.66 sq.km. inundation during 07-09th August.

Noticeably, the flood extent remained classified under high areal inundation during during 17-19th July, 20–22nd July, 23–25th July, 01-03rd August and 04-06th August. The area was predicted as 574.22, 484.41, 416.86, 481.89 and 433.89 sq.km, respectively. Out of 19 districts, there were 14 districts which were unfavorably affected due to floods and these districts were Katihar, Khagaria, Darbhanga, Araria, Purnea, Madhubani, Madhepura, Saharsa, Samastipur, Sheohar, Sitamarhi, Vaishali, Supaul, Paschim and Purbi Champaran. Results estimated the highest flood inundation happened during 17-19th July which engulfed around 574.22 sq.km area of Northern Bihar.

It can further affirm that the districts located on the river bank of Kosi, Burhi Gandak, and Ganga are the most salient reason of being most engulfed apart from the excessive rainfall throughout the seasonal channels (called as "dhars"). There are not bunds available to stop flood-water accompanying the Kosi river inundation from Katihar district. It was also reported that such districts were distinguished under very high waterlogging hazard, as 40–60% of regions were affected by waterlogging [19]. It remarked that flood engulfed regions mostly influenced agricultural fields as it is an agrarian dominated state. Thus, whenever the flooding event occurs, it affects the standing crops, such as rice, maize, sugarcane, banana, barley, vegetables, and fruits (banana, litchi, watermelon, coriander) etc.

Figure 4. Flood extent on 19th July, 31st July and 12th August derived based on Sentinel-1A (SAR) data

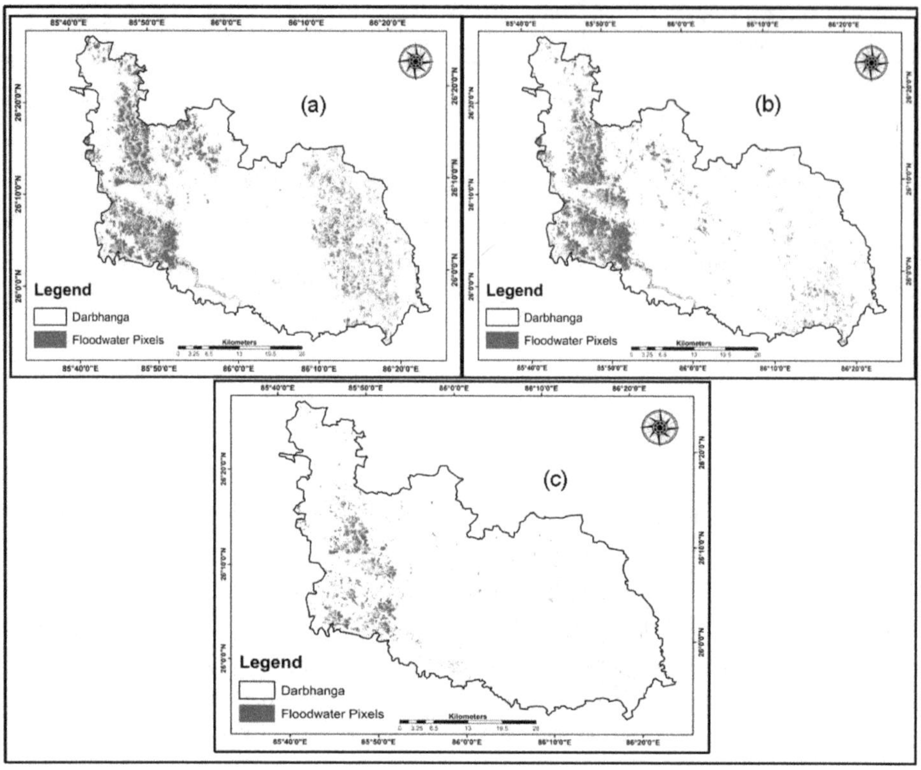

4.2 Sentinel 1-A Based Flood Inundation Assessment Over Darbhanga District of North Bihar

Regression and progression of the flooding over the Darbhanga district based on S-1A data during July-August 2019 were shown in Figure 4. SAR data were considered as more accurate and effective in flood mapping with compare to the optical datasets. Several literatures suggested considering the VH band to delineate flood pixels. The inundation maps during 19th July presented the flood induced areas which were estimated as ~448 sq.km across the district. Flooding extent reflects the effect of rainfall induced flooding in Figure 4. Due to the data gap, the next flood extent was estimated as ~363 sq.km on 31st July 2019 (Table 3). Flood extent started shrinking day by day followed by the precipitation. Until then, it got submerged the eastern and western parts of the district. The standing crops in the agricultural fields were got inundated and affected at a larger scale as it is agricultural dominated area. At multiple location such as vegetations, settlements and fallow land were inundated. On 12th August, flood inundated areas were reduced by 202.52 sq.km and estimated as ~160 sq.km. Flood extent under high areal inundation was estimated on 19th July as 447.04 sq.km. Based on several studies, it can be stated that almost every year flooding events lead to uprooting of standing crops in kharif season (July-December) across the state.

Table 3. Sentinel-1A based Flood Inundation of 19th and 31st July and 12th August with % inundation (Total region of Darbhanga district is 2289 sq.km) over Darbhanga district (Bold marking indicates dominant inundation)

Flooding duration	Flood inundated area (in sq.km)	Flood inundated area (in %)
	in VH polarization	in VH polarization
19th July	**447.04**	**19.52**
31st July	362.8	15.84
12th August	160.28	7.00

4.3 Comparison of Flood Extent of MODIS NRT Flood data with Sentinel 1-A and Validation with SENTINEL-2A Data

Over the Darbhanga district, flood inundation area retrieved from MODIS NRT Flood data has validated and compared with the Sentinel-1 SAR and S2-A data. The Darbhanga district was considered for validation because it is one of the prominent flood-affected regions. The flood extent was spatially distributed and comparable to each other (Figure 5 (a), (b), (c)), and optical False Color Composite (FCC) image presented flooding pixels (Figure 5c). The flood affected area was 556.91 sq.km in MODIS NRT flood data during 17-19th July (i.e., mean of 3 days composite), while 447.04 sq.km in SAR data on 19th July. The flood affected extent was overestimated in MODIS NRT and mostly attributed to sensor's resolution (250 m) capability (coarser) than the SAR (10 m) data. The error probability in Sentinel-1 data was less in comparison to MODIS NRT Flood product. During flood events, flood inundation extent produced by MODIS NRT data may obtain an overestimated area in different districts.

Figure 5. Flood extent map of MODIS NRT (a) and Sentinel-1A SAR (b) data over the entire Darbhanga district was shown during 16–19th August and 19th August, respectively. The S2-A image (a) of 17th July, 2019 was also used to verify flooding extent.

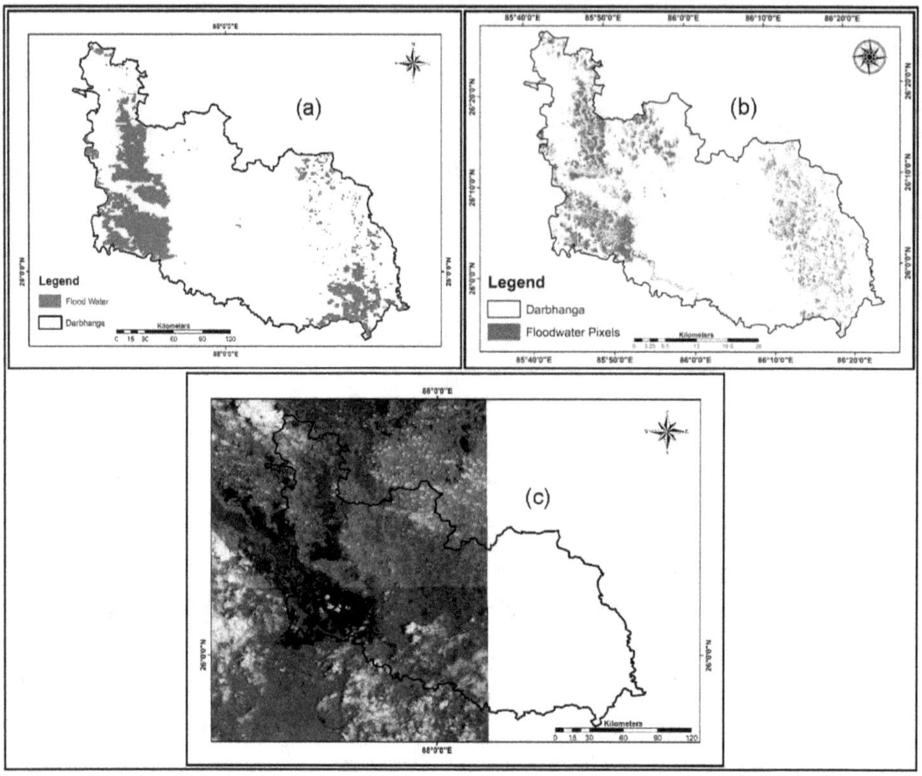

5. SUMMARY AND FUTURE PERSPECTIVES

India is a flood-prone country and thus under constant threat of flooding. Every year, floods destroy lives, livestock, and infrastructure brings an enormous financial toll. During disasters, obtaining the reliable and desired information is crucial. As a part of the operational methods, flood inundation was validated using Sentinel 2-A images. Unfortunately, due to the dominance of cloudy conditions, no (cloud-free) Sentinel 2-A images were available during the peak flood period. Apart from wall-to-wall comparisons with optical or field-based studies, a determination of the extent of flooding in India may serve disaster management purposes such as relief operations. Because, the monitoring of flood and its detection is an essential phase to grant support for civilians and other authorities' protection. In these circumstances,

EO data can perform a significant role, primarily the SAR images which permits continuous and efficient inspections of Earth.

In this chapter, relevancy of radar remote sensing data for flood inundation mapping was demonstrated as they do not depend on the availability of cloud. The presence of a MODIS-based NRT flood product (composite data) developed by NASA/DFO has proven advantageous to observe and estimate the flood-affected inundation area across a broader geographical space. It was witnessed from the case study that the MODIS-based NRT flood data can be efficiently utilized to evaluate flood progression as well as regression across the Northern Bihar district. However, it was worth noticing that the inundation spectrum procured from MODIS NRT data provided an overestimated region under flood still maintained spatial pattern. The MODIS-based NRT flood data can achieve the spatial variability of temporal rainfall that may lead to the progressive and regressive flooding over an area. Though the adopted methodology was quite easy but can still be helpful for the non-expert society to create NRT inundation maps in sparse areas to promote flood risk control. Nevertheless, the SAR data and hydrological models can easily portray the progressive and regressive nature of spatio-temporal flood mapping. It can be concluded from the NRT flood product that most of the regions during July 2019 were affected by the floods. During July-August 2017, a major Northern region of Bihar was affected by flooding. Flood maps were also validated with the supplementary dataset and field reports. The areal space throughout the flooding event exhibited the spatial pattern of the flooded area. The high spatial resolution of Sentinel-1 (SAR) images contributed additional precision to the inundated area than any other present dataset (Matgen et al. 2007; Mason et al. 2014) (Figure 5b). However, there were some uncertainties of flood maps due to the floating vegetated areas (Tsyganskaya et al. 2018). The Sentinel-1 images of cultivated land such as rice plantation were appeared to be the inundated areas and to overcome these uncertainties, local knowledge is crucial. Alternatively, unmanned aerial vehicles (UAVs) are an emerging tool for observing real time floods in disaster management. In general, SNAP tools take a longer time for preprocessing Sentinel-1 images. There is also a lag time between the image availability in ESA Sentinel hub and in GEE. If time lag is reduced in the future, the method could be implemented in GEE environment for a more rapid production of flood extent.

Based on the outcomes of this case study, we can conclude that earth observation and geospatial technologies provide an immediate report for efficient conclusions for comprehensive flood disaster supervision in the Northern Bihar region. Due to the predominance of severe weather conditions during flooding time, freely available and regularly sampled Sentinel-1 SAR earth observation data has great potential in producing flood information with high accuracy and high spatial resolution in a twelve-day interval. The open accessibility of freely available data is beneficial especially for developing countries. Natural flood disasters are common and cannot be stopped. However, efficient tools for mapping flood inundation and its assessment can be quite useful for emergency response and disaster management. Future perspectives of this case study is to incorporate 3-D hydrological modeling to map, predict the flood risk zones for such devastating events and to incorporate machine/deep learning technique for reducing the error as well as generating computations on cloud-based environments (GEE platform) for the operational users involved in planning, mitigation exercises, and flood-related emergency responses.

The shortcomings of this study are the availability of the satellite data and hydraulic parameters throughout the entire flooding event. During the flooding period, optical data obtained from passive sensors, have restrictions for flood mapping due to the presence of cloud. SAR sensors also have a limitation in their temporal resolution at regular intervals as it is not always necessary to have the availability of SAR data during the flood occurrences because SAR data is more sensitive towards moisture content present in the object, near-by river channels, and canals etc. The hydrological principles and machine learning algorithms along with the satellite data can be used in flood mapping progression and regression throughout (Alsdorf et al. 2007b; Matgen et al. 2011; Schlaffer et al. 2015b; Ajadi et al. 2016; Teng et al. 2017c; Afshari et al. 2018c; Speckhann et al. 2018b; McGrath et al. 2018b). The hydraulic parameters such as the roughness of the surface, discharge, slope, and run-off coefficients should remain set for flood mapping according to land use as well as the soil type distribution.

ACKNOWLEDGMENT

We are sincerely thankful to NASA GES/DISC, USGS and ASF for providing MODIS NRT Flood product, Sentinel-1A, and Sentnel-2A satellite data.

Conflicts of Interest: Authors declare no conflict of interest.

REFERENCES

Afshari, S., Tavakoly, A. A., & Rajib, M. A. (2018a). Comparison of new generation low-complexity flood inundation mapping tools with a hydrodynamic model. *Journal of Hydrology (Amsterdam)*, *556*, 539–556. doi:10.1016/j.jhydrol.2017.11.036

Afshari, S., Tavakoly, A. A., Rajib, M. A., Zheng, X., Follum, M. L., Omranian, E., & Fekete, B. M. (2018b). Comparison of new generation low-complexity flood inundation mapping tools with a hydrodynamic model. *Journal of Hydrology (Amsterdam), 556*, 539–556. doi:10.1016/j.jhydrol.2017.11.036

Ajadi, O., Meyer, F., & Webley, P. (2016). Change Detection in Synthetic Aperture Radar Images Using a Multiscale-Driven Approach. *Remote Sensing, 8*(6), 482. doi:10.3390/rs8060482

Alsdorf, D. E., Rodríguez, E., & Lettenmaier, D. P. (2007a). Measuring surface water from space. *Reviews of Geophysics, 45*(2), RG2002. Advance online publication. doi:10.1029/2006RG000197

Brivio, P. A., Colombo, R., Maggi, M., & Tomasoni, R. (2002). Integration of remote sensing data and GIS for accurate mapping of flooded areas. *International Journal of Remote Sensing, 23*, 429–441. doi:10.1080/01431160010014729

Costache, R. (2019). Flood Susceptibility Assessment by Using Bivariate Statistics and Machine Learning Models - A Useful Tool for Flood Risk Management. *Water Resources Management, 33*, 3239–3256. doi:10.100711269-019-02301-z

Gan, T. Y., Zunic, F., Kuo, C.-C., & Strobl, T. (2012). Flood mapping of Danube River at Romania using single and multi-date ERS2-SAR images. *International Journal of Applied Earth Observation and Geoinformation, 18*, 69–81. doi:10.1016/j.jag.2012.01.012

Grimaldi, S., Li, Y., Pauwels, V. R. N., & Walker, J. P. (2016). Remote Sensing-Derived Water Extent and Level to Constrain Hydraulic Flood Forecasting Models: Opportunities and Challenges. *Surveys in Geophysics, 37*(5), 977–1034. doi:10.100710712-016-9378-y

Kumar, R., Jain, V., Prasad Babu, G., & Sinha, R. (2014). Connectivity structure of the Kosi megafan and role of rail-road transport network. *Geomorphology, 227*, 73–86. doi:10.1016/j.geomorph.2014.04.031

Lian, J.-J., Guo, X.-Y., Ma, C., & Xu, K. (2019). Optimal Reservoir Flood Control Operation Using a Hedging Model and Considering the Near-Field Vibrations Induced by Flood Release. *Water Resources Management, 33*(8), 2645–2663. doi:10.100711269-019-02231-w

Macchione, F., Costabile, P., Costanzo, C., & De Santis, R. (2019). Moving to 3-D flood hazard maps for enhancing risk communication. *Environmental Modelling & Software, 111*, 510–522. doi:10.1016/j.envsoft.2018.11.005

Manjusree, P., Prasanna Kumar, L., & Bhatt, C. M. (2012). Optimization of threshold ranges for rapid flood inundation mapping by evaluating backscatter profiles of high incidence angle SAR images. *International Journal of Disaster Risk Science*, *3*, 113–122. doi:10.100713753-012-0011-5

Martinis, S., Kersten, J., & Twele, A. (2015). A fully automated TerraSAR-X based flood service. *ISPRS Journal of Photogrammetry and Remote Sensing*, *104*, 203–212. doi:10.1016/j.isprsjprs.2014.07.014

Mason, D. C., Giustarini, L., Garcia-Pintado, J., & Cloke, H. L. (2014). Detection of flooded urban areas in high resolution Synthetic Aperture Radar images using double scattering. *International Journal of Applied Earth Observation and Geoinformation*, *28*, 150–159. doi:10.1016/j.jag.2013.12.002

Matgen, P., Hostache, R., Schumann, G., Pfister, L., Hoffmann, L., & Savenije, H. H. G. (2011). Towards an automated SAR-based flood monitoring system: Lessons learned from two case studies. *Physics and Chemistry of the Earth Parts A/B/C*, *36*(7-8), 241–252. doi:10.1016/j.pce.2010.12.009

Matgen, P., Schumann, G., & Henry, J.-B. (2007). Integration of SAR-derived river inundation areas, high-precision topographic data and a river flow model toward near real-time flood management. *International Journal of Applied Earth Observation and Geoinformation*, *9*, 247–263. doi:10.1016/j.jag.2006.03.003

McGrath, H., Bourgon, J.-F., Proulx-Bourque, J.-S., Nastev, M., & Abo El Ezz, A. (2018). A comparison of simplified conceptual models for rapid web-based flood inundation mapping. *Natural Hazards*, *93*(2), 905–920. doi:10.100711069-018-3331-y

Modrick, T. M., & Georgakakos, K. P. (2015). The character and causes of flash flood occurrence changes in mountainous small basins of Southern California under projected climatic change. *Journal of Hydrology. Regional Studies*, *3*, 312–336. doi:10.1016/j.ejrh.2015.02.003

Mure-Ravaud, M., Binet, G., Bracq, M., Perarnaud, J.-J., Fradin, A., & Litrico, X. (2016). A web based tool for operational real-time flood forecasting using data assimilation to update hydraulic states. *Environmental Modelling & Software*, *84*, 35–49. doi:10.1016/j.envsoft.2016.06.002

Pandey, A. C., Singh, S. K., & Nathawat, M. S. (2010). Waterlogging and flood hazards vulnerability and risk assessment in Indo Gangetic plain. *Natural Hazards*, *55*(2), 273–289. doi:10.100711069-010-9525-6

Pandey, R. K., Crétaux, J.-F., Bergé-Nguyen, M., Tiwari, V. M., Drolon, V., Papa, F., & Calmant, S. (2014). Water level estimation by remote sensing for the 2008 flooding of the Kosi River. *International Journal of Remote Sensing, 35*(2), 424–440. doi:10.1080/01431161.2013.870678

Sakamoto, T., Van Nguyen, N., & Kotera, A. (2007). Detecting temporal changes in the extent of annual flooding within the Cambodia and the Vietnamese Mekong Delta from MODIS time-series imagery. *Remote Sensing of Environment, 109*, 295–313. doi:10.1016/j.rse.2007.01.011

Schlaffer, S., Matgen, P., Hollaus, M., & Wagner, W. (2015a). Flood detection from multi-temporal SAR data using harmonic analysis and change detection. *International Journal of Applied Earth Observation and Geoinformation, 38*, 15–24. doi:10.1016/j.jag.2014.12.001

Schlaffer, S., Matgen, P., Hollaus, M., & Wagner, W. (2015b). Flood detection from multi-temporal SAR data using harmonic analysis and change detection. *International Journal of Applied Earth Observation and Geoinformation, 38*, 15–24. doi:10.1016/j.jag.2014.12.001

Schumann, G. J.-P., & Domeneghetti, A. (2016). Exploiting the proliferation of current and future satellite observations of rivers: Invited Commentary. *Hydrological Processes, 30*(16), 2891–2896. doi:10.1002/hyp.10825

Schumann, G. J.-P., Frye, S., Wells, G., Adler, R., Brakenridge, R., Bolten, J., Murray, J., Slayback, D., Policelli, F., Kirschbaum, D., Wu, H., Cappelaere, P., Howard, T., Flamig, Z., Clark, R., Stough, T., Chini, M., Matgen, P., Green, D., & Jones, B. (2016). Unlocking the full potential of Earth observation during the 2015 Texas flood disaster: EO imagery assisting the texas flood disaster response. *Water Resources Research, 52*(5), 3288–3293. doi:10.1002/2015WR018428

Shakya, A., & Kumar, A. (2019). Noise Clustering-Based Hypertangent Kernel Classifier for Satellite Imaging Analysis. *Photonirvachak (Dehra Dun), 47*, 2009–2025. doi:10.100712524-019-01044-z

Singh, S. K., Pandey, A. C., & Nathawat, M. S. (2011). Rainfall Variability and Spatio Temporal Dynamics of Flood Inundation during the 2008 Kosi Flood in Bihar State, India. *Asian Journal of Earth Sciences, 4*, 9–19. doi:10.3923/ajes.2011.9.19

Sinha, R., Bapalu, G. V., Singh, L. K., & Rath, B. (2008). Flood risk analysis in the Kosi river basin, north Bihar using multi-parametric approach of analytical hierarchy process (AHP). *Photonirvachak (Dehra Dun), 36*(4), 335–349. doi:10.100712524-008-0034-y

Speckhann, G. A., Borges Chaffe, P. L., Fabris Goerl, R., Abreu, J. J., & Altamirano Flores, J. A. (2018). Flood hazard mapping in Southern Brazil: A combination of flow frequency analysis and the HAND model. *Hydrological Sciences Journal, 63*(1), 87–100. doi:10.1080/02626667.2017.1409896

Teng, J., Jakeman, A. J., Vaze, J., Croke, B. F. W., Dutta, D., & Kim, S. (2017). Flood inundation modelling: A review of methods, recent advances and uncertainty analysis. *Environmental Modelling & Software, 90*, 201–216. doi:10.1016/j.envsoft.2017.01.006

Tripathi, P., Parida, B. R., & Pandey, A. C. (2019). Spatio-Temporal Rainfall Variability and Flood Prognosis Analysis using Satellite Data over North Bihar during the August 2017 Flood Event. *Hydrology, 6*(2), 38. doi:10.3390/hydrology6020038

Tsyganskaya, V., Martinis, S., Marzahn, P., & Ludwig, R. (2018). SAR-based detection of flooded vegetation – a review of characteristics and approaches. *International Journal of Remote Sensing, 39*, 2255–2293. doi:10.1080/01431161.2017.1420938

Twele, A., Cao, W., Plank, S., & Martinis, S. (2016). Sentinel-1-based flood mapping: A fully automated processing chain. *International Journal of Remote Sensing, 37*(13), 2990–3004. doi:10.1080/01431161.2016.1192304

Uddin, M. (1581). Meyer (2019) Operational Flood Mapping Using Multi-Temporal Sentinel-1 SAR Images: A Case Study from Bangladesh. *Remote Sensing, 11*. Advance online publication. doi:10.3390/rs11131581

Van Dijk, A., Mueller, N., & Lymburner, L. (2011). *MODIS-based standing water detection for flood and large reservoir mapping: Algorithm development and applications for the Australian continent*. CSIRO. doi:10.4225/08/58518bd176131

Wahab, A. M., & Muhamad Ludin, A. N. (2018) Flood vulnerability assessment using artificial neural networks in Muar Region, Johor Malaysia. *IOP Conference Series: Earth and Environmental Science, 169*. 10.1088/1755-1315/169/1/012056

Wan, X., Yang, Q., Jiang, P., & Zhong, P. (2019). A Hybrid Model for Real-Time Probabilistic Flood Forecasting Using Elman Neural Network with Heterogeneity of Error Distributions. *Water Resources Management, 33*(11), 4027–4050. doi:10.100711269-019-02351-3

Chapter 9
Inferring Relationship of Landslides, Tectonics, and Climate:
Tons Valley, NW Himalaya

Imlirenla Jamir
https://orcid.org/0000-0002-6658-9466
Department of Geology, Nagaland University, India

Pranaya Diwate
Centre for Climate Change and Water Research, Suresh Gyan Vihar University, India

Vipin Kumar
https://orcid.org/0000-0001-7982-6806
Department of Geology, University of Liege, Belgium

Gambhir Singh Chauhan
https://orcid.org/0000-0001-6225-7678
Department of Geology, H.N.B. Garhwal University, India

ABSTRACT

Landslides, despite being the surficial impression of climate-tectonic-erosion linkage, are rarely explored in this context in Himalaya. The need for such study becomes more crucial in the evaluation of the regional hillslope denudation budget. We are of the understanding that the distributional pattern of landslides can reveal the relative significance of tectonic and climate. To test this hypothesis, ~ 55 landslides of the Tons River valley, Himalaya along with the tectonic and climate proxies are used in the present study. Steepness index and valley floor width to valley height ratio are

DOI: 10.4018/978-1-7998-5027-4.ch009

used to infer the tectonic regime whereas; Tropical Rainfall Measurement Mission based daily rainfall data and swath profile of Normalized Difference Vegetation Index are used to deduce spatial variability in climate. The study revealed the possible existence of a positive feedback system in the Higher Himalaya Crystalline and the simultaneous role of tectonic-climate in the Lesser Himalaya Crystalline. The LHS is found to possess a zone of landslide cluster, possibly due to local fault.

ABBREVIATIONS

HHC, Higher Himalayan Crystalline; LHC, Lesser Himalayan Crystalline; LHS, Lesser Himalayan Sequence; MCT, Main Central Thrust; MT, Munsiari Thrust

1.0 INTRODUCTION

Landslides are the primary mass wasting process in the tectonically active hilly terrain (Hovius et al., 1997; Ballantyne, 2002; Sanchez et al., 2010; Gupta et al., 2016; 2017; Kumar et al., 2018, 2019a, 2019b; Jamir et al., 2017, 2019). Along with tectonics, the climate is also considered to control the distribution of landslides (Borgatti and Soldati, 2010; Kumar et al., 2019b). Though lithology and structures have also been noted to control the occurrence and distribution of landslides, tectonics and climate dominate at the regional scale (Guzzetti et al., 1996; Sanchez et al., 2010). Nonetheless, inter-relationship of climate and tectonics has rarely been associated with spatially varying landslide distribution patterns. Though the spatial distribution of landslides can be achieved easily using high-resolution satellite imagery and subsequent ground-truthing, there are many limitations to infer temporal distribution owing to the delineation of individual failure events, loss of landslide scarp, vegetation growth and dating constraints (Lang et al., 1999; Kumar et al., 2019b).

The Tons River valley, NW Himalaya is an appropriate region to infer such inter-relationship among landslide, tectonics, and climate considering the varying litho-tectonic and climate conditions of the valley (Uniyal and Prasad, 2006; Raman and Punia, 2011). The valley is endowed with numerous landslides in the structurally and lithologically contrasted rock mass. Steepness index (k_s) and valley floor width to valley height ratio (V_f) are used to infer tectonic variability whereas, TRMM based daily rainfall data of the last 19 years (2000-2018) and the NDVI are used to represent spatial variability in climate. This study will help to find out the relative influence of tectonics and climate on the landslide distribution and dimension, which may also be used further for the hazard assessment.

2.0 STUDY AREA

The NE-SW flowing Tons River valley originates from the Bandarpunch glacier, NW Himalaya and finally joins Yamuna River valley as a tributary. The study area covers ~110 km long stretch along the Tons River valley covering parts of the Higher Himalayan Crystalline (HHC), Lesser Himalayan Crystalline (LHC) and Lesser Himalaya Sequence (LHS) (**Fig. 1**). The lithological setting of the study area is dominated by the gneissic and granitic rocks belonging to the HHC, LHC and LHS litho-tectonic divisions of the Himalaya.

Figure 1. Map of study area

STD, MCT, and MT are South Tibetan Detachment, Main Central Thrust, and Munsiari Thrust, respectively. Litho-units and regional faults are based on **Kumar et al., 2019b**. *The red quadrangular in the inset highlights the location of study area. Red circles at the edge of Tons River valley represent the extent of study area.*

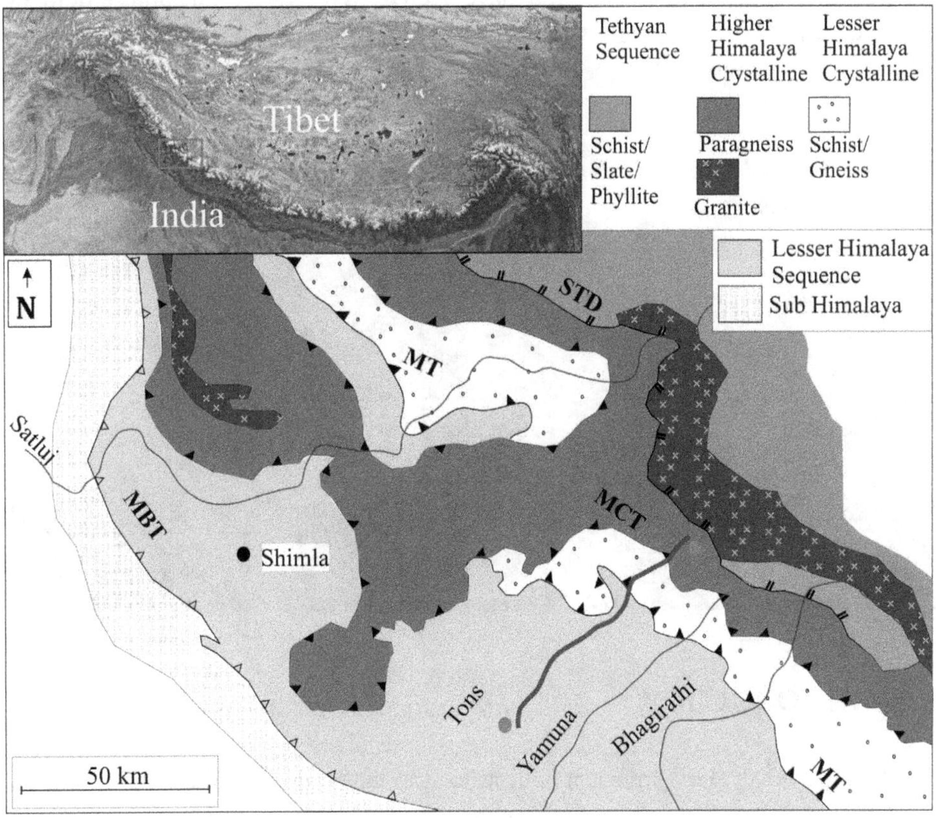

The HHC is separated from the LHC by the Main Central Thrust (MCT) (Locally termed as Vaikrita thrust). Further SW, the LHC is separated from the LHS by the Munsiari Thrust (Fuchs, 1982). The HHC comprises crystalline nappes whereas, the LHC comprises mostly quartzite, slate, and schist. The LHS in the region comprises limestone, slate, and quartzite. The climate of the valley is humid temperate during summer and cold and dry in the winter. The area receives rainfall mostly in the SW monsoon month (June- September) (Uniyal and Prasad, 2006; Raman and Punia, 2012). The study area lies in the western extent of the NW Himalaya and is supposed to possess relatively less exhumation and/ or tectonic deformation in the HHC region than the eastern counterpart of the NW Himalaya i.e., Kali River valley (Eugster et al., 2018). The Indian Summer Monsoon that precipitates along the Himalaya in an ENE to NW direction has also been considered to deplete in this part of the NW Himalaya (Bookhagen and Burbank, 2006). The present study aims to explore such tectonic and climate implications in the study area and their relations with the landslide distribution. Further, the Tons River valley in the LHC and LHS region comprises ~3-4 levels of terraces that have been used for agriculture and settlement purposes (Fig. 2). The straight top and linear alignment of these terraces might be related to lineaments and/or local faults in the region.

Figure 2. Field photographs showing relatively wide valley setting in the LHC (left) and LHS (right)

3.0 METHODOLOGY

The Google Earth (GE) Imagery is used for the landslide mapping owing to high resolution, free, easy, quick, and wide coverage. In view of the mapping limitations of the GE Imagery, ground truthing is also performed by comparing the known distances on ground with the GE measurement. The known distances were measured during the field work. A difference of 1.99% was noted between the known distances and measured ones in the Google Earth Imagery.

3.1 Steepness Index (k_s) and Valley Floor Width to Valley Height Ratio (V_f)

Steepness index (k_s) has been used widely as a proxy to infer tectonic regime in different fluvial settings (Flint, 1974; Howard and Kerby, 1983; Wobus et al., 2006). Usually, higher k_s represent high uplift region and low k_s to low uplift zones (Synder et al., 2000). However, anomaly in the k_s pattern may also occur owing to lithological variation and/or tributary junction. In order to minimize such uncertainty and strengthen the inference, another proxy i.e., valley floor width to valley height ratio was also used. The k_s was computed using the following equation;

$$\mathbf{S=k_s\, A^{-\theta}} \tag{1}$$

Here, S is river channel gradient (m/m); k_s is steepness index (mm^{-1}); A is an upstream drainage area (m^2) and θ is the concavity index. The θ was measured by the power regression of slope and upstream drainage area log-log plot. The slope and upstream drainage area values were determined at 500 m interval along the Tons River valley using the ALOS_PALSAR DEM having a pixel size of 12.5 m. The slope and upstream drainage area log-log plot gave a concavity index (θ) of 0.65. This concavity index was used to determine the steepness index profile along the study area.

To understand the relation of valley morphology and landslides, the ratio of valley floor width to valley height (V_f) (Bull and McFadden, 1977) was also determined using the following equation;

$$\mathbf{V_f= 2V_{fw} / [(E_{ld}-E_{sc}) + (E_{rd}-E_{sc})]} \tag{2}$$

Here, V_{fw} is valley floor width; E_{sc} is valley floor elevation; E_{ld} and E_{rd} are left and right valley divides, respectively. These parameters were extracted from ALOS_PALSAR DEM at 500 m interval. Generally, higher V_f characterizes the tectonically tranquil region, and lower V_f characterizes rapid uplift and valley incision (Keller and Pinter, 2002).

3.2 Rainfall and Normalized Difference Vegetation Index (NDVI)

To understand the spatial variability of the climate, TRMM (3B42) based daily rainfall data (2000-2018) and NDVI swath profile was used. The NDVI (Deering,

1978; Jackson and Huete, 1991) was used as an indirect proxy of climate. It was derived using following equation;

NDVI= (NIR-R)/(NIR+R) (Equation 3)

Here, NIR and R correspond to Near Infrared and Red bands of the electromagnetic spectrum. The data of these bands was extracted from the Landsat 8 imagery using ArcGIS 10.5 software, ESRI. The NDVI swath profile was prepared using swath width of 4 km and sampling interval of 30 m along the valley.

4.0 RESULTS AND DISCUSSION

High values of the k_s are observed at the Main Central Thrust (MCT), Munsiari Thrust (MT) and in the LHS, possibly due to existence of local fault (Fig. 3). However, the high value of k_s may also occur due to tributary junction and/or lithological contrast that result in a relatively more slope gradient in the bedrock channel. Therefore, in order to ascertain the most probable cause, another proxy i.e., V_f was used to support the inferences made through the k_s. The V_f ratio resulted in lower values at the Main Central Thrust (MCT), Munsiari Thrust (MT) and in the localized zone in the LHS. It is noted that the high values of the steepness index (k_s) and the low values of valley floor width to valley height ratio (V_f) are generally associated with the narrow/deep gorges whereas, low k_s and high V_f generally coexist with the wide valley settings (Kumar et al., 2019b; Jamir et al., 2019). It is to note that the variations in the k_s are attributed to the tectonic uplift and/or bedrock erosion owing to the underlying principle of the $k_{s.}$ The k_s has been considered to be an approximate equivalent to the ratio of bedrock uplift and bedrock erosion. Since the bedrock erosion is a highly complex process owing to diverse controlling factors and resultant bedrock uplift also remains complex to infer (Kirby and Whipple, 2001). Therefore, to infer the controlling factor, the valley was supposed to be in tectonic equilibrium (and let the climate be a variable) i.e., bedrock uplift » bedrock erosion and variation in k_s generally attribute to the tectonic disturbance.

Further, the HHC and LHC region comprises relatively more rockfalls that can be ascribed to the relatively more bedrock incision by the fluvial action in this region that in turn may be related to rapid uplift. Various researchers have observed relatively high exhumation rates in the HHC in the NW Himalaya that in turn can be understood as a counter mechanism to the relatively high hillslope erosion in this region (Thiede et al., 2009; Kumar et al., 2019b; Jamir et al., 2019). Apart from the tectonic disturbances i.e., exhumation rate, climate also acts vitally in the HHC region that acts as the orographic barrier to the saturated winds and receives high

Figure 3. Tectonic and precipitation regimes in the Tons River valley
1, 2, 3, 4 and 5 refer to different rainfall data (TRMM based) regions, respectively. The k_s and V_f refer to steepness index and valley width to valley height ratio.

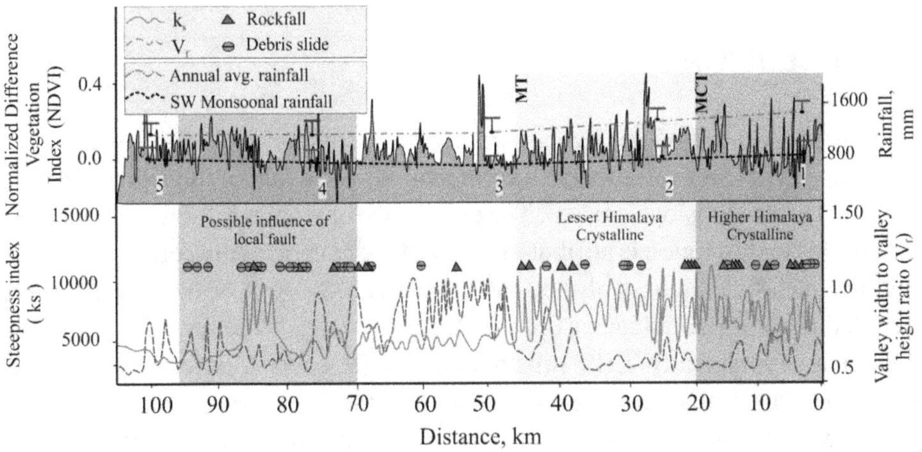

precipitation. The relative increase in the rainfall in the HHC region is supposed to be one of the primary reasons for enhanced hillslope erosion (Fig. 3).

The LHC comprises almost equal distribution of debris slides and rockfalls depicting widespread hillslope erosion, though relatively less rainfall than the HHC, as inferred from the rainfall variation (Fig. 3). It is to note that relatively wider fluvial channel in this region and the existence of paleo-fluvial terraces also implies relatively less tectonic disturbances (Fig. 2). Thus, in the absence of tectonically supported bedrock uplift, rainfall-induced hillslope erosion is supposed to enhance. The River valley widens and hillslope toe erosion prevails. The vegetation density, inferred from the NDVI, remains almost the same along the valley, except few peaks at the MCT. Though the effect of the vegetation is not found to be prominent on the hillslopes, sudden decrease beyond the MCT in the HHC can be considered to result in less shear strength of hillslope material (Kumar et al., 2019b; Jamir et al., 2019).

Similar to the LHC, the influence of the tectonics is found to be relatively less in the LHS, as inferred from the k_s and V_f regime. However, a sudden increase in the k_s at a localized region and cluster of debris slides in this region is also observed. We are of understanding that it could be due to the influence of local fault. This finding in the LHS opens new prospects for the future study. Further, wide valley setting in this region highlights the tendency of lateral fluvial erosion (Fig. 2). Such lateral erosion is observed to prevail in the settings where thick overburden on the fluvial channel restricts the bedrock incision (Bull, 2009; Kumar et al., 2019b; Jamir et al., 2019). The regional faults; Main Central Thrust (MCT) and Munsiari Thrust (MT) are noted to spatially co-exist with the cluster of landslides. This spatial co-existence is attributed to litho-structural contrast at fault that results in the rock mass shearing and hence enhanced hillslope erosion.

5.0 CONCLUSION

The gneissic rockmass of the Higher Himalayan Crystalline i.e., hanging wall of the MCT, is observed to possess the high tectonic activity and relatively high precipitation that result in more hillslope erosion. The dominance of relatively more rockfall type of landslides in this region is attributed to relatively more tectonic perturbations. In the schist/gneissic rockmass of the LHC, however, the effect of the tectonics and precipitation appear simultaneous in view of more debris slide type landslides. In the quartzitic (dominantly) LHS, hillslopes are subjected to slope toe erosion in the wider valley floors that are supposed to support the lateral fluvial erosion. Such lateral fluvial erosion through toe cutting results in more debris into the valley floor and further supports the lateral erosion process. The k_s and V_f indices, meant to infer tectonic influence, showed sudden variation in localized region in the LHS that is attributed to the influence of local fault.

ACKNOWLEDGMENT

VK and IJ acknowledge Prof. Michael J. Crozier and Prof. Mauri McSaveney for the constructive discussion during 9th ICG, 2017 (India) and iRALL school, 2018 (Chengdu, China), respectively. IJ, VK and GSC are also thankful to Prof. Onkar Dikshit (IIT Kanpur) for the landslide training program (2017) that helped in the landslide data extraction & interpretation.

REFERENCES

Ballantyne, C. K. (2002). Paraglacial geomorphology. *Quaternary Science Reviews*, *21*(18), 1935–2017. doi:10.1016/S0277-3791(02)00005-7

Bookhagen, B., & Burbank, D. W. (2006). Topography, relief, and TRMM-derived rainfall variations along the Himalaya. *Geophysical Research Letters*, *33*(8), L08405. doi:10.1029/2006GL026037

Borgatti, L., & Soldati, M. (2010). Landslides as a geomorphological proxy for climate change: A record from the Dolomites (northern Italy). *Geomorphology*, *120*(1), 56–64. doi:10.1016/j.geomorph.2009.09.015

Bull, W. B. (2009). *Tectonically active landscapes* (1st ed.). John Wiley and Sons. doi:10.1002/9781444312003

Bull, W. B., & McFadden, L. D. (1977). Tectonic geomorphology north and south of the Garlock fault, California. Geomorphology in arid regions. *Proceedings of the eighth annual geomorphology symposium*, 115-138.

Deering, D. W. (1978). *Rangeland reflectance characteristics measured by aircraft and spacecraft sensors* (Ph.D. Dissertation). Texas A&M University, College Station, TX.

Eugster, P., Thiede, R. C., Scherler, D., Stübner, K., Sobel, E. R., & Strecker, M. R. (2018). Segmentation of the Main Himalayan Thrust Revealed by Low-Temperature Thermochronometry in the Western Indian Himalaya. *Tectonics*, *37*(8), 2710–2726. doi:10.1029/2017TC004752

Flint, J. J. (1974). Stream gradient as a function of order, magnitude, and discharge. *Water Resources Research*, *10*(5), 969–973. doi:10.1029/WR010i005p00969

Fuchs, G. R. (1982). *Explanations of the geologic-tectonic map of the Himalaya; Geol.* Surv.

Gupta, V., Jamir, I., Kumar, V., & Devi, M. (2017). Geomechanical characterization of slopes for assessing rockfall hazards in the Upper Yamuna Valley, Northwest Higher Himalaya, India. *Himalayan Geology*, *38*(2), 156–170.

Gupta, V., Nautiyal, H., Kumar, V., Jamir, I., & Tandon, R. (2016). Landslide hazards around Uttarkashi township, Garhwal Himalaya, after the tragic flash flood in June 2013. *Natural Hazards*, *80*(3), 1689–1707. doi:10.100711069-015-2048-4

Guzzetti, F., Cardinali, M., & Reichenbach, P. (1996). The influence of structural setting and lithology on landslide type and pattern. *Environmental & Engineering Geoscience*, *2*(4), 531–555. doi:10.2113/gseegeosci.II.4.531

Hovius, N., Stark, C. P., & Allen, P. A. (1997). Sediment flux from a mountain belt derived by landslide mapping. *Geology*, *25*(3), 231–234. doi:10.1130/0091-7613(1997)025<0231:SFFAMB>2.3.CO;2

Howard, A. D., & Kerby, G. (1983). Channel changes in badlands. *Geological Society of America Bulletin*, *94*(6), 739–752. doi:10.1130/0016-7606(1983)94<739:CCIB>2.0.CO;2

Jackson, R. D., & Huete, A. R. (1991). Interpreting vegetation indices. *Preventive Veterinary Medicine*, *11*(3-4), 185–200. doi:10.1016/S0167-5877(05)80004-2

Jamir, I., Gupta, V., Kumar, V., & Thong, G. T. (2017). Evaluation of potential surface instability using finite element method in Kharsali Village, Yamuna Valley, Northwest Himalaya. *Journal of Mountain Science*, *14*(8), 1666–1676. doi:10.100711629-017-4410-3

Jamir, I., Gupta, V., Thong, G. T., & Kumar, V. (2019). Litho-tectonic and precipitation implications on landslides, Yamuna valley, NW Himalaya. *Physical Geography*, 1–24. Advance online publication. doi:10.1080/02723646.2019.1672024

Keller, E. A., & Pinter, N. (2002). *Active Tectonics: Earthquakes, Uplift, and Landscape* (2nd ed.). Prentice Hall.

Kirby, E., & Whipple, K. (2001). Quantifying differential rock-uplift rates via stream profile analysis. *Geology*, *29*(5), 415–418. doi:10.1130/0091-7613(2001)029<0415:QDRURV>2.0.CO;2

Kumar, V., Gupta, V., & Jamir, I. (2018). Hazard evaluation of progressive Pawari landslide zone, Satluj valley, Himachal Pradesh, India. *Natural Hazards*, *93*(2), 1029–1047. doi:10.100711069-018-3339-3

Kumar, V., Gupta, V., Jamir, I., & Chattoraj, S. L. (2019a). Evaluation of potential landslide damming: Case study of Urni landslide, Kinnaur, Satluj valley, India. *Geoscience Frontiers*, *10*(2), 753–767. doi:10.1016/j.gsf.2018.05.004

Kumar, V., Gupta, V., & Sundriyal, Y. P. (2019b). Spatial interrelationship of landslides, litho-tectonics, and climate regime, Satluj valley, Northwest Himalaya. *Geological Journal*, *54*(1), 537–551. doi:10.1002/gj.3204

Lang, A., Moya, J., Corominas, J., Schrott, L., & Dikau, R. (1999). Classic and new dating methods for assessing the temporal occurrence of mass movements. *Geomorphology*, *30*(1), 33–52. doi:10.1016/S0169-555X(99)00043-4

Raman, R., & Punia, M. (2012). The application of GIS-based bivariate statistical methods for landslide hazards assessment in the upper Tons river valley, Western Himalaya, India. Georisk. *Assessment and Management of Risk for Engineered Systems and Geohazards*, *6*(3), 145–161. doi:10.1080/17499518.2011.637504

Sanchez, G., Rolland, Y., Corsini, M., Braucher, R., Bourlès, D., Arnold, M., & Aumaître, G. (2010). Relationships between tectonics, slope instability and climate change: Cosmic ray exposure dating of active faults, landslides and glacial surfaces in the SW Alps. *Geomorphology*, *117*(1), 1–13. doi:10.1016/j.geomorph.2009.10.019

Snyder, N. P., Whipple, K. X., Tucker, G. E., & Merritts, D. J. (2000). Landscape response to tectonic forcing: Digital elevation model analysis of stream profiles in the Mendocino triple junction region, northern California. *Geological Society of America Bulletin, 112*(8), 1250–1263. doi:10.1130/0016-7606(2000)112<1250:LRTTFD>2.0.CO;2

Srivastava, V., Srivastava, H. B., & Lakhera, R. C. (2010). Fuzzy gamma based geomatic modelling for landslide hazard susceptibility in a part of Tons river valley, northwest Himalaya, India. *Geomatics, Natural Hazards & Risk, 1*(3), 225–242. doi:10.1080/19475705.2010.490103

Thiede, R. C., Ehlers, T. A., Bookhagen, B., & Strecker, M. R. (2009). Erosional variability along the northwest Himalaya. *Journal of Geophysical Research. Earth Surface, 114*, F101015.

Uniyal, A., & Prasad, C. (2006). Disaster management strategy for mass wasting hazard prone Naitwar Bazar and surrounding areas in Upper Tons valley in Uttarkashi district, Uttaranchal (India). *Disaster Prevention and Management: An International Journal, 15*(5), 821–837. doi:10.1108/09653560610712766

Wobus, C., Whipple, K. X., Kirby, E., Snyder, N., Johnson, J., Spyropolou, K., Crosby, B., & Sheehan, D. (2006). *Tectonics from topography: Procedures, promise, and pitfalls* (Vol. 398). Geological Society of America Special Papers.

Chapter 10
Sentinel SAR Data and In-Situ-Based High-Resolution Above-Ground Carbon Stocks Estimation Within the Open Forests of Ramgarh District

Akash Anand
https://orcid.org/0000-0003-1086-5676
Department of Geoinformatics, Central University of Jharkhand, India

ABSTRACT

The present study deals with an approach to estimate the above ground biomass (AGB) to assess the total carbon stock of forest cover present in Ramgarh district using remote sensing and GIS techniques. Due to the fact that biomass estimation is one of the most influential biophysical parameters in traditional carbon sequestration techniques, satellite remote sensing plays an important role in AGB and carbon stock estimation. Presently, AGB is estimated using Sentinel1A SAR data in conjunction with in-situ field data, which is conducted in 20 different sites within the forest area. Biomass is calculated for each plot, and a correlation analysis is performed with the backscatter value obtained from SAR data to generate an allometric equation that is used to calculate the AGB and carbon stock for the entire forest cover. Both Polarization VV and VH are correlated with field data in which cross-polarized backscatter value shown a stronger correlation of 0.75 (R2 Value). C-band is proved to be the best band for the estimation of biomass and carbon stock in tropical mixed forests.

DOI: 10.4018/978-1-7998-5027-4.ch010

1. INTRODUCTION

Forest on the earth covers nearly one-third of total surface area and plays a vital role in the global carbon cycle changes (Franklin, 2001). Forest biomass which is also expressed as dry weight per unit area is one of the most influencing factors of the forest ecosystem, accurate estimation of Above Ground Biomass (AGB) is a critical component for carbon stock quantification, suitability analysis, analyzing fuels for forest fires etc. (Bhatt, Bisht, Rawal, & Dhar, 2007; Gorte, 2009). The estimated standing biomass in India is 8375 million tons in year 1986 whereas the carbon stock is estimated to be 4178 million tons (Dadhwal, Sehgal, Singh, & Rajak, 2003). The first ever woody biomass estimation is done in India by the Forest Survey of India (FSI) in 1995 in which they have used field inventory data, forest cover and various thematic maps as input, afterwards they carried out another survey to estimate the total carbon stock of Indian forests using remote sensing data and sample plot data during 2008-2010 (Gilbert, 2012). Presently, the scenario is at an alarming rate due to the deforestation and declining condition of mixed forest within the growing cities and it is continuously increasing (Pandey, Anand, & Srivastava, 2019; Ranjan, Anand, Vallisree, & Singh, 2016).

The most common method for biomass estimation is through allometric equations which is applied to the field inventory data to estimate biomass and carbon stock of forests. Many generalized allometric equations are generated by a number of researchers for different tree species, in a generalized allometric equation the relation between various physical parameters such as height of the tree, diameter at breast height, wood density, crown diameter, tree species etc. are established (Huang, Ziniti, Torbick, & Ducey, 2018; Montes, Gauquelin, Badri, Bertaudiere, & Zaoui, 2000; Navar, 2009; Nelson, 1999; Xiao & Ceulemans, 2004). (Kim et al., 2011), focused their study towards the site specific allometric equations and predicted forest biomass for a regional level whereas (Vieilledent et al., 2012) stated a universal approach towards biomass and carbon stock estimation in which a simple allometry of height and diameter is required from field inventories which is applicable for given forest sites where biomass allometric models are not available.

Carbon stock means the total carbon stored in living biomass that are present in the ecosystem, the carbon sequestration by tree in the forest are the source of carbon in the forest whose quantity varies with the forest type (Lucht, Schaphoff, Erbrecht, Heyder, & Cramer, 2006; Winjum, Dixon, & Schroeder, 1992). According to (Eggleston, Buendia, Miwa, Ngara, & Tanabe, 2006), there are five carbon pools in the ecosystem that include above ground, below ground, litters, woody debris, and organic matters present in soil, in which the carbon present above the ground are dominant. (S. Brown, 1997; Lugo & Brown, 1992)stated that the total carbon sequestration is approximately half of the forest biomass whereas the total carbon sequestration of the forest is inferred from its total biomass accumulation. Numerous studies are published on biomass and carbon sequestration that mainly focuses on boreal and temperate forests (Le Toan et al., 2011; Santos, Lacruz, Araujo, & Keil, 2002) however study on tropical forests are limited due to the complex species composition and structure (Sinha, Jeganathan, Sharma, & Nathawat, 2015).

Due to its ability of measure the reflectance at broad spectral bands having precise information, the multispectral images are widely used to perform classifications. There are several image classification techniques used for the classification of land use land cover which are broadly of two types, supervised classification technique and unsupervised classification techniques. The supervised classification technique is broadly consists of decision tree (Friedl & Brodley, 1997), support vector machine (Furey et al., 2000), neural nets (Weiss, Kapouleas, & Shavlik, 1990), maximum likelihood classification (Paola & Schowengerdt, 1995) etc. whereas the unsupervised classification techniques consists of k-means (Rollet, Benie, Li, Wang, & Boucher, 1998), spectral clustering (Kamvar et al., 2003), ISODATA (Irvin, Ventura, & Slater, 1997) etc. In present study, the maximum likelihood classification technique is used to do the classifications.

The role of remote sensing technology for the assessment of biomass and carbon sequestration has been found as a better alternative against the conventional methods (Anand, Singh, & Kanga, 2018). It has been observed that, optical remote sensing are widely used for biomass estimation but for the tropical forests it is recommended to use the radar data products as they are weather resistant and has an ability to penetrate into the forest canopy and it is sensitive to surface roughness. As per the empirical relationships between biomass and radar backscattering it is observed that C and L band saturates at different levels dependent upon incident angles and polarization (Kurvonen, Pulliainen, & Hallikainen, 1999; Lucas et al., 2006; Luckman, Baker, Kuplich, Yanasse, & Frery, 1997).

BACKGROUND

The biomass of any forests is a very useful way of assessing of the quantity and the components of the forest. Estimating the total above ground biomass of open forests which is also known as biomass density when articulated as dry weight per unit area is a suitable way of calculating the amount of resource available for different kinds of ecological uses and most importantly the change in the open forest diversity and the corresponding changes to the regional climate caused by its above ground carbon sequestration (Miehle, Grote, Battaglia, Feikema, & Arndt, 2010; Shrestha & Dhillion, 2003). The total above ground biomass of forests estimations is directly proportional to the carbon pools in forest species because carbon is roughly about

47% of the total stored biomass. Assessment of the biomass density of open forests using different empirical models have been derived by the scientific community that also measures the contribution of open forest deforestation and carbon burning that led to the increase in atmospheric carbon dioxide, GHG's and other trace gases (S. Brown, 2002; Hall & Uhlig, 1991; Hawbaker et al., 2009; Kuhlbusch & Crutzen, 1996) . Besides, above ground biomass density estimations of open forests are extremely significant for analyzing other global ecological as well as biogeochemical cycles, because the amount of other nutrient present in open forests are also linked with the quantity of above ground biomass present in the system.

Carbon stock states the total amount of carbon present or stored in the forest ecosystem, mostly in living biomass such as trees, organisms etc. and soil, but also to an extent also in dead wood and litter there is a few amount of carbon present. Carbon sequestration by the plant and trees in the forest ecosystem is the major source of the carbon stock within any type of forest and its amount varies with forest type (Bradford, Jensen, Domke, & D'Amato, 2013; Schaphoff et al., 2013).

As per (Roy & Ravan, 1996), estimation of above ground biomass can broadly be divided into two groups that is, using the in-situ based approach using destructive method and estimation of above ground biomass using a remote sensing methodology. In in-situ based approach, tree measurements within the sample size can be converted to biomass using either by predefined biomass expansion factors or allometric regression models. As accumulating in-situ data for the assessment of below ground biomass is difficult hence most of the research on biomass estimation focused on the above ground biomass. In the remote sensing approach, the above ground carbon is calculated by establishing a relation between fields based measurements and the satellite backscattering values to map the overall biomass and carbon stock of a particular region.

The AGB and carbon approximation using C band sentinel data has been achieved a good result in mixed vegetation covers with lesser biomass as well. Radar backscattering values at longer wavelengths like L and P bands is lower than as compared to C band for mixed vegetation biomass sites, especially grassland, shrub land, forest regeneration sites, and small distributed plantations. Land surface covered with grass, bushes, or small plantations have very low backscattering value at P band since the scattering components of grass and small plants are very less as compared with the P band's wavelength to provide significant backscattering values. The same land surface, however, will be very rough at C band so a much robust backscattering may occur at C band. The major scattering element for the C band, are basically the leaves and the branches, it generally saturates at 10 kg/m^2, while the tropical forests with a much difficult structure and mixed plant/tree species, the L and P band saturates commonly at 100 t/ha (Imhoff, 1993; Saini et al., 2004). In the areas with simple structure and one or two dominant species the saturation level is about 250 t/ha. The best bands for biomass estimation in mixed forests are C and L band with cross polarization (K. J. Ranson, Saatchi, & Sun, 1995; S. Ranson & Stewart, 1994). The backscattering of the L band has shown more penetration in vegetation canopy than C band so L band is more related to the tree trunk and major branches and less sensitive to the regional environmental conditions so it is also appreciated in the biomass and carbon estimation. (Hoekman & Djankov, 1997) stated that using a combination of C and L band could show a better result than using of one band.

2. STUDY AREA

Ramgarh is one of the youngest district of Jharkhand state, it is established on 12th September 2007. The total area of Ramgarh district is 1341 km^2 in which the forest area covers 487.93 km^2, the district is a part of Chotanagpur plateau and contains a large stock of coal from Lower Gondwana age (Chandramouli, 2011). Damodar River is the major river flows in the district having number of small tributaries which flows throughout the district. The study area falls in the sub-humid climatic zone where the day temperature generally rises up to 40^0C in summer and 10^0C in winters with an annual rainfall of 1251.2 mm (Bhanja et al., 2017). The present study area is mostly tropical dry deciduous forest and dry mixed deciduous forest having *Shorea robusta* whereas *Annona reticulate, Annona squamosa, Azadirachta indica, Adansonia digitata, Caesalpinia pulcherima, Delonix regia,Dalbergia sissoo Swietenia mahagoni, Saraca asoka, Syzygium cumini etc,* are also found during the investigation. These observed species are present in a heterogeneous way throughout the study area and it is observed that the hilly areas have the higher forest density and the lowlands are mostly open forest.

3. MATERIALS AND METHODOLOGY

3.1. Data Used

In present study, Landsat ETM+ satellite image is acquired from the United Stat Geological Survey (USGS) to calculate the LULC also the near-infrared band and red band is further used to derive above ground biomass and carbon stock (RANJAN et al., 2017). Sentinel 1A C-band data is also used in present study to get the sigma naught value to further correlate it with the in-situ measurements. A field survey is conducted within the study area to measure the AGB of 15 different sites to establish a correlation with the sigma naught value obtained from the SAR backscattering

Figure 1. Study Area

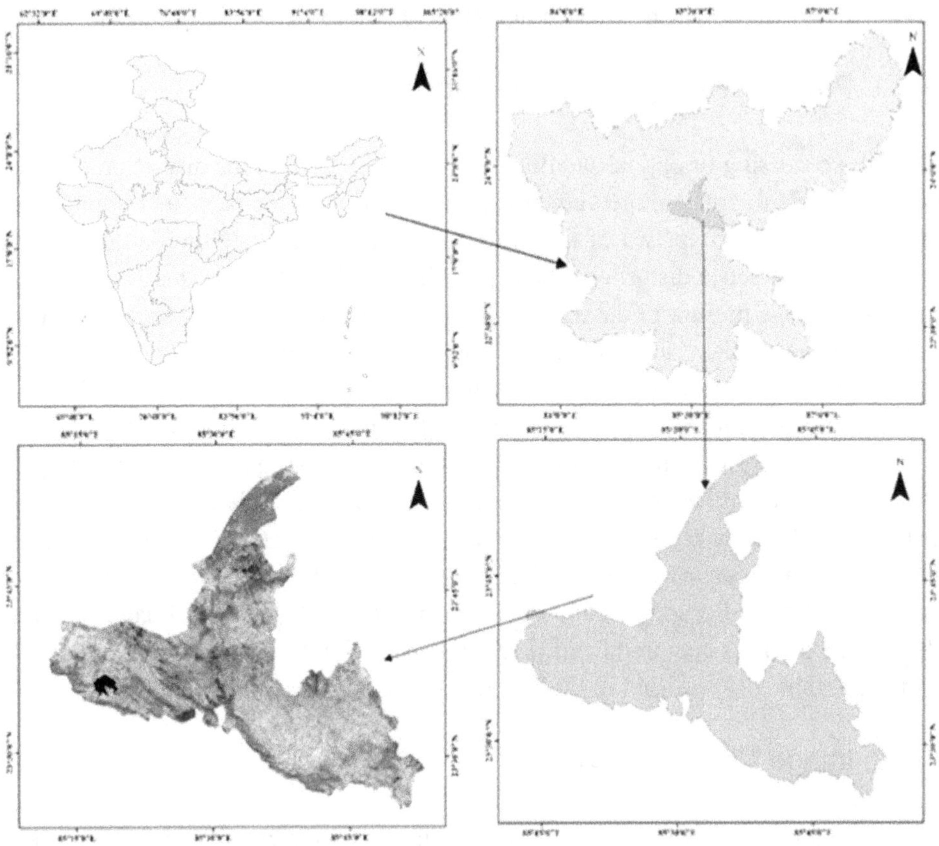

values (Attema et al., 2008; Freeman & Saatchi, 2004; Patel, Srivastava, Panigrahy, & Parihar, 2006).

3.2. Methodology

Landsat ETM+ image is acquired from the USGS Earth Explorer website (https://earthexplorer.usgs.gov/) and the preprocessing is done using ERDAS Imagine software that include layer stacking, study area masking, noise removal and top of atmospheric reflectance retrieval. Landsat ETM+ image is used to generate the land use land cover of the present study area using Maximum Likelihood Classification (MLC) method to calculate the spatial coverage of particular feature class (Ranjan, Anand, Kumar, Verma, & Murmu, 2018). MLC is a supervised classification technique in which a training set is made using signature collected manually for each feature class. The accuracy assessment is also done by calculating the overall accuracy and kappa coefficient using the confusion matrix (Thompson & Walter, 1988).

3.2.1. Pre-Processing of Optical Data

The pre-processing of optical satellite data targets to correct the distortion and the quality of data to create more authentic representation of the particular sight. This characteristically comprises of the initial handling of raw satellite data to firstly correct the geometric distortions then to rectify the data radiometrically and lastly to remove noise present in the data (T. Lillesand, Kiefer, & Chipman, 2015). The radiance measured by any given classification over a given entity is generally influenced by such a element as changes in image radiance, atmospheric conditions, viewing angle and instrument characteristics. Digital elevation correction and earth to sun distance correction was also carried out for mosaicking more than one images which are taken at different intervals or learning the temporal changes in reflectance of the ground feature at altered location. Atmospheric correction is meant to reduces the interference caused by atmospheric gases or particles to solar illumination variation (T. M. Lillesand, Kiefer, & Chipman, 2000). The Landsat image used for the present analysis acquired from USGS Earth Explorer which was pre-processed and geographically rectified to UTM projection Zone 45.

3.2.2. Image Classification

There are a number of classification techniques used for classifying the land use land cover using multi-temporal datasets, but Maximum Likelihood Classification (MLC) technique was adopted mostly and it is utilized in present study to analyse the variation in LU/LC of Ramgarh District using Landsat data. Basically, the maximum likelihood classifier takes up the pixels of an unknown class and decides certain possibility of going to a specific set (Ismail, Mutanga, Kumar, & Bob, 2008), the possibilities in MLC technique are the same for all the unknown classes and the input signatures in each multi-spectral band follow the Gaussian distribution function to classify the given image (Goodman, 1963; T. M. Lillesand et al., 2000). The supervised classification implicates training set for every class and the training set is used to prompt spectral reflectance signature of each land use category. The collected signatures would then be taken by the classifier to group the pixels into a certain class which has the similar spectral pattern.

3.2.3. Sentinel 1 Pre-Processing

The pre-processing of Sentinel 1 data is done with the Sentinel 1 Toolbox (S1TBX) inside SNAP software (Dostálová et al., 2016). The metadata information as well as the velocity information is taken from the orbit file of Sentinel 1 data. The radiometric correction is then performed to remove the distortion caused by sun's azimuth and atmospheric conditions which affects the reflected energy coming from earth's surface. Mathematically the radiometric calibration of Sentinel 1 data is shown as:

$$\mathbf{Value(i) = |DN_i|^2/A_i}$$

where, value(i) is the radiometrically corrected value, A_i is a calibration coefficient of ith pixel which is provided in SAR metadata as a grid format that is a two-dimensional array, DN_i is the digital number of ith pixel (El Hajj, Baghdadi, Zribi, & Angelliaume, 2016). Image noise is then calculated to remove it from the radiometrically calibrated value to get a noise free radiometrically corrected value. Sentinel 1 provides a noise LUT (Look Up Table) for each SAR data which can be used to derive a calibrated noise profile that is matched with the radiometrically calibrated data to get the pixel-wise noise value (Cong & Ding, 1994; Francis, Rose, & Vranesic, 1991). The noise can be calculated as,

$$\mathbf{noise(i) = \eta_i/A_i^2}$$

where noise(i) is the calibrated noise profile and η_i is noise LUT of ith pixel. Now the final noise removed radiometrically corrected value (σ^o) is calculated by subtracting the noise from the radiometrically corrected image.

$$\sigma^o = \text{value(i)} - \text{noise(i)}$$

$$\sigma^o = (DN_i^2 - \eta_i)\ /A_i^2$$

After noise removal, speckle filtering is done using Lee Adaptive filters to remove the distortion caused by the speckles that corrupts the information reflected back from the ground (Tyo, Goldstein, Chenault, & Shaw, 2006). Lastly, the terrain correction is performed to correct the geographical errors caused by the side looking aperture radar.

3.2.4. Measurement of Tree Biomass

Presently, the method involved in the measurement of tree biomass is done without cutting or harming the trees, the method included three variables namely Height of the tree (H), Diameter at Breast Height (DBH) and Wood Density (WD). DBH is measured at a breast height (h) of 1.3 meters from the ground and only the diameter is considered for the measurement if diameter >10 cm (Iizuka, Yonehara, Itoh, & Kosugi, 2018).

Figure 2. Tree Height Measurement (Chavan & Rasal, 2010)

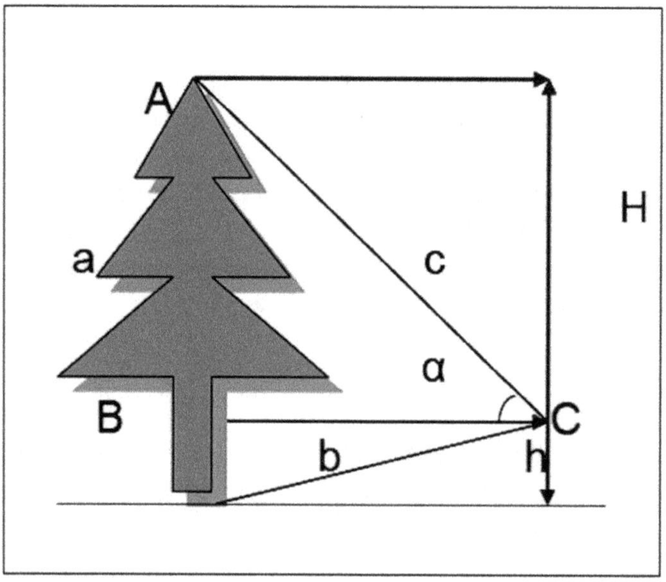

The height of the tree is calculated using the Equation 1, where h is the breast height, b is the distance between tree and observer and α is the angle between tree top and the eye view of observer at DBH. Here a is the height of the tree above h, b is given as the distance between tree and the observer and c is the slope between tree and the observers eye piece.

$$H = h + b^{*}\tan\alpha \quad (1)$$

There are many mathematical allometric equation that has been developed and used by the researchers for biomass and carbon stock estimation of trees (S. Brown, 1996; Negi, Rikhari, & Singh, 1992; Pandey et al., 2019) but these equations are species specific, particularly in the tropical regions. So, to overcome this issue a general equation is developed by modifying the species-wise allometric equations which is more general in nature (Alves et al., 1997; S. L. Brown, Schroeder, & Kern, 1999; Schroeder, Brown, Mo, Birdsey, & Cieszewski, 1997) and also relevant in field. The general allometric model developed by to estimate above ground biomass and carbon stock has been used in present study (S. Brown & Iverson, 1992; Ketterings, Coe, van Noordwijk, & Palm, 2001). After calculating the height of the tree another equation is used to determine the AGB given by which is mostly suitable for the tropics and generated for mixed forest (Warran & Patwardhan, 2008). The mathematical equation for the calculation of AGB is given in Equation 2,

$$\text{Bio-volume} = 0.4 \times (D/\pi)^{2} \times H \tag{2}$$

$$\text{AGB} = S \times \text{Bio-volume} \tag{3}$$

where Y is the AGB (t. /Pixel), D is the diameter at breast height (cm) and S is the wood density (gm/cm^2). The standard average value for the wood density of tropical forests is derived to be 0.6 gm/cm^3 (Warran & Patwardhan, 2008).

3.2.5. Tree Sampling and Data Collection

Field sampling was carried out manually in 20 different sample plots during spring season of the year 2018 for measuring different parameters namely tree height, and Diameter at Breast Height (DBH) of different trees present in the study area (Gregoire, Valentine, & Furnival, 1995; Hairiah, Sitompul, van Noordwijk, & Palm, 2001; Keller, Palace, & Hurtt, 2001). Among 20 sample plots quarter 75% plots are utilized in the generation of allometric equation whereas 25% plots are used for validation of the results. Quadrat in total 15 plots with size of size of 30m x 30 m were laid down to measure the average tree heights and average diameter at breast height of trees within the plot. The plots are further subdivided into 9 equally distributed plots of 10m X 10m to match the cell sixe of sentinel 1A data. This was majorly carried out for performing the models and generating the relations between fields based data and backscattering value and testing the accuracy of the AGB estimations (Luckman et al., 1997). The tree height is measured using Clinometer and a hand held GPS is used for recording the geological locations of the plot (Penman et al., 2003). The measured field value is used to calculate the mean Bio-volume of the plot and then to derive the Above Ground Biomass using Equation 3 for which the specific wood density is taken as 0.6 gm/cm^3.

Table 1. Field Measurements

Plot No.	Height (m)	Diameter (cm)	Bio- Volume (m^3/pixel)	Above Ground Biomass (t./pixel)
1	17.236	26	472.6981	28.36189
2	20.931	30	764.246	45.85476
3	18.257	31	711.7926	42.70756
4	10.438	25	264.6659	15.87995
5	10.981	26	301.1544	18.06927
6	12.417	31	484.1063	29.04638
7	11.005	25	279.0428	16.74257
8	19.261	24	450.0927	27.00556
9	10.867	24	253.941	15.23646
10	17.57	26	481.8581	28.91149
11	11.349	30	414.3819	24.86292
12	12.549	28	399.1406	23.94843
13	11.686	27	345.6162	20.73697
14	11.523	32	478.7031	28.72218
15	10.674	27	315.6861	18.94116

3.2.6. Estimating Above Ground Biomass and Carbon Stock

Correlation analysis is performed to assess the degree of association between two variables in the form of a linear equation (Husch, Beers, & Kershaw Jr, 2002). In present study, a correlation is established between the AGB of each plot and their respective sigma naught value extracted from the Sentinel SAR data. A linear curve fitting tool is used to generate correlation equation and the correlation coefficient (R^2) is observed to see the accuracy of the equation. (Stein, 1999). The observed regression equation is used to calculate the biomass of the overall forest within the study area. Carbon stock is calculated using estimated above ground biomass and the carbon fraction which is assumed 0.47 as per (Cruz-Barahona, 2019).

4. RESULT AND DISCUSSION

The present study illustrates the biomass and carbon estimation using the in-situ field inventory data and satellite images. Sentinel C band sigma naught value is acquired to generate a linear relation with the estimated biomass value to map the overall biomass and carbon stock of the forest within the study area. A relation is also established between the tree height, girth diameter and their product with the tree biomass to estimate the dependency of tree biomass on both the parameters. Being a mineral abundant region, Ramgarh is losing its forest area due to extensive mining activities, so present analysis is very useful for calculating the remaining carbo stock within the study area.

4.1. Land Use Land Cover Classification

Supervised land use land cover classification is done to classify the different land use classes and to the see the spatial distribution of seven different classes namely forest, cropland, waterbody, settelments, wastelands, coal mining areas and other.

Figure 3. Land Use Land Cover of the study area

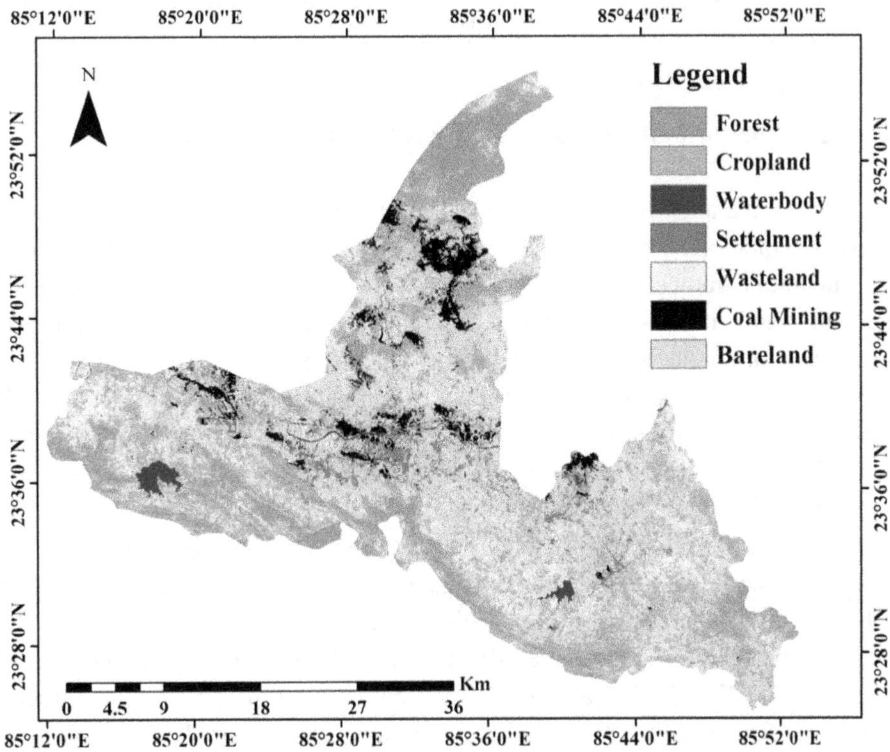

Table 2 is showing the area wise distribution of each land use feature, in which it is shown that the forest area is 167.39 Km^2 of the total 1383.68 km^2 of the whole study are. As is whole forest area is mainly the open forest, a single allometric equation is applied for calculating above ground biomass and carbon stock. The forest class in land use land cover is used to mask the overall forest for further calculations.

Table 2. Land Use Land Cover statistics

S. No.	Feature Class	Area(Km^2)
1	Forest	167.39
2	Cropland	339.39
3	Waterbody	19.82
4	Settlement	40.63
5	Wasteland	122.25
6	Coal Mining	57.96
7	Others	596.24

4.2. Assessment of Volume and Biomass Equation

Tree height and diameter at breast height (DBH) have been measured. Furthermore, allometric equations were applied to estimate tree volume and biomass, hence converted into carbon stocks. A small attempt was made to check the suitability of tree volume and biomass equation for the estimation of carbon stock. This equation instead of relating single parameter viz. tree height or DBH, uses the product of tree height and DBH. Linear empirical relationships AGB vs tree height and AGB vs DBH when established for measuring AGB showed positive relationships with R^2 0.6637 and 0.4115 respectively (Figure 4 & 5). Relationship between AGB and product of tree height and DBH when tested gave very high R^2 value (0.931) (Figure 6) demonstrating high the potential of the product in the estimation of AGB when compared to parameter when correlated solely. The linear relation has been established in all the cases, as the relation between biomass and tree height has given $Y = 2.0013X - 1.9154$, the relation between biomass and tree diameter given a relation $2.1599X - 33.657$ and the relation between biomass with the product of height and diameter of the tree has been estimated as $Y = 0.0779X - 3.9092$ with $P<0.05$ in all the cases. Hence, the tree volume and biomass equation was found suitable for estimating the carbon stock.

Figure 4. Scatter plot of the correlation between tree biomass and tree biomass

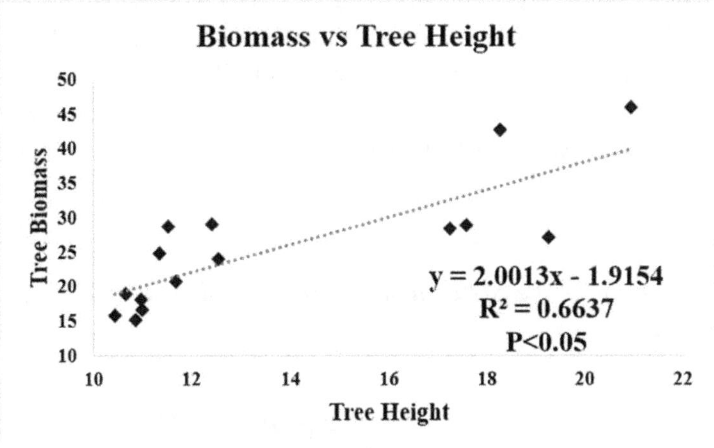

Figure 5. Scatter plot of the correlation between tree biomass and tree diameter

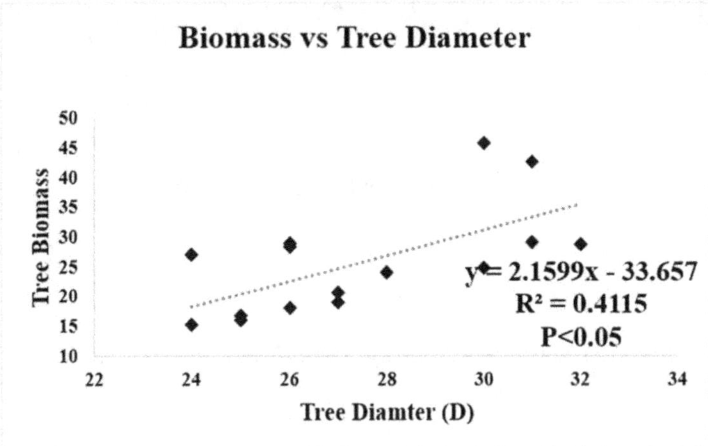

4.2.1. Correlation Between AGB and Radar Backscatter

With the suitability checking of tree height and diameter product, the above ground biomass for the forests in the 15 selected plots were calculated. The relationship between calculated AGB and backscatter coefficients of VV and VH of Sentinel data analyzed based on the correlation coefficient for the forests of Ramgarh District and it is found that VV is showing better relation as compared to VH polarization. Linear empirical model was generated for measuring tree biomass from backscatter data extracted from Sentinel imageries. The results show that there is a positive relationship between backscatter coefficient and AGB with the larger sensitivity to tree biomass found at VV when compared to VH. Thus AGB vs VV relationship was used for estimating biomass in the present study. The plot biomass and C-band VV backscatter coefficient (sigma naught) of the forest trees are closely linked, and the evaluation results were satisfactory ($R^2 = 0.7573$) with a linear equation $Y = 21.29X + 6.1638$ having $P<0.05$.

Figure 6. Scatter plot of the correlation between tree biomass and the Tree Height Diameter Product

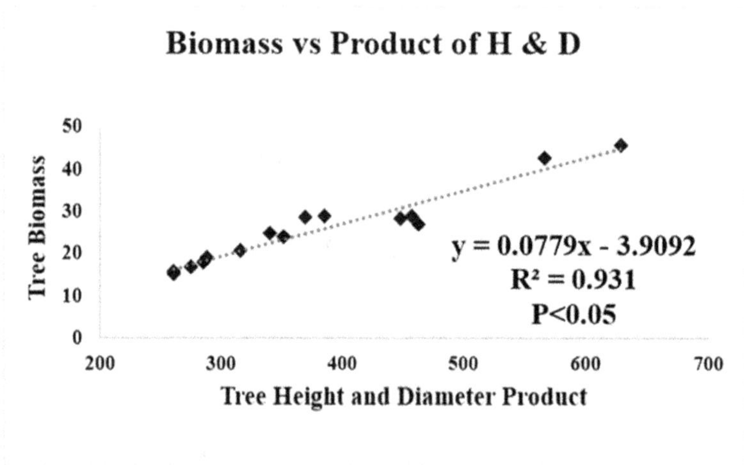

Figure 7. Scatter plot of the correlation between the VH cross-polarized backscatter and the Above Ground Biomass

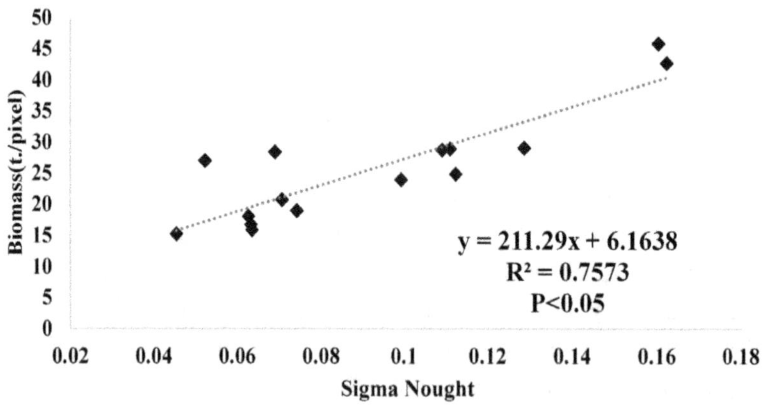

4.2.2. Biomass and Carbon Stock Estimation

The in-situ biomass values are used to generate the equation using linear regression model as illustrated in Figure 7. Sentinel C band backscattering values are used to establish the relationship between satellite imagery and ground based data. During the analysis, some of the high biomass values and highly negative sigma naught values are considered as outliers and removed from the model.

After establishing a relation between satellite imagery and ground value, an above ground biomass and carbon stock map is generated for the forest area within the region. It is calculated that the above ground biomass is varying from 6.47081 t. /pixel to 224.382 t. /pixel and the above ground carbon stock is varying from 3.04128 t. C. /pixel to 105.459 t. C. /pixel.

Figure 8. Above Ground Biomass in Ramgarh District

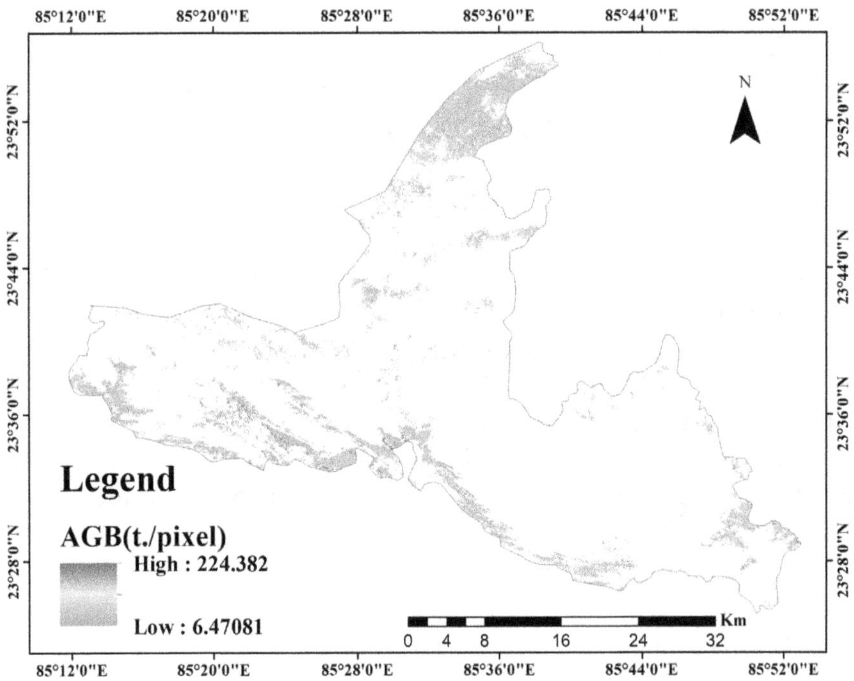

Figure 9. Estimated carbon in Ramgarh District

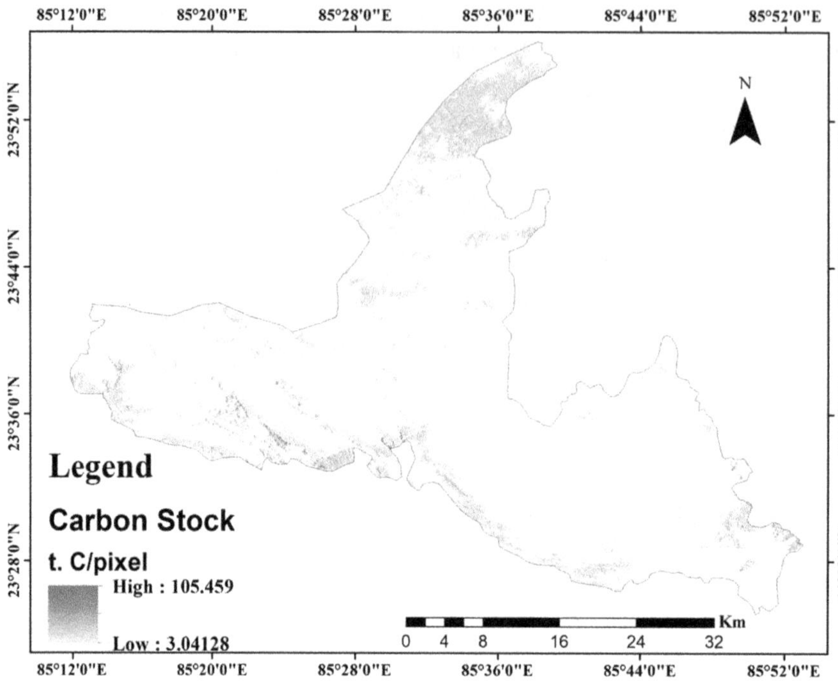

The estimated biomass and observed biomass is compared with each other and approximately 1/3rd of the ground data are used for validating the linear model. The validation of above ground biomass and carbon stock was performed in order to assess the result accuracy, the validation showed a R^2 value of 0.822 at a significance of $P<0.05$. The spatial distribution of forest has been assessed mostly at the borders of the study area as the center part is mostly the mining area and significantly increasing its diameter, it is a major source of deforestation in this region causing drastic loss in carbon stock.

5. CONCLUSION

The results demonstrates that the in-situ field inventory data in conjunction with the satellite imagery especially the C-band data is an effective way of assessing forest biomass and carbon stock at a very high spatial resolution of 10 meters. The study concluded the role of geospatial technology in assessing the relation between C-band sigma naught value with the ground based biomass and mapping the relation within the study area. Presently, an accurate estimation of forest biomass is performed at each pixel and has been validated to the measured forest biomass which shown a R^2 of 0.822. This study also established a relation of tree biomass (calculated from destructive method) with the height and diameter of the tree to show the dependency of biomass with its height and diameter.

Future work at this study area includes the identification of different species present in the mixed forest region and estimate the species wise above ground biomass and carbon stock. As the area is covered with the mining activities and continuously affecting the forest biomass, a decadal assessment is also needed to be done to see the trends of degradation and predicting the future scenarios if the mining continuous with the same pace. Moreover, it is very essential to monitor the changes taking place to the regional land cover to protect the forest area and provide restoration facilities.

ACKNOWLEDGMENT

Author would like to thank USGS Earth-Explorer for providing Landsat and Sentinel 1A data for free and Ramgarh Forest Administration for allowing us to do field sampling within the forest region.

REFERENCES

Alves, D., Soares, J. V., Amaral, S., Mello, E., Almeida, S., da Silva, O. F., & Silveira, A. (1997). Biomass of primary and secondary vegetation in Rondônia, Western Brazilian Amazon. *Global Change Biology*, *3*(5), 451–461. doi:10.1046/j.1365-2486.1997.00081.x

Anand, A., Singh, S. K., & Kanga, S. (2018). Estimating the change in Forest Cover Density and Predicting NDVI for West Singhbhum using Linear Regression. *ESSENCE Int. J. Env. Rehab. Conserv.*, *IX*(1), 193–203.

Attema, E., Davidson, M., Floury, N., Levrini, G., Rosich, B., Rommen, B., & Snoeij, P. (2008). *Sentinel-1 ESA's new European radar observatory.* Paper presented at the 7th European Conference on Synthetic Aperture Radar.

Bhanja, S. N., Mukherjee, A., Rodell, M., Wada, Y., Chattopadhyay, S., Velicogna, I., Pangaluru, K., & Famiglietti, J. S. (2017). Groundwater rejuvenation in parts of India influenced by water-policy change implementation. *Scientific Reports*, *7*(1), 7453. doi:10.103841598-017-07058-2 PMID:28785088

Bhatt, A., Bisht, A. K., Rawal, R., & Dhar, U. (2007). Assessment of status and biomass of Swertia angustifolia: a high value Himalayan medicinal plant. *Afr. J. Plant Sci, 1*(1), 1-6.

Bradford, J. B., Jensen, N. R., Domke, G. M., & D'Amato, A. W. (2013). Potential increases in natural disturbance rates could offset forest management impacts on ecosystem carbon stocks. *Forest Ecology and Management*, *308*, 178–187. doi:10.1016/j.foreco.2013.07.042

Brown, S. (1996). *Tropical forests and the global carbon cycle: estimating state and change in biomass density. In Forest ecosystems, forest management and the global carbon cycle*. Springer.

Brown, S. (1997). *Estimating biomass and biomass change of tropical forests: a primer* (Vol. 134). Food & Agriculture Org.

Brown, S. (2002). Measuring carbon in forests: Current status and future challenges. *Environmental Pollution*, *116*(3), 363–372. doi:10.1016/S0269-7491(01)00212-3 PMID:11822714

Brown, S., & Iverson, L. R. (1992). Biomass estimates for tropical forests. *World Resource Review, 4*(3), 366-384.

Brown, S. L., Schroeder, P., & Kern, J. S. (1999). Spatial distribution of biomass in forests of the eastern USA. *Forest Ecology and Management*, *123*(1), 81–90. doi:10.1016/S0378-1127(99)00017-1

Chandramouli, C. (2011). Census of India 2011: rural urban distribution of population. *Registrar General and Census Commissioner, Ministry of Home Affairs.* http:// censusindia. gov. in/2011-prov-results/paper2/data_files/india/Rural_Urban_2011.pdf

Chavan, B., & Rasal, G. (2010). Sequestered standing carbon stock in selective tree species grown in University campus at Aurangabad, Maharashtra, India. *International Journal of Engineering Science and Technology*, *2*(7), 3003–3007.

Cong, J., & Ding, Y. (1994). FlowMap: An optimal technology mapping algorithm for delay optimization in lookup-table based FPGA designs. *IEEE Transactions on Computer-Aided Design of Integrated Circuits and Systems, 13*(1), 1–12. doi:10.1109/43.273754

Cruz-Barahona, C. (2019). *Metodología para calcular costos y beneficios ambientales del Centro de Transferencia y Transformación de Materiales.*

Dadhwal, V., Sehgal, V., Singh, R., & Rajak, D. (2003). Wheat yield modelling using satellite remote sensing with weather data: Recent Indian experience. *Mausam (New Delhi), 54*(1), 253–262.

Dostálová, A., Naeimi, V., Wagner, W., Elefante, S., Cao, S., & Persson, H. (2016). *Geocoding uncertainty analysis for the automated processing of Sentinel-1 data using Sentinel-1 Toolbox software.* Paper presented at the Image and Signal Processing for Remote Sensing XXII.

Eggleston, S., Buendia, L., Miwa, K., Ngara, T., & Tanabe, K. (2006). *2006 IPCC guidelines for national greenhouse gas inventories* (Vol. 5). Institute for Global Environmental Strategies Hayama.

El Hajj, M., Baghdadi, N., Zribi, M., & Angelliaume, S. (2016). Analysis of Sentinel-1 radiometric stability and quality for land surface applications. *Remote Sensing, 8*(5), 406. doi:10.3390/rs8050406

Francis, R. J., Rose, J., & Vranesic, Z. (1991). *Technology mapping of lookup table-based FPGAs for performance.* Paper presented at the 1991 IEEE International Conference on Computer-Aided Design Digest of Technical Papers. 10.1109/ICCAD.1991.185334

Franklin, S. E. (2001). *Remote sensing for sustainable forest management.* CRC press. doi:10.1201/9781420032857

Freeman, A., & Saatchi, S. S. (2004). On the detection of Faraday rotation in linearly polarized L-band SAR backscatter signatures. *IEEE Transactions on Geoscience and Remote Sensing, 42*(8), 1607–1616. doi:10.1109/TGRS.2004.830163

Friedl, M. A., & Brodley, C. E. (1997). Decision tree classification of land cover from remotely sensed data. *Remote Sensing of Environment, 61*(3), 399–409. doi:10.1016/S0034-4257(97)00049-7

Furey, T. S., Cristianini, N., Duffy, N., Bednarski, D. W., Schummer, M., & Haussler, D. (2000). Support vector machine classification and validation of cancer tissue samples using microarray expression data. *Bioinformatics (Oxford, England)*, *16*(10), 906–914. doi:10.1093/bioinformatics/16.10.906 PMID:11120680

Gilbert, N. (2012). India's forest area in doubt. *NATNews*, *489*(7414), 14. PMID:22962698

Goodman, N. R. (1963). Statistical analysis based on a certain multivariate complex Gaussian distribution (an introduction). *Annals of Mathematical Statistics*, *34*(1), 152–177. doi:10.1214/aoms/1177704250

Gorte, R. W. (2009). *US tree planting for carbon sequestration*. Academic Press.

Gregoire, T. G., Valentine, H. T., & Furnival, G. M. (1995). Sampling methods to estimate foliage and other characteristics of individual trees. *Ecology*, *76*(4), 1181–1194. doi:10.2307/1940925

Hairiah, K., Sitompul, S., van Noordwijk, M., & Palm, C. (2001). *Methods for sampling carbon stocks above and below ground*. ICRAF Bogoi.

Hall, C. A., & Uhlig, J. (1991). Refining estimates of carbon released from tropical land-use change. *Canadian Journal of Forest Research*, *21*(1), 118–131. doi:10.1139/x91-016

Hawbaker, T. J., Keuler, N. S., Lesak, A. A., Gobakken, T., Contrucci, K., & Radeloff, V. C. (2009). Improved estimates of forest vegetation structure and biomass with a LiDAR-optimized sampling design. *Journal of Geophysical Research. Biogeosciences*, *114*(G2).

Hoekman, B., & Djankov, S. (1997). Determinants of the export structure of countries in Central and Eastern Europe. *The World Bank Economic Review*, *11*(3), 471–487. doi:10.1093/wber/11.3.471

Huang, X., Ziniti, B., Torbick, N., & Ducey, M. J. (2018). Assessment of forest above ground biomass estimation using multi-temporal C-band sentinel-1 and polarimetric L-band PALSAR-2 data. *Remote Sensing*, *10*(9), 1424. doi:10.3390/rs10091424

Husch, B., Beers, T. W., & Kershaw, J. A. Jr. (2002). *Forest mensuration*. John Wiley & Sons.

Iizuka, K., Yonehara, T., Itoh, M., & Kosugi, Y. (2018). Estimating Tree Height and Diameter at Breast Height (DBH) from Digital surface models and orthophotos obtained with an unmanned aerial system for a Japanese Cypress (Chamaecyparis obtusa) Forest. *Remote Sensing*, *10*(1), 13.

Imhoff, M. L. (1993). Radar backscatter/biomass saturation: Observations and implications for global biomass assessment. *Proceedings of IGARSS'93-IEEE International Geoscience and Remote Sensing Symposium.*

Irvin, B. J., Ventura, S. J., & Slater, B. K. (1997). Fuzzy and isodata classification of landform elements from digital terrain data in Pleasant Valley, Wisconsin. *Geoderma, 77*(2-4), 137–154. doi:10.1016/S0016-7061(97)00019-0

Ismail, R., Mutanga, O., Kumar, L., & Bob, U. (2008). Determining the optimal spatial resolution of remotely sensed data for the detection of sirex noctilio infestations in pine plantations in kwazulu-natal, south africa. *The South African Geographical Journal, 90*(1), 22–31. doi:10.1080/03736245.2008.9725308

Kamvar, K., Sepandar, S., Klein, K., Dan, D., Manning, M., & Christopher, C. (2003). *Spectral learning.* Paper presented at the International Joint Conference of Artificial Intelligence.

Keller, M., Palace, M., & Hurtt, G. (2001). Biomass estimation in the Tapajos National Forest, Brazil: Examination of sampling and allometric uncertainties. *Forest Ecology and Management, 154*(3), 371–382. doi:10.1016/S0378-1127(01)00509-6

Ketterings, Q. M., Coe, R., van Noordwijk, M., & Palm, C. A. (2001). Reducing uncertainty in the use of allometric biomass equations for predicting above-ground tree biomass in mixed secondary forests. *Forest Ecology and Management, 146*(1-3), 199–209. doi:10.1016/S0378-1127(00)00460-6

Kim, C., Jeong, J., Kim, R.-H., Son, Y.-M., Lee, K. H., Kim, J.-S., & Park, I.-H. (2011). Allometric equations and biomass expansion factors of Japanese red pine on the local level. *Landscape and Ecological Engineering, 7*(2), 283–289. doi:10.100711355-010-0131-2

Kuhlbusch, T. A., & Crutzen, P. J. (1996). Black carbon, the global carbon cycle, and atmospheric carbon dioxide. *Biomass Burning and Global Change, 1*, 160-169.

Kurvonen, L., Pulliainen, J., & Hallikainen, M. (1999). Retrieval of biomass in boreal forests from multitemporal ERS-1 and JERS-1 SAR images. *IEEE Transactions on Geoscience and Remote Sensing, 37*(1), 198–205. doi:10.1109/36.739154

Le Toan, T., Quegan, S., Davidson, M., Balzter, H., Paillou, P., Papathanassiou, K., ... Shugart, H. (2011). The BIOMASS mission: Mapping global forest biomass to better understand the terrestrial carbon cycle. *Remote Sensing of Environment, 115*(11), 2850–2860. doi:10.1016/j.rse.2011.03.020

Lillesand, T., Kiefer, R. W., & Chipman, J. (2015). *Remote sensing and image interpretation*. John Wiley & Sons.

Lillesand, T. M., Kiefer, R. W., & Chipman, J. (2000). *Remote sensing and image interpretation*. John Willey & Sons.

Lucas, R. M., Cronin, N., Lee, A., Moghaddam, M., Witte, C., & Tickle, P. (2006). Empirical relationships between AIRSAR backscatter and LiDAR-derived forest biomass, Queensland, Australia. *Remote Sensing of Environment*, *100*(3), 407–425. doi:10.1016/j.rse.2005.10.019

Lucht, W., Schaphoff, S., Erbrecht, T., Heyder, U., & Cramer, W. (2006). Terrestrial vegetation redistribution and carbon balance under climate change. *Carbon Balance and Management*, *1*(1), 6. doi:10.1186/1750-0680-1-6 PMID:16930462

Luckman, A., Baker, J., Kuplich, T. M., Yanasse, C. C. F., & Frery, A. C. (1997). A study of the relationship between radar backscatter and regenerating tropical forest biomass for spaceborne SAR instruments. *Remote Sensing of Environment*, *60*(1), 1–13. doi:10.1016/S0034-4257(96)00121-6

Lugo, A. E., & Brown, S. (1992). Tropical forests as sinks of atmospheric carbon. *Forest Ecology and Management*, *54*(1-4), 239–255. doi:10.1016/0378-1127(92)90016-3

Miehle, P., Grote, R., Battaglia, M., Feikema, P. M., & Arndt, S. K. (2010). Evaluation of a process-based ecosystem model for long-term biomass and stand development of Eucalyptus globulus plantations. *European Journal of Forest Research*, *129*(3), 377–391. doi:10.100710342-009-0343-x

Montes, N., Gauquelin, T., Badri, W., Bertaudiere, V., & Zaoui, E. H. (2000). A non-destructive method for estimating above-ground forest biomass in threatened woodlands. *Forest Ecology and Management*, *130*(1-3), 37–46. doi:10.1016/S0378-1127(99)00188-7

Navar, J. (2009). Allometric equations for tree species and carbon stocks for forests of northwestern Mexico. *Forest Ecology and Management*, *257*(2), 427–434. doi:10.1016/j.foreco.2008.09.028

Negi, G., Rikhari, H., & Singh, S. (1992). Phenological features in relation to growth forms and biomass accumulation in an alpine meadow of the Central Himalaya. *Vegetatio*, *101*(2), 161–170.

Nelson, J. C. (1999). A review of the efficacy of serotonergic and noradrenergic reuptake inhibitors for treatment of major depression. *Biological Psychiatry*, *46*(9), 1301–1308. doi:10.1016/S0006-3223(99)00173-0 PMID:10560035

Pandey, P. C., Anand, A., & Srivastava, P. K. (2019). Spatial distribution of mangrove forest species and biomass assessment using field inventory and earth observation hyperspectral data. *Biodiversity and Conservation*, 1–20.

Paola, J. D., & Schowengerdt, R. A. (1995). A detailed comparison of backpropagation neural network and maximum-likelihood classifiers for urban land use classification. *IEEE Transactions on Geoscience and Remote Sensing*, *33*(4), 981–996. doi:10.1109/36.406684

Patel, P., Srivastava, H. S., Panigrahy, S., & Parihar, J. S. (2006). Comparative evaluation of the sensitivity of multi-polarized multi-frequency SAR backscatter to plant density. *International Journal of Remote Sensing*, *27*(2), 293–305. doi:10.1080/01431160500214050

Penman, J., Gytarsky, M., Hiraishi, T., Krug, T., Kruger, D., Pipatti, R., . . . Tanabe, K. (2003). *Good practice guidance for land use, land-use change and forestry*. IPCC National Greenhouse Gas Inventories Programme and Institute for Global Environmental Strategies (IGES), Kanagawa, Japan. Intergovernmental Panel on Climate Change. www. ipcc-nggip. iges. or. j p/public/gpglulucf/gpglulucf. htm

Ranjan, A. K., Prakash, S., Anand, A., Verma, S. K., Murmu, L., & Kumar, P. B. (2017). Ground water prospect variability analysis with spatio temporal changes in Ranchi City, Jharkhand India using Geospatial Technology. *International Journal of Earth Sciences and Engineering*, *10*(3), 616–625.

Ranjan, A. K., Anand, A., Kumar, P. B. S., Verma, S. K., & Murmu, L. (2018). Prediction of Land Surface Temperature within Sun City Jodhpur (Rajasthan) in India Using Integration of Artificial Neural Network and Geoinformatics Technology. *Asian Journal of Geoinformatics*, *17*(3).

Ranjan, A. K., Anand, A., Vallisree, S., & Singh, R. K. (2016). LU/LC change detection and forest degradation analysis in Dalma Wildlife Sanctuary using 3S technology: A case study in Jamshedpur-India. *Aims Geosciences*, *2*(4), 273–285. doi:10.3934/geosci.2016.4.273

Ranson, K. J., Saatchi, S., & Sun, G. (1995). Boreal forest ecosystem characterization with SIR-C/XSAR. *IEEE Transactions on Geoscience and Remote Sensing*, *33*(4), 867–876. doi:10.1109/36.406673

Ranson, S., & Stewart, J. (1994). *Management for the public domain: enabling the learning society*. Macmillan International Higher Education. doi:10.1007/978-1-349-23787-6

Rollet, R., Benie, G., Li, W., Wang, S., & Boucher, J. (1998). Image classification algorithm based on the RBF neural network and K-means. *International Journal of Remote Sensing*, *19*(15), 3003–3009. doi:10.1080/014311698214398

Roy, P., & Ravan, S. A. (1996). Biomass estimation using satellite remote sensing data—An investigation on possible approaches for natural forest. *Journal of Biosciences*, *21*(4), 535–561. doi:10.1007/BF02703218

Saini, S., Bowser, J., Enck, R., Luciani, V., Heim, P., & Dagenais, M. (2004). *A semiconductor optical amplifier with high saturation power, low noise figure and low polarization dependent gain over the C-band.* Paper presented at the 17th Annual Meeting of the IEEE Lasers and Electro-Optics Society. 10.1109/LEOS.2004.1363131

Santos, J., Lacruz, M. P., Araujo, L., & Keil, M. (2002). Savanna and tropical rainforest biomass estimation and spatialization using JERS-1 data. *International Journal of Remote Sensing*, *23*(7), 1217–1229. doi:10.1080/01431160110092867

Schaphoff, S., Heyder, U., Ostberg, S., Gerten, D., Heinke, J., & Lucht, W. (2013). Contribution of permafrost soils to the global carbon budget. *Environmental Research Letters*, *8*(1), 014026. doi:10.1088/1748-9326/8/1/014026

Schroeder, P., Brown, S., Mo, J., Birdsey, R., & Cieszewski, C. (1997). Biomass estimation for temperate broadleaf forests of the United States using inventory data. *Forest Science*, *43*(3), 424–434.

Shrestha, P. M., & Dhillion, S. S. (2003). Medicinal plant diversity and use in the highlands of Dolakha district, Nepal. *Journal of Ethnopharmacology*, *86*(1), 81–96. doi:10.1016/S0378-8741(03)00051-5 PMID:12686446

Sinha, S., Jeganathan, C., Sharma, L., & Nathawat, M. (2015). A review of radar remote sensing for biomass estimation. *International Journal of Environmental Science and Technology*, *12*(5), 1779–1792. doi:10.100713762-015-0750-0

Stein, S. E. (1999). An integrated method for spectrum extraction and compound identification from gas chromatography/mass spectrometry data. *Journal of the American Society for Mass Spectrometry*, *10*(8), 770–781. doi:10.1016/S1044-0305(99)00047-1

Thompson, W. D., & Walter, S. D. (1988). A reappraisal of the kappa coefficient. *Journal of Clinical Epidemiology*, *41*(10), 949–958. doi:10.1016/0895-4356(88)90031-5 PMID:3057117

Tyo, J. S., Goldstein, D. L., Chenault, D. B., & Shaw, J. A. (2006). Review of passive imaging polarimetry for remote sensing applications. *Applied Optics*, *45*(22), 5453–5469. doi:10.1364/AO.45.005453 PMID:16855644

Vieilledent, G., Vaudry, R., Andriamanohisoa, S. F., Rakotonarivo, O. S., Randrianasolo, H. Z., Razafindrabe, H. N., Rakotoarivony, C. B., Ebeling, J., & Rasamoelina, M. (2012). A universal approach to estimate biomass and carbon stock in tropical forests using generic allometric models. *Ecological Applications*, *22*(2), 572–583. doi:10.1890/11-0039.1 PMID:22611855

Warran, A., & Patwardhan, A. (2008). Carbon sequestration potential of trees in and around pune city. Case study Department of Environmental Sciences, University of Pune.

Weiss, S. M., Kapouleas, I., & Shavlik, J. (1990). An empirical comparison of pattern recognition, neural nets and machine learning classification methods. *Readings in Machine Learning*, 177-183.

Winjum, J. K., Dixon, R. K., & Schroeder, P. E. (1992). Estimating the global potential of forest and agroforest management practices to sequester carbon. *Water, Air, and Soil Pollution*, *64*(1-2), 213–227. doi:10.1007/BF00477103

Xiao, C.-W., & Ceulemans, R. (2004). Allometric relationships for below-and aboveground biomass of young Scots pines. *Forest Ecology and Management*, *203*(1-3), 177–186. doi:10.1016/j.foreco.2004.07.062

Chapter 11
Status and Opportunities for Forest Resources Management Using Geospatial Technologies in Northeast India

Kasturi Chakraborty
North Eastern Space Applications Centre, India

Thota Sivasankar
https://orcid.org/0000-0003-2422-8731
NIIT University, India

Junaid Mushtaq Lone
https://orcid.org/0000-0001-6417-1392
Geoinformatics Programme, Department of Earth Sciences, University of Kashmir, India

K. K. Sarma
North Eastern Space Applications Centre, India

P. L. N. Raju
North Eastern Space Applications Centre, India

ABSTRACT

The forest resource of North East Region (NER) of India is a store house of several unique, endangered, endemic, medicinal plant, bamboo, etc. species in diverse forest type and high forest density. Several authors and organizations have contributed to the study of the richness and diversity distributed in different forest types and forest density. This chapter attempts to highlight the uniqueness of the forest of

DOI: 10.4018/978-1-7998-5027-4.ch011

NER and the role of geospatial technology and presents various interesting studies pertaining to the region as an input to forest resource assessment. Remote sensing and GIS have an important role in NER forest resource assessment, management, and conservation. Various studies carried out with the help of remote sensing and GIS technology have highlighted the ongoing forest degradation and deforestation taking place in this region due to developmental activity and economic benefits. There is continuous improvement in the forest estimates from coarse resolution satellite data to unmanned aerial vehicles (UAV) in the recent times.

INTRODUCTION

Forests provide a variety of environmental, economic, social and ecosystem services which are of immense importance for mankind. India is among the world's top ten nations in terms of forest area with nearly 712,249 km^2 that is about 21.67% of India's total geographical area (GFRA, 2015; ISFR, 2019). The Northeastern states of India cover roughly one-fourth of the forest area of the country accounting for over 17 million hectares (Dikshit and Dikshit, 2014). A range of ecological variations are witnessed due to the abrupt topographic variations in altitude ranging from 200-5000m. This consists of deep valleys and undulating hills giving rise to different climatic zones in different regions which eventually determine the vegetative pattern of this region. A varied vegetative pattern is seen here consisting of grasslands, meadows, scrub-forests, marshes, swamps, tropical forests, temperate forests and alpine vegetation. Moreover, this region is also bestowed with a number of sacred grove forests which have cultural and spiritual significance for the people in this region leading to rich biodiversity of this region. Due to its immensely rich flora and fauna, this region is also regarded as one of the world's biodiversity hotspots (Nayar, 1996; Dikshit and Dikshit, 2014).

According to the 16th forest cover assessment by Forest Survey of India (FSI), the Northeastern states of India consists total forest cover of 1,70,541 km^2. The total forest cover of this region is highly distributed in states of Arunchal Pradesh (39.10%) and Assam (16.61%) followed by Mizoram (10.56%), Meghalaya (10.04%), Manipur (9.88%), Nagaland (7.32%), Tripura (4.53%) and Sikkim (1.96%). The information of geographical area of forest cover classes, such as, very dense forest, moderately dense forest and open forest in Northeast India sates is given in Table 1. It is observed that there is a significant decrease in forest cover in this region. All the top 5 states in India showing a decrease in forest cover are a part of Northeast India namely, Mizoram (531 sq km), Nagaland (450 sq km) and Arunachal Pradesh (190 sq km) followed by Tripura (164 sq km) and Meghalaya (116 sq km) between 2015-2017 (ISFR, 2017). Thus, conservation of these forests is highly critical.

Table 1. Forest cover and scrub in Northeast India states (Source: ISFR, 2019)

State	Very Dense Forest (in sq. km)	Moderately Dense Forest (in sq. km)	Open Forest (in sq. km)	Scrub (in sq. km)
Arunachal Pradesh	21095	30557	15036	229
Assam	2795	10279	15253	173
Manipur	905	6386	9556	1181
Meghalaya	489	9267	7363	600
Mizoram	157	5801	12048	1
Nagaland	1273	4534	6679	635
Sikkim	1102	1552	688	307
Tripura	654	5236	1836	29

The most recent assessment (ISFR, 2019) shows decrease in forest cover to an extent of 765 sq km in most of the northeastern states. Approximately 30% of Northeastern forests of India face immense pressure due to rapid changes in land use land cover especially due to the practice of shifting cultivation, which is very prevalent in this region (Kumar, 2011). Extensive shifting cultivation, compounded by increase in population and demand for agricultural land has reduced the jhum cycle to less than 5 years as compared to 10 years practiced earlier (NITI Aayog, 2018, Kumar, 2011). This radical change in jhum cycle does not allow the soil to recover and forests to regenerate and is one of the primary factors for the reduction in forest cover and increase in deforestation and degradation. Instances of anthropogenically induced forest fires, mining and logging activities are also witnessed in northeast region. An instance of deforestation in upper Assam and the likely impact on the wildlife habitat and ecosystems has been pointed by Khushwaha (2018).

Geospatial technology has a promising role to play in the conservation and management of forest of this region. The ground truth collection has become easy, efficient and precise with the use of GPS. Good quality GPS devices can give location information under the forest canopy and have been used extensively for land use land cover and forest area mapping (Kushwaha, 2005). The Indian Regional Navigation Satellite System (IRNSS), with an operational name of NAVIC (NAVigation with Indian Constellation), is an independent regional navigation satellite system developed by India. It is designed to provide accurate position information service to users in India as well as the region extending up to 1500 km from its boundary, which is its primary service area (IRNSS: NAVIC). NAVIC provide standard positioning service to all users with a position accuracy of about 5 m (https://earth.esa.int/web/eoportal/satellite-missions/i/irnss) which may improve the accuracy of forest boundary mapping.

In comparison to the conventional ground based approaches, Earth observation data acquired from various sensors onboard airborne or satellite have shown immense potentials. Remote sensing together with GIS (Geographical Information System) is an important tool, for monitoring changes in land cover by identifying the potential hotspots for conservation; extent and rate of deforestation and forest cover loss (Kumar, 2011).The advanced algorithms and availability of high spatial and spectral resolution datasets are enabling for improved forest resources mapping and monitoring. Increase in free and open remotely sensed data from various space agencies have significantly increased its use for various applications including forest studies in a big way. In addition, data provided from Indian remote sensing (IRS) satellites with improved spatial extent and repetivity has further aided in forest studies of this region. Few important past, present and future IRS satellites details which focus on land applications are given in Table 2. Improved access to open data, inter-operability of platforms, and analysis-ready satellite data are becoming a standard, enabling the wider remote sensing community novel uses of these datasets.

The chapter summarizes the status of remote sensing and geospatial technologies in support of forest resource management in Northeast India and the need for geospatial technology based studies in this region.

Table 2. Few Indian remote sensing satellites focus on land applications

Satellite	Year	Sensors	Wavelength (μm)	Spatial Resolution (m)	Swath (km)
IRS-1A	1988 - 1996	LISS-I, and LISS-II A/B (3 sensors)	0.45-0.52 0.52-0.59 0.62-0.68 0.77-0.86	72.5 (LISS-I) 36 (LISS-II)z	148
IRS-1B	1991 - 2003	Satellite and instruments are identical to those of IRS-1A			
IRS-P2	1994 - 1997	LISS-II M	0.45-0.52 0.52-0.59 0.62-0.68 0.77-0.86	32 x 37	131
IRS-1C	1995 - 2007	LISS-III	0.52-0.59 0.62-0.68 0.77-0.86 1.55-1.70	23.5 23.5 23.5 70	142 142 142 148
		PAN	0.50-0.75	5.8	70
		WiFS	0.62-0.68 0.77-0.86	188	804
IRS-1P3	1996 - 2004	WiFS	0.62-0.68 0.77-0.86 1.55-1.70	188	804
IRS-1D	1997 - 2010	Satellite and instruments are identical to those of IRS-1C			

continues on following page

Table 2. Continued

Satellite	Year	Sensors	Wavelength (μm)	Spatial Resolution (m)	Swath (km)
IRS-P6 (Resourcesat-1)	2003 - 2013	LISS-IV	0.52-0.59 0.62-0.68 0.77-0.86	5.8 5.8 5.8	70
		LISS-III	0.52-0.59 0.62-0.68 0.77-0.86 1.55-1.70	23.5 23.5 23.5 23.5	140
		AWiFS	0.62-0.68 0.77-0.86 1.55-1.70	70 70 70	740
IRS-P5 (Cartosat-1)	2005 - present	PAN-F PAN-A	0.50-0.75 0.50-0.75	2.5 2.5	30 30
CartoSat-2	2007 - present	PAN camera	0.50-0.85	< 1	9.6
IRS-P6 (Resourcesat-2)	2011- present	LISS-IV	Same as Resourcesat-1		
		LISS-III	Same as Resourcesat-1		
		AWiFS	0.52-0.59 0.62-0.68 0.77-0.86 1.55-1.70	70 70 70 70	740
Resourcesat-2A	2016 - present	Satellite and instruments are identical to those of Resourcesat-2			
RISAT-1	2012 - 2017	SAR Instrument	(C-band) (5.60 cm)	< 2 - 50	100 - 600
CartoSat-3	2019 - present	PAN camera Multispectral camera	0.5 to 0.75 0.52-0.59 0.62-0.68 0.77-0.86	0.28 1.14 1.14 1.14	16 16 16 16
RISAT-1b	2021 (upcoming mission)	Satellite and instruments are identical to those of RISAT-1			
NISAR	Late 2021 (upcoming mission)	SAR Instrument	L-band (24 cm) S-band (12 cm)	3-48	>240

FOREST RESOURCES IN NORTHEAST INDIA

The northeastern region of India serves as a rich repository of flora and fauna due to its wide range of physiographic and eco-climatic conditions. The region is rich in medicinal plants and many other rare and endangered trees (see Table 3). The region has at least 7,500 flowering plants, 700 orchids, 58 bamboos, 64 citrus, 28 conifers, 500 mosses, 700 ferns and 728 lichen species. Bamboo is one of the fastest growing plants playing important role in carbon sequestration. As per ISFR

(2017) maximum pure bamboo brakes are observed in Arunachal Pradesh (137 sq km) followed by Manipur (95 (sq km) and Nagaland (57 sq km). More than 50% of the bamboo species occur in North Eastern India (Loushambam, 2017). The highest diversity in terms of floral species richness among northeastern states is of the order of; Arunachal Pradesh 5000, Sikkim ± 4500, Meghalaya ± 3500, Assam ± 3010, Manipur ± 2500, Nagaland ± 2250, Mizoram ± 2200, Tripura ± 1600 (Chatterjee *et al.,* 2006). According to Red data book published by the Botanical Survey of India, of the 1500 endangered floral species, 800 are reported from North East India. Important tree species in different forest types across Northeast India are given in Table 4.

Table 3. Forest resources of Northeast India (Source: ICAR, 2007)

Forest	17.11 (million ha)
Orchids	600 (175 rare spp.)
Medicinal and aromatic plants including	5000 flowering species
Bamboo resources	50% of the country
Indigenous crop germplasm	4500 Nos.

Table 4. Distribution of forest types with their representative tree species across Northeast India

Forest type	Tree species	Area (km^2)
Tropical Wet evergreen forest	*Dipterocarpus turbinatus, Dipterocarpus retusus, Litsea monopetala, Artocarpus chaplasha, Garcinia pedunculata*	14714
Tropical Semi evergreen forest	*Terminalia myriocarpa, Tetrameles nudiflora, Mesua ferrea, Dillenia indica, Duabanga sonneratioides*	13921
Tropical Moist deciduous forest	*Terminalia myriocarpa, Shorea robusta, Lagerstroemia parviflora, Sterculia villosa, Callicarpaarborea*	51149
Tropical Dry deciduous forest	*Anthocephalus chinensis, Gmelina arborea, Albizia chinensis, Albizia procera, Sterculia villosa*	176
Subtropical Broadleaved hill forest	*Schima wallichii, Rhus javanica, Castanopsis kurzii, Quercus griffithii, Symplocos glomerata*	4896
Subtropical Pine forest	*Pinus kesiya, Acer oblongum, Magnolia campbelli, Schima wallichii, Quercus serrata*	640
Montane Wet Temperate forest	*Michelia doltsopa, Michelia champaca, Betula alnoides, Actinodaphne macroptera, Gordonia obtusa*	1480
Himalayan Moist Temperate forest	*Lithocarpus fenestratus, Castanopsis kurzii, Myrica esculenta, Rhododendron arboreum, Engelhardtia spicata*	18

(Source: Reddy et al., 2015)

APPLICATIONS OF GEOSPATIAL TECHNOLOGY

Satellite remote sensing, in conjunction with geographic information systems (GIS), has been widely applied and been recognized as a powerful and effective tool in detecting land use and land cover change (Ehlers *et al.*, 1990;Westmoreland and Stow, 1992; Yeh and Li, 1996; Weng, 2001).The predominant causes for dwindling forest wealth in NER have been identified as over-exploitation, overgrazing, illegal encroachments, unsustainable practices, forest fires, and an indiscriminate sitting of development projects in the forest areas (GoI, 1999). The management of the forest has suffered in the recent past due to pressure on land, decreasing cycle of shifting cultivation, illegal exploitation of forest for timber and lack of scientific management strategy. Based on the 2005-2007 data published by FSI, significant scale forest degradation is occurring in the northeastern states with 599 sq km of moderately dense forests declining to lower canopy density classes. This area could be considered as potential for "reducing degradation" option under REDD (Murthy *et al.*, 2013).Increase in human andlivestock population, increased extraction of fuel wood,lack of land ownership rights, shortening of jhum cycle,conversion of natural forests into plantations for horticulturalcrops, mining, overgrazing, and forest fire are themajor causes of deforestation in North East India (Bhatt and Sachan,2004).

The categorization of forests into different forest types allows for scientific analysis in diverse applications such as forest and wildlife management, silviculture research, resource assessment, environmental impact assessment, carbon stock assessment, climate change studies etc. (ISFR 2015). Forest type classification is very important from the point of view of policy, management and conservation priority (ISFR 2015). In India, NRSA for the first time carried out a countrywide mapping of forest cover using Landsat MSS data in 1972–75 and later in 1980–82 (National Remote Sensing Agency 1983). Forest Survey of India (FSI) carried out forest type mapping in grid dimensions of 2 ½' and 2 ½' based on IRS 1D LISS III data of 2002 and forest type maps at 1:1 M scale of 1982 (ISFR, 2009).FSI started providing forest cover maps at a scale of 1:250,000 after every two years since the year 1987 and later at an improved scale of 1:50,000 (Lele and Joshi, 2009).The first digital forest cover type classification using satellite data was attempted in 1978 for Nagaland (Madhavan, 1979). First biome level vegetation map of India was prepared by Indian Institute of Remote Sensing using IRS-1C WiFS data on 1:1 M scale have identified 22 vegetation cover types (Roy *et al.,* 2006; Joshi *et al.,* 2006). A total of 100 vegetation classes have been mapped (Roy *et al.,* 2012; 2015). Considerably high-resolution data of IRS LISS III and Landsat TM/ETM+ satellites have been used to generate information on forest types in various case studies carried out in India (Reddy *et al.,* 2008; 2011;Roy *et al.,* 2012). 'Atlas of Forest Types of India' was published by FSI in the year 2011 at 1:50,000 scale that contains maps of 178 forest types out of 200 forest types described by authors in India based on Champion and Seth (1968) Classification (ISFR 2015).

This landuse information from the available maps and records does not represent the current situation and the LULC changes. Natural Resources Census programme of ISRO has scope for generation of maps of Land Use Land Cover and seasonal agricultural information during kharif, rabi and zaid seasons in near-real time at 1: 250000 scale. AWiFS sensor data onboard Resourcesat-1, 2 and 2A is extensively used for this project and annual maps are generated every year since 2004-05 crop year. Land Use Land Cover information in 1: 50,000 scale is generated in the country at 5 years interval using IRS LISS-III data of 2005-06, 2011-12 and 2016-17. Level-III classification system is used while mapping the land use / land cover classes (NRSC). Area under wasteland and its change is available under the Wasteland mapping and monitoring program. Realizing the importance of reliable database on wastelands, the Department of Land Resources in collaboration with National Remote Sensing Centre, Department of Space had published Wastelands Atlases of India in the years 2000, 2005, 2010, 2011 and 2019. According to these Wasteland Atlases of India, the extent of wastelands in the country has been estimated to be 63.85 million hectares (Mha) in Atlas - 2000, 55.64 Mha in Atlas - 2005, 47.23 Mha in Atlas – 2010 and 46.70 Mha in Atlas – 2011 and 55.76 Mha in Atlas - 2019. North Eastern Space Applications Centre (NESAC) with the help of State Remote Sensing Centers of NER is implementing actively participates in the national level programs to generate datasets pertaining to NER. In addition to Forest and LULC mapping, several studies have focused on the bamboo and bamboo flowering mapping in this region (Das, 2012;Goswami, 2010; Lallianthanga and Sailo, 2012). Although large amount of datasets are available but often there are limitations of data availability and identification of feature (Nongkynrih, 2018) which require decision with respect to cost and the information extraction.

Deforestation alone accounts for approximately 20% of anthropogenic emissions (FAO 2006; Stern, 2006).Northeast India, one of the recognized global biodiversity hotspots, approximately 40% of total forest cover is under pressure of rapid land-use changes and also found that the forest patches greater than 500 km^2 are relatively least affected by anthropogenic activities due to their physiographic location and inaccessibility to these areas(Lele and Joshi, 2009).Study using geospatial technology by Reddy *et al.*, (2013) shows the net deforestation rateis as high as ~4.82 in North Eastern Region. The major deforestation sites (>1km^2) are found to be mostly located in the north eastern region according to the study by Reddy et al., (2016). Area analysis shows that 97,875 sq. km (40%) of the forest in NE India falls in the dynamic areas where rapid forest cover changes have been taking place since 1972. The analysis reported that overall dynamic areas are highest in Tripura (56%), Meghalaya (54%), Mizoram (48%), Nagaland (52%), Manipur (38%) and

Assam (33%).A very high net deforestation rate of ~5.29 was reported in parts of North Tripura district, which is the deforestation hotspot in the whole of Barak basin (Reddy *et al.*, 2013). A time series analysis of satellite imagery of 1994 to 2001 was carried out in Sonitpur district, Assam, which highlights extensive deforestation in the moist deciduous and other forest areas (Srivastatva *et al.*, 2002).

The altered pattern of land use, particularly shifting cultivation was identified as the cause of forest cover change (Chakraborty, 2009). Kushwaha *et al.,* (2018) examined the forest cover depletion in parts of Assam and Arunachal Pradesh over an area of 42,375 km^2 in an elephant landscape falling in the Lesser Himalaya, North East India and observed the total loss of 7590 km^2 in forest cover from 1924 to 2009. They identified the increasing human population and subsequent demand on the land for cultivation as major reasons for forest cover depletion.The total area of 44,000 km^2 is assumed to be affected by shifting cultivation that has changed the natural landscapes and lead to poor species composition (Roy and Tomar, 2000; Roy and Joshi, 2002;Behera *et al.,* 2005).In the process of development of decadal (1985-1995-2005) land use and land cover database (Roy *et al.,*2015), observed a steady decrease in the forest area recorded in both central India and parts of northeast India between 1985 and 2005.Quantification and monitoring of deforestation in India over eight decades (1930-2013) showed an increasing trend in deforestation over northeast India as given in Figure 1 (Reddy *et al.,* 2016).

Joshi et al. (2006) analyzed the spatio-temporal changes in north-east India using Shannon's entropy index. The role of topography and forest change has been observed in several studies carried out in north east region. One of such study by Chakraborty et al., (2009) analyzed the land use/ land cover dynamics during 1970-2005 with respectto elevation and direction of slope in Umngot watershed in Meghalaya. It was observed that dense and open evergreen forests areas are fewer disturbances.

Figure 1. (a) Areal extent of forest cover change in Northeast region; (b) Net rate of deforestation in Northeast region
(Source: Reddy et al., 2016)

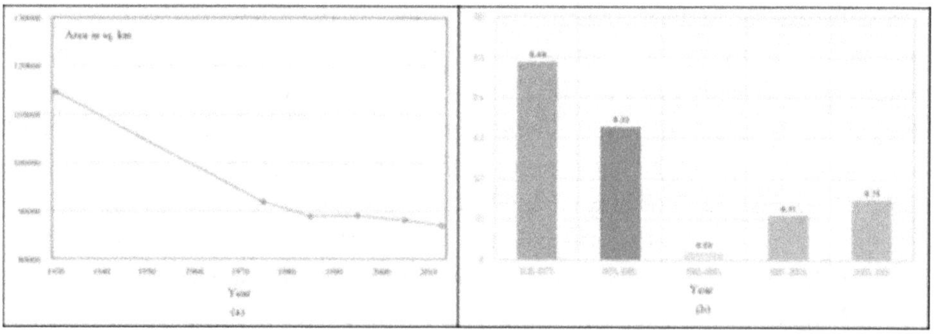

Also, Bamboo areas are identified increasing trend due to its regeneration capacity. The dynamics are largely found particularly in 0-30% slope and southern aspect of the topography.

According to the FSI State of Forest Report 2017, the states with the thickest forests have also lost the most trees due to development projects like those for irrigation, mining and infrastructure: Arunachal lost 190 km^2, Meghalaya 116 km^2 and Mizoram 531 km^2 between 2015 and 2017.Unscientific mining of minerals poses a serious threat to the environment, resulting in a reduction of forest cover and loss of biodiversity, soil erosion and pollution of air, water and land (Chabukdhara and Singh 2016). The primitive and unscientific 'rat-hole' method of mining adopted by private operators and related activities have led to large-scale environmental degradation and severe ecosystem destruction including denudation of forest cover, scarcity of water, air and water pollution, degradation of soil and agricultural lands, land subsidence, haphazard dumping of coal and overburden in Meghalaya (Swer and Singh 2003,2004). Coal mining has adversely affected the vegetation and the density of trees, shrubs and herbs in mined areas in the Nokrek Biosphere Reserve in Meghalaya, India, (Sarma and Barik 2011).Chakraborty et al., (2014a) analyzed the expansion of cement manufacturing units and mining areas in Lumshnong, Jaintia hills and its impact on surrounding forest cover. It was reported that these activities lead to significant loss of forest cover of 1265.36 ha during 2005-2011, particularly within the radius of 5 km.

Huge quantity of carbon and environmental services may get negatively affected when natural forests are substituted by plantations but technically by definition they remain forests. Thus, the loss of forest and associated biodiversity extend far beyond the simple statistics of deforestation due to inclusion of plantations and orchards (Jha *et al.,* 2000; Romijn *et al.,* 2013).According to FAO, it was observed that the plantations increased significantly in India about 15,400 km^2 per annum during 1995-2005 period (Puyravaud et al., 2010). The increasing demand for natural rubber has adverse effect on forests in northeast region of India. North Eastern Development Finance Corporation (NEDFi) reported that rubber plantation expanded positively in Northeastern India states. For instance, it is increased by 69,721 ha during 1976-2015 in Tripura, about 32,500 ha during 2006-2014 in Assam, 1,302 ha during 2000-2007 in Meghalaya, 271 ha during 2000-2007 in Manipur, 18 ha during 2005-2007 in Mizoram, 1,901 ha during 2000-2007 in Nagaland (http://databank.nedfi.com/content/ner-databank). Chakraborty et al., (2018) observed that monoculture rubber plantation (Hevea brasiliensis) expanded approximately 23.95 km^2 in area during 1997-2013 in two sub-watersheds, viz. Singla and Longai. The analyses were carried from multi-temporal IRS LISS III data of 23.5 m resolution acquired during March 1997, 2003, 2005, 2010 and 2013.

Forest fire is a global concern and is considered as a proximate cause of land use/land cover change leading to loss of forests. Larger swath MODIS and Visible Infrared Imaging Radiometer Suite (VIIRS) datasets have shown significant potentials for operational active forest fire mapping and monitoring. At regional level, North Eastern Space Applications Centre (NESAC) is providing the forest fire alerts on daily basis during fire season and also assesses burnt area. Similar information also been provided by National Remote Sensing Centre (NRSC) and Forest Survey of India (FSI) on national level whereas National Aeronautics and Space Administration (NASA) on global level. Forest fire is a predominant phenomenon in several parts of Northeast India changing the vegetation structure and composition (Chakraborty *et al.,* 2014b). Historical data analysis of fire counts in North Eastern Region (NER) of India shows approx. 10,000 fire incidents each year (Chakraborty *et al.,* 2017).The slash and burn shifting cultivation, or locally known as jhum the predominant form of agriculture in the hill tracts of North East India is found to be the major cause of forest fire in the region. The fires in the northeast India mostly pertain to slash and burn agriculture (Ramakrishnan, 1988; Majumder *et al.,* 2011). Chakraborty et al. (2014b) did the zonation of fire proneness in the Northeastern region (NER) of India using the historical data of fire incidents from January 2001 to April 2013, fuel characteristics, such as, forest type, forest density and other land use types, topography information, such as, elevation, slope and aspect as well as proximity details, such

Figure 2. Forest fire hazard map of Northeast India

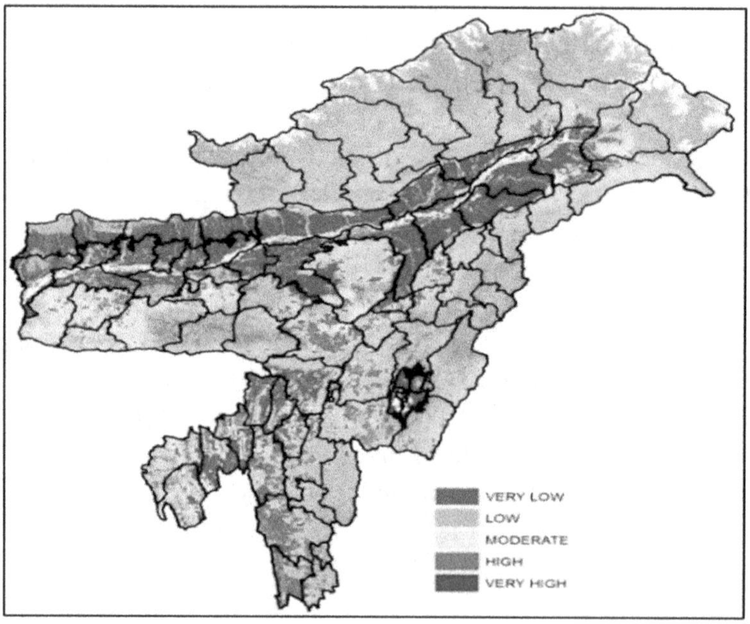

as distance to road and settlement. The generated forest fire hazard map is given in Figure 2. They found that large areas in NER are prone to forest fire. Five of the seven states have significant proportion of area under high risk zones. Tripura has approx. 52% followed by Nagaland 42%, Mizoram 36%, Manipur 35% and Meghalaya 32% area under highly prone to forest fire hazard. Whereas, Arunachal Pradesh has maximum area under moderate hazard zone (44%) and Assam has maximum low hazard zones. Mizoram has above 40% area under moderate vulnerability followed by Arunachal Pradesh, Manipur, Meghalaya and Nagaland.

The ongoing change in LULC and the forest loss has several implications. It is observed that largest portions of elephant habitat persist in northeast region, an estimate shows a depletion 10,188 sq km elephant habitat in this region in past eight decades (Padalia, 2019). LULC and forest dynamics have detrimental effect on prevailing climate of this region. Deforestation and land use change is estimated to contribute to 17% of the global CO2 emissions. Ravindranath *et al.,* 2011, in their study on the climate change vulnerability profiles for North East India has opined that vulnerabilities related to human activities will not change much in future, but climate change in NER will contribute an additional vulnerability. Several villages in this region derive the water supply from the spring water since the forested watershed of northeast region are the source of clean portable water and the ongoing forest change has led to the launch of several spring rejuvenation programs.

The rainfall data analysis for 110 years (1901-2010) by Deka *et al.,* (2013), analysed the monthly, seasonal and annual trends in the Brahmaputra and Barak basins of Assam. The analysis revealed significant decreasing trend of rainfall during monsoon and post-monsoon seasons in the Barak basin during 1901-2010. In another study carried out by Saha *et al.,*(2015) on the spatial variability in temporal variability trends of precipitation and its impact on agricultural scenario of Mizoram, observed the long term monthly rainfall observations (1986-2014) for 12 rain gauge stations. According to their findings, there is significant reduction in post monsoon and winter rainfall recorded for most of the sites. Increasing dryness during the winter months will intensify in Mizoram leading to acute water shortage and suggested for adopting suitable water management strategies.

To understand the response to climate change and to estimate carbon pools, the accurate estimation of forest biomass is imperative. Remotely sensed data acquired from optical, radar and LiDAR sensors has become an important tool for estimating and monitoring forest biomass over large areas. Particularly, spaceborne Synthetic Aperture Radar (SAR) data has been widely exploited for forest above-ground biomass (AGB) estimation due to its higher saturation limits and unique sensitivity towards various components of vegetation. Kumar et al., (2019) and Sivasankar et al. (2018) demonstrated the potentials of ALOS PALSAR and Sentinel-1 SAR datasets for forest AGB estimation in parts of northeast India and reported that the highest coefficient of determination (R^2) can be obtained from L-band HV backscatter as compared to C-band backscatter. However, it is well known that SAR backscatter is a composite effect of forest AGB as well as topographic characteristics. Sivasankar et al. 2019 have given furthermore details on the influence of aspect and degree of slope on SAR backscatter for AGB estimation. It is observed that the regression approach without consideration of the topographic characteristics for forest AGB may result in over or underestimation, depending on the topography aspect with respect to the SAR sensor look direction (Lone et al. 2017). Moreover, it is observed that the value of backscatter increases with increase in the degree of slope over aspects facing the SAR sensor due to a decrease in local incidence angle. Similarly, a decrease in the value of backscatter is observed over back-slope regions due to an increase in local incidence angle. Thus, Sivasankar et al. (2019) proposed an approach by considering the topographic characteristics along with SAR backscatter as inputs to support vector machine (SVM) for forest AGB estimation. The study reported significant improvement in the accuracy with higher R^2 of 0.91 and lower root mean square error (RMSE) of 18.21 tons/ha with L-band dual-polarization (HH and HV) backscatter in Nongkhyllem wildlife sanctuary and reserve forest, Meghalaya. In addition, advanced remote sensing inputs using high-resolution data acquired from Unmanned Aerial Vehicle (UAV) have also been analyzed for forest biometric parameter extraction in northeast India (Chakraborty et al. 2019).

CONCLUSION

Forests in Northeast India is known for its diverse and unique species composition. Several studies referred brings out the loss of forests, reasons of deforestation, possible impacts etc. on the valuable ecosystem, habitat destruction and its impact on climate of this region. Geospatial technology has shown immense potential in forest resources assessment and monitoring. Significant research has already been done in north east India with respect to forest cover mapping and monitoring, fire alerts, burnt area assessment and forest biomass estimation etc. Most of the previous studies have used satellite based optical images for the above analysis. However, recent use of high resolution UAV data, Synthetic Aperture Radar (SAR) data have opened up new research arena and contributed significantly in forest resources monitoring and management. Last few years have witnessed extensive use of Lidar and hyperspectral datasets in forestry. It is observed from the various studies cited here that the region lacks and definitely requires more number of works to be carried out in the advanced field of remote sensing technique which can bring out pertinent results in critical areas. Since each sensor data has its own pros and cons for a specific application, multi-sensor data may be a better alternative. Thus, multi-sensor data based studies can be initiated for forestry applications. The penetration of this technology in the various research works carried out in colleges and universities can further aid in better understanding of the issues and forest management strategies. More important is to have a synchronization of this technology with the local ground based knowledge for a holistic forest management design.

REFERENCES

NITI Aayog. (2018). *Report of Working Group III, Shifting Cultivation: Towards a Transformational Approach*. Author.

Behera, M. D., Kushwaha, S. P. S., & Roy, P. S. (2005). Rapid assessment of biological richness in a part of Eastern Himalaya: An integrated three-tier approach. *Forest Ecology and Management*, *207*(3), 363–384. doi:10.1016/j.foreco.2004.10.070

Chabukdhara, M., & Singh, O. P. (2016). Coal mining in northeast India: An overview of environmental issues and treatment approaches. *International Journal of Coal Science & Technology*, *3*(2), 87–96. doi:10.100740789-016-0126-1

Chakraborty, K., Chabukdhara, M., Mondal, P. P., Borah, S. B., Sarma, K. K., Sudhakar, S., & Raju, P. L. N. (2017). Status of forest fire – A case study from northeast India. *Asian Conference on Remote Sensing*.

Chakraborty, K., Joshi, P. K., & Sarma, K. K. (2009). Land use/land cover dynamics in Umngot watershed of Meghalaya using geospatial tools. *Journal of the Indian Society of Remote Sensing*, *37*(1), 99–106. doi:10.100712524-009-0015-9

Chakraborty, K., Mondal, P. P., Chabukdhara, M., & Sudhakar, S. (2014b). Forest fire scenario and challenges of mitigation during fire season in North East India. *The International Archives of the Photogrammetry, Remote Sensing and Spatial Information Sciences*, *40*(8), 27–33. doi:10.5194/isprsarchives-XL-8-27-2014

Chakraborty, K., Saikom, V., Borah, S. B., Kalita, M., Gupta, C., Meitei, L. R., ... Raju, P. L. N. (2019). Forest biometric parameter extraction using unmanned aerial vehicle to aid in forest inventory data collection. *Current Science, 117*(7).

Chakraborty, K., & Sudhakar, S. (2014a). The expansion and impact of cement manufacturing units and mining areas in Lumshnong, Jaintia hills, Meghalaya. *Current Science*, 997–1000.

Chakraborty, K., Sudhakar, S., Sarma, K. K., Raju, P. L. N., & Das, A. K. (2018). Recognizing the rapid expansion of rubber plantation–a threat to native forest in parts of northeast India. *Current Science*, *114*(1), 207. doi:10.18520/cs/v114/i01/207-213

Champion, S. H., & Seth, S. K. (1968). *A revised survey of the forest types of India.* A revised survey of the forest types of India.

Chatterjee, S., Saikia, A., Dutta, P., Ghosh, D., Pangging, G., & Goswami, A. K. (2006). *Background paper on biodiversity significance of North East India, for the study on natural resources, water and environment nexus for development and growth in North Eastern India*. Forest Conservation Programme.

Das, D. J. (2012). *A study on impact of shifting cultivation on Landuse/landcoverin Mizoram, using Geospatial tools* (Ph.D. thesis). FRI (deemed) University.

Deka, R. L., Mahanta, C., Pathak, H., Nath, K. K., & Das, S. (2013). Trends and fluctuations of rainfall regime in the Brahmaputra and Barak basins of Assam, India. *Theoretical and Applied Climatology*, *114*(1-2), 61–71. doi:10.100700704-012-0820-x

Dikshit, K. R., & Dikshit, J. K. (2014). Natural Vegetation: Forests and Grasslands of North-East India. In *North-East India: Land, People and Economy* (pp. 213–255). Springer. doi:10.1007/978-94-007-7055-3_9

GlobalFRA. (GFRA). (2015). How are the world's forests changing? (2nd ed.). Food and Agriculture Organization of the United Nations.

GoI. (1999). *Economic survey*. Economic Division, Ministry of Finance, Government of India.

Goswami, J., Tajo, L., & Sarma, K. K. (2010). Bamboo resources mapping using satellite technology. *Current Science*, 650–653.

ICAR. Meghalaya. (2007). *Vision 2025*. Retrieved from: http://www.kiran.nic.in/pdf/reports/Vision_2025_ICARNEH.pdf

Indian State Forest Report (ISFR). (2009). *Forest Survey of India*. Retrieved from http://fsi.nic.in/forest-report-2009

Indian State Forest Report (ISFR). (2015). *Forest Survey of India*. Retrieved from http://fsi.nic.in/forest-report-2015

Indian State Forest Report (ISFR). (2017). *Forest Survey of India*. Retrieved from http://fsi.nic.in/forest-report-2017

Indian State Forest Report (ISFR). (2019). *Forest Survey of India*. Retrieved from http://fsi.nic.in/forest-report-2019

Jadkowski, M. A., Howard, R. R., & Brostuen, D. E. (1990). Application of SPOT data for regional growth analysis and local planning. *Photogrammetric Engineering and Remote Sensing*, *56*(2), 175–180.

Jha, C. S., Dutt, C. B. S., & Bawa, K. S. (2000). Deforestation and land use changes in Western Ghats, India. *Current Science*, 231–238.

Joshi, P. K., Lele, N., & Agarwal, S. P. (2006). Entropy as an indicator of fragmented landscape. *Current Science*, *91*(3), 276–278.

Joshi, P. K., Roy, P. S., Singh, S., Agrawal, S., & Yadav, D. (2006). Vegetation cover mapping in India using multi-temporal IRS Wide Field Sensor (WiFS) data. *Remote Sensing of Environment*, *103*(2), 190–202. doi:10.1016/j.rse.2006.04.010

Kumar, A., Kishore, B. S. P. C., Saikia, P., Deka, J., Bharali, S., Singha, L. B., Tripathi, O. P., & Khan, M. L. (2019). Tree diversity assessment and above ground forests biomass estimation using SAR remote sensing: A case study of higher altitude vegetation of North-East Himalayas, India. *Physics and Chemistry of the Earth Parts A/B/C*, *111*, 53–64. doi:10.1016/j.pce.2019.03.007

Kumar, D. (2011). Monitoring forest cover changes using remote sensing and GIS: A global prospective. *Research Journal of Environmental Sciences*, *5*(2), 105–123. doi:10.3923/rjes.2011.105.123

Kushwaha, S. P. S. (2005). Geospatial Technology for Forest Resources Management. *International Journal of Ecology and Environmental Sciences*, *31*(1), 9–20.

Kushwaha, S. P. S., Nandy, S., Shah, M. A., Agarwal, R., & Mukhopadhyay, S. (2018). Forest cover monitoring and prediction in a Lesser Himalayan elephant landscape. *Current Science, 115*(3).

Lallianthanga, R. K., & Sailo, R. L. (2012). Monitoring of bamboo flowering using satellite remote sensing and GIS techniques in Mizoram, India. *Science Vision*, *12*(4), 147.

Lele, N., & Joshi, P. K. (2009). Analyzing deforestation rates, spatial forest cover changes and identifying critical areas of forest cover changes in North-East India during 1972–1999. *Environmental Monitoring and Assessment*, *156*(1-4), 159–170. doi:10.100710661-008-0472-6 PMID:18677545

Lone, J. M., Sivasankar, T., Sarma, K. K., Qadir, A., & Raju, P. L. N. (2017). Influence of slope aspect on above ground biomass estimation using ALOS-2 data. *International Journal of Scientific Research*, *6*(6), 1422–1428.

Loushambam, R. S., Singh, N. R., Taloh, A., & Mayanglambam, S. (2017). Bamboo in North East India. *Indian Journal of Hill Farming*, *30*(2), 181–185.

Madhavan Unni, N. V. (1979). *Satellite remote sensing survey of natural resources of Nagaland. Report*. National Remote Sensing Agency.

Majumder, M., Shukla, A. K., & Arunachalam, A. (2010). Agricultural practices in Northeast India and options for sustainable management. In *Biodiversity, Biofuels, Agroforestry and Conservation Agriculture* (pp. 287–315). Springer. doi:10.1007/978-90-481-9513-8_10

Murthy, I. K., Sharma, N., & Ravindranath, N. H. (2013). Harnessing REDD+ opportunities for forest conservation and carbon stock enhancement in the Northeastern States of India. *Nature and Science*, *5*(3), 349–358.

Nayar, M. P. (2002). Hot spots of endemic plants of India, Nepal and Bhutan. Tropical Botanical Garden Research Institute.

Nongkynrih, J. M., Pohshna, C., & Sarma, K. K. (2018). Dynamics of shifting cultivation in relation to slope and elevation in parts of Nagaland, India. *Current Science*, *114*(5), 1094. doi:10.18520/cs/v114/i05/1094-1099

Padalia, H., Ghosh, S., Reddy, C. S., Nandy, S., Singh, S., & Kumar, S. A. (2019). Assessment of historical forest cover loss and fragmentation in Asian elephant ranges in India. *Environmental Monitoring and Assessment*, *191*(3), 802. doi:10.100710661-019-7696-5 PMID:31989279

Prasad, P. R. C., Reddy, C. S., Rajasekhar, G., & Dutt, C. B. S. (2008). Mapping and analyzing vegetation types of North Andaman Islands, India. *GIS Development*.

Puyravaud, J. P., Davidar, P., & Laurance, W. F. (2010). Cryptic destruction of India's native forests. *Conservation Letters*, *3*(6), 390–394. doi:10.1111/j.1755-263X.2010.00141.x

Ramakrishnan, P. S. (1998). Sustainable development, climate change and tropical rain forest landscape. *Climatic Change*, *39*(2-3), 583–600. doi:10.1023/A:1005398606125

Ravindranath, N. H., Rao, S., Sharma, N., Nair, M., Gopalakrishnan, R., Rao, A., ... Krishna, N. (2011). Climate change vulnerability profiles for North East India. *Current Science*, *101*(3).

Reddy, C. S., Dutta, K., & Jha, C. S. (2013). Analysing the gross and net deforestation rates in India. *Current Science*, *105*(11), 1492–1500.

Reddy, C. S., Hari Krishna, P., & Ravikiran, A. (2011). Mapping the vegetation types of Rajasthan, India using remote sensing data. *Journal of Environmental Research and Management*, *2*(1), 1–9.

Reddy, C. S., Jha, C. S., Dadhwal, V. K., Krishna, P. H., Pasha, S. V., Satish, K. V., ... Diwakar, P. G. (2016). Quantification and monitoring of deforestation in India over eight decades (1930–2013). *Biodiversity and Conservation*, *25*(1), 93–116. doi:10.100710531-015-1033-2

Reddy, C. S., Jha, C. S., Diwakar, P. G., & Dadhwal, V. K. (2015). Nationwide classification of forest types of India using remote sensing and GIS. *Environmental Monitoring and Assessment*, *187*(12), 777. doi:10.100710661-015-4990-8 PMID:26615560

Romijn, E., Ainembabazi, J. H., Wijaya, A., Herold, M., Angelsen, A., Verchot, L., & Murdiyarso, D. (2013). Exploring different forest definitions and their impact on developing REDD+ reference emission levels: A case study for Indonesia. *Environmental Science & Policy*, *33*, 246–259. doi:10.1016/j.envsci.2013.06.002

Roy, P., Roy, A., Joshi, P., Kale, M., Srivastava, V., Srivastava, S., ... Sharma, Y. (2015). Development of decadal (1985–1995–2005) land use and land cover database for India. *Remote Sensing*, *7*(3), 2401–2430. doi:10.3390/rs70302401

Roy, P. S., Behera, M. D., Murthy, M. S. R., Roy, A., Singh, S., Kushwaha, S. P. S., ... Gupta, S. (2015). New vegetation type map of India prepared using satellite remote sensing: Comparison with global vegetation maps and utilities. *International Journal of Applied Earth Observation and Geoinformation*, *39*, 142–159. doi:10.1016/j.jag.2015.03.003

Roy, P. S., & Joshi, P. K. (2002). Landscape fragmentation and biodiversity conservation. *Map India*, 281-285.

Roy, P. S., Joshi, P. K., Singh, S., Agarwal, S., Yadav, D., & Jegannathan, C. (2006). Biome mapping in India using vegetation type map derived using temporal satellite data and environmental parameters. *Ecological Modelling*, *197*(1-2), 148–158. doi:10.1016/j.ecolmodel.2006.02.045

Roy, P. S., Kushwaha, S. P. S., Murthy, M. S. R., Roy, A., Kushwaha, D., Reddy, C. S., . . . Singh, S. (2012). Biodiversity characterisation at landscape level: national assessment. Indian institute of remote sensing, Dehradun, India, 140.

Roy, P. S., & Tomar, S. (2000). Biodiversity characterization at landscape level using geospatial modelling technique. *Biological Conservation*, *95*(1), 95–109. doi:10.1016/S0006-3207(99)00151-2

Saha, S., Chakraborty, D., Choudhury, B. U., Singh, S. B., Chinza, N., Lalzarliana, C., ... Singh, A. R. (2015). Spatial variability in temporal trends of precipitation and its impact on the agricultural scenario of Mizoram. *Current Science*, *109*(12), 2278–2282. doi:10.18520/cs/v109/i12/2278-2282

Sarma, K., & Barik, S. K. (2011). Coal mining impact on vegetation of the Nokrek Biosphere Reserve, Meghalaya, India. *Biodiversity (Nepean)*, *12*(3), 154–164. doi:10.1080/14888386.2011.629779

Sivasankar, T., Lone, J. M., Sarma, K. K., Qadir, A., & Raju, P. L. N. (2018). The potential of multi-frequency multi-polarized ALOS-2/PALSAR-2 and Sentinel-1 SAR data for aboveground forest biomass estimation. *International Journal of Engineering and Technology*, *10*(3), 797–802. doi:10.21817/ijet/2018/v10i3/181003095

Sivasankar, T., Lone, J. M., Sarma, K. K., Qadir, A., & Raju, P. L. N. (2019). Estimation of Above Ground Biomass Using Support Vector Machines and ALOS/PALSAR data. *Vietnam Journal of Earth Sciences*, *41*(2), 95–104. doi:10.15625/0866-7187/41/2/13690

Srivastava, S., Singh, T. P., Singh, H., Kushwaha, S. P. S., & Roy, P. S. (2002). Assessment of large-scale deforestation in Sonitpur district of Assam. *Current Science*, 1479–1484.

Stern, N. H. (2006). *The Economics of Climate Change: The Stern Review*. Cambridge University Press.

Swer, S., & Singh, O. P. (2003). Coal mining impacting water quality and aquatic biodiversity in Jaintia Hills district of Meghalaya. *Himalayan Ecology*, *11*(2), 29.

Swer, S., & Singh, O. P. (2004). Water quality, availability and aquatic life affected by coal mining in ecologically sensitive areas of Meghalaya. In *Proceedings of the 3rd National Seminar on Inland Water Resources and Environment*. University of Kerala.

Weng, Q. (2001). A remote sensing? GIS evaluation of urban expansion and its impact on surface temperature in the Zhujiang Delta, China. *International Journal of Remote Sensing*, *22*(10), 1999–2014.

Westmoreland, S., & Stow, D. A. (1992). Category identification of changed land-use polygons in an integrated image processing/geographic information system. *Photogrammetric Engineering and Remote Sensing*, *58*, 1593–1599.

Yeh, A. G. O., & Li, X. (1996). Urban growth management in the Pearl River delta – an integrated remote sensing and GIS approach. *ITC Journal*, *1*, 77–85.

Chapter 12
Statistical Analysis of Major Flood Events During 1980-2015 in Middle Ganga Plain, Ganga River Basin, India

Aman Arora
https://orcid.org/0000-0001-9396-8720
Department of Geography, Faculty of Natural Sciences, Jamia Millia Islamia, New Delhi, India

Masood Ahsan Siddiqui
Department of Geography, Faculty of Natural Sciences, Jamia Millia Islamia, New Delhi, India

Manish Pandey
https://orcid.org/0000-0001-8291-2043
University Center for Research and Development (UCRD), Chandigarh University, Mohali, India

ABSTRACT

To understand the vicious nature of extreme flood events for the most flood prone region of Ganga River Basin, this study uses 36 years (1980-2015) of flood records from Dartmouth Flood Observatory (DFO) and the Centre for Research on the Epidemiology of Disasters (CRED) Emergency Events Database (EM-DAT). Further, the Water Level (WL) data collected from Central Water Commission (CWC) for same period are utilized to compare with the data of DFO and EM-DAT to identify the major flood events recorded in the Middle Ganga Plain (MGP). The final dataset comprises of 15 attributes (parameters) and is prepared of identified 99 flood

DOI: 10.4018/978-1-7998-5027-4.ch012

instances for statistical analysis. The descriptive statistical analysis is performed for the following parameters: severity class, flood duration in days, affected flood area, flood magnitude, total number of deaths, and total count of displaced people. The graphical representation of all selected parameters provides an insight of common flood events, which lie between ±95% confidence level and exclude the major events as outliers.

1. INTRODUCTION

There is a long list of disasters, particularly natural disasters which impact society and economy significantly. The disasters are on rise. The Special Reports on Extreme Events and Disasters (SREX) of Intergovernmental Panel on Climate Change (IPCC) predicted further increase of disasters i.e. floods, cyclones, droughts, heatwaves, etc. The report further quoted that the magnitude and frequency of extreme temperature are liable to increase in the 21st century while the cold days will decrease significantly. With the extreme temperature, the heavy precipitation is also expected to increase in the coming years (IPCC 2012). Floods being among the most pervasive phenomena, they are being further monitored and assessed by international institutions like United Nations, World Bank, etc. (UNDP Emergency Analyst 2008). They (floods) alone have caused property damage at the scale of > $50 million in 2018 only (EMDAT 2018, 2019; World Bank – World Development Indicators 2019). As reported by EM-DAT, 2019 (Centre for Research on the Epidemiology of Disasters (CRED) EM-DAT Team n.d.), there have been 4960 reported flood events worldwide. There has been reported loss of 23037130 people due to all types of natural disasters; out of which 30% (6960299 persons) have lost their lives due to floods only (Arora et al. 2019). Since 1953 to 2016, there has been estimated loss of life and property (including damage to the structure, agricultural loss, and cattle) has been reported to be huge (₹ 3475812.01 million) (CWC 2018). The year 2015 has brought the most severe loss to the country with a total loss estimated to be ₹ 572910.98 million. Flood has a special significance as major part of India is affected frequently. Impacts of floods include direct mortality and indisposition and indirect displacement and widespread damage of crops, infrastructure and property. The analysis of precipitation data during 1950-2000 reveals that the rainfall intensity will increase in coming decades, so the flood events (Vittal et al. 2016). As per report of National Disaster Management Authority (NDMA) about 15% area of India is prone to floods. Majority of such areas occur in a linear form along the major rivers of the rivers like Ghaghara, Ganga, and Gandak in Uttar Pradesh and almost entire Bihar.

The current study aims to identify the most severe and devastating flood events for the study area during the study period by accomplishing the statistical analysis of major flood events. To do so, the identification of major flood events by systematic records historical flood events being recorded by EM-DAT and DFO have been done and then it is verified by the water level and discharge data for the duration of 1980-2015 provided by CWC.

2. STUDY AREA

Although not a definite physical unit, the Middle Ganga Plain lies between at latitudinal extend of 24°30'N-27°50'N and on a margin of longitudinal spread of 81°47'E- 87°50'E. The MGP is a huge physical area of 1,44,409 km^2 and with the immense human, cultural and economic significance that makes it the heart region of India. It covers the Eastern Uttar Pradesh, Bihar Plains and some parts of Jharkhand state (Figure 1). The complete plain is lying within the Himalayan region in the north and the peninsular embankments in the south, respectively. The Ganga, having major tributaries- Ghaghara, Gandak, Son etc., is the main river which divides this plain from the middle and gives a clear cut North and South Middle Ganga Plain respectively. The western and eastern flanks of the MGP are wide open. The central part of the east-west continuum of the enormous isotropic Plain-there is no physical periphery as the plain gradually opens up in the west at the margin of the Upper Ganga Plain and so intangibility becomes invisible in the east bordering the Lower Ganga Plain. In conclusion, it is a par excellence transitional region, an essential part of the Ganga Valley, which has having all kinds of human, physical, cultural and economic properties, which makes it most significant expanse of the country (Singh 1971).

3. HISTORICAL RECORDS OF FLOODS IN MGP

The historical record analysis of floods is largely based on government records, damage reports, newspapers and unpublished records, etc. The other sources of information are verbal communication from flood victims (Benito et al. 2004; Salas, Wohl, and Jarrett 1994). Though there is no clear-cut boundary of MGP to explain the historical flood records yet as per available studies nearly 70% area of MGP lies in Bihar while remaining 30% comprises of Uttar Pradesh. Hence, the explanation of large and devastating flood events are narrowed down while studying the government reports such as flood reports issued by Flood Management Information System (FMIS) of Bihar (http://fmis.bih.nic.in/) and Uttar Pradesh (http://fmiscup.in/), annual report published by CWC, reports issued by Bihar State Disaster Management Authority (http://bsdma.org) and Disaster Management Department, Govt. of Uttar Pradesh (https://rahat.up.nic.in/). Apart from reports, DFO and CRED-EM-DAT for the duration of 1985-2015 and 1980-2015 respectively for India have been consulted. DFO is an online system for global river discharge and runoff measurement. The agency provides data and flood maps of different flood events occurred worldwide with brief report (Brakenridge 2010). Whereas, the Centre for CRED EM-DAT, launched in 1988, is an Emergency Events Database. It was established with the help of the World Health Organization (WHO) and the Government of Belgium (CRED 2011). EM-DAT provides essential core data of disasters, worldwide from 1900 till recent. It has a rich database of more than 22000 events (Becerra, Cavallo, and Noy 2014; Brakenridge et al. 2003; Fallis 2013; Hoyois and Guha-sapir 2003; Kundzewicz, Pińskwar, and Brakenridge 2013; Magilligan and Goldstein 2001; Schmidt-Thomé et al. 2006; Sodhi and Tang 2014).

The flooded regions are commonly called by locals as *'Khadir Land'* in the Western Districts and *'Diara Land'* in the eastern parts of the State. *'Diara'* areas in Uttar Pradesh and Bihar, is spreading over about 15.41 lakh hectares of along the rivers like Ganga, Ghaghra, Sarda, and Gomti and the maximum stretch of it found in the Ballia district. The river bed *'Diara'* is fine sand and deposits alongside rivers and used for cultivation during the months of January to June, until monsoon hits in late June. The *'Diara Land'* of eastern UP and Bihar are most affected parts by floods.

Figure 1. Location map of Study Area and selected Hydrological Stations

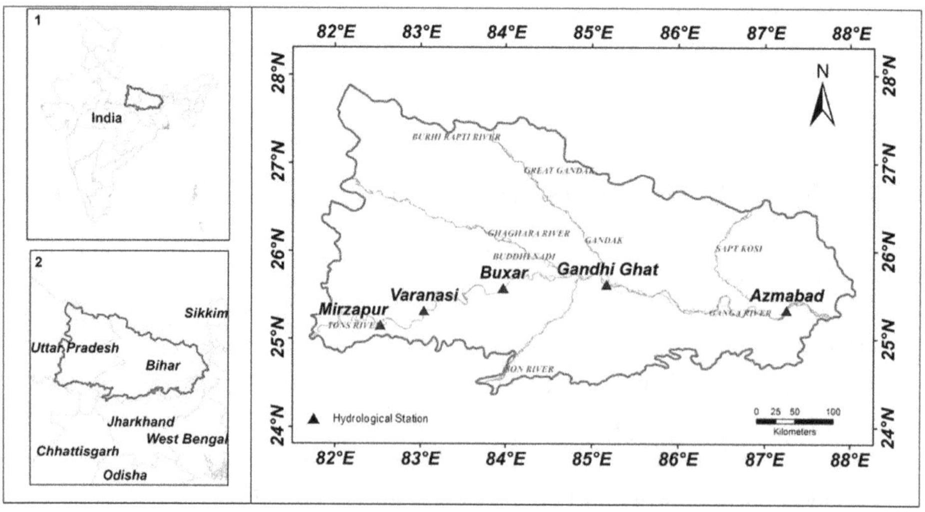

The analysis of historical records of major floods reveals that most of the flood occurs during the months of August and September and the northern region of Bihar and Eastern Uttar Pradesh and partially south Bihar are adversely affected. The reasons for floods besides rainfall may be summarized as: (i) Recurrent spillages in the snow-fed rivers of North Bihar coming from the mountain regions; and (ii) releasing of heavy amount of water from the dams on Indo-Nepal border brings havoc situation, event there is no sign of heavy rainfall in Bihar or Eastern Uttar Pradesh (iii) Substantial silting of some of these rivers caused frequent avulsions and breaches in embankments. There are also blockages in the north and south bank tributaries of Ganga in high floods situation and drainage congestion in low lying flood areas happen frequently.

4. METHODOLOGY FOR IDENTIFICATION AND VERIFICATION OF MAJOR EVENTS

The following steps have been followed to accomplish the statistical analysis of flood events:

4.1. Data Collection and Selection

For the statistical analysis, the event lists obtained from reports and both databases of EM-DAT and DFO have been combined into a single list after removing redundancy. First the combined list was filtered out by states- Uttar Pradesh and Bihar, as the major portion of Eastern Uttar Pradesh and Bihar cover the study area, i.e. MGP. The list of major events for the selected region was further refined by using the coordinates of the events. Finally, total Ninety-Nine (99) major events were found for the study area from 1980 to 2015 period.

It may be noted that the flood occurs at the same time at different places or districts or state (not recognize any political boundary) and the DFO & EM-DAT databases recorded the data of occurred flood event for any year on the basis of its expansion or affected areas, either it was within the boundary of UP and Bihar or added up some information from nearby political division, i.e. West Bengal, Jharkhand, etc. It has not been collected or recorded by state or district boundary. It provides the factual data of affected people or fatalities or affected areas. Therefore, the numbers presented in the databases are not confined to Uttar Pradesh or Bihar or both only.

4.2. Data Collection and Selection

The unwanted attributes such as glide number, reference number, location details, etc. have been removed from the EM-DAT and DFO databases. The final database has the following attributes for events-reference id, affected regions, start date, end date, duration in days, year of event, nos. of death, total affected population, total affected area, total damage, cause of flood, severity index, flood magnitude, longitude & latitude (Table 1).

Table 1. Format of input data

S. No.	Attribute	Data Type	Description
1.	Reference Id	Text	Flood Id
2.	Affected Regions	Text	Major places affected in UP and Bihar
3.	Start Date	Numeric	Begin date of the flood
4.	End Date	Numeric	End date of the flood
5.	Duration in Days	Numeric	Number of days flood occurred (Days)
6.	Nos. of death	Numeric	Number of dead people
7.	Total Affected Population (in Nos.)	Numeric	Number of displaced persons
8.	Total Damage (in US$)	Numeric	The total damage due to flood
9.	Cause of Flood	Text	Torrential rain or flash flood etc.
10.	Severity Class	Numeric	Severity of flood is related to flood frequency, calculated on scale of 1-2
11.	Flood Affected Area (in Sq. Km.)	Numeric	Geographic area affected (in km2)
12.	Magnitude (M)	Numeric	Magnitude or frequency of floods
13.	Longitude	Numeric	Longitude value of the location coordinates of the flood event
14.	Latitude	Numeric	Latitude value of the location coordinates of the flood event
15.	Description / Remarks	Text	Highlights of flood

Source: EM-DAT and DFO Databases

4.3. Systematic Identification and Verification of Major Flood Events

Systematic flood records include directly and continuously measured flood discharge and water level (Salas et al. 1994). The collected hydrological in-situ water level observations from CWC for the period of 1980-2015 have been taken as a source for listing the systematic records. Out of collected data from seventeen (17) hydrological stations, five (05) stations for Ganga River at Mirzapur, Varanasi, in Uttar Pradesh and Buxar, Gandhi Ghat and Azmabad in Bihar have been selected. Rests of the stations are located at the tributaries of the Ganga River (Figure 1). From the collected records for the study period of water level data the cross examination of input data, collected from DFO and EM-DAT databases, has been completed. The corresponding data measurement which provide the unprecedented situation, i.e. the Warning Level (WL), Danger Level (DL) and Highest Flood Level (HFL) for respective hydrological stations are collected from CWC website (http://cwc.gov.in/). Based on situation, the CWC has categorized floods as (i) Normal Flood, the water level is either below WL or close to WL but below to DL (ii) severe flood, if the water level crosses WL by one meter and close to danger level but below HFL (iii) Extreme flood situation arises when the water level crossed the HFL (CWC 1973; CWC Handbook for Flood Protection 2012; Flood Forecast Monitoring Directorate 2012).

Table 2. Recorded highest water level on hydrological stations during 1980-2015

Level	Stations on Ganga River (All measurements in Meter)				
	Azamabad	Buxar	Gandhi Ghat	Mirzapur	Varanasi
WL	*NA*	*59.32*	*47.6*	*76.724*	*70.262*
DL	*NA*	*60.32*	*48.6*	*77.724*	*71.262*
HFL	*32.41*	*61.43*	*50.52*	*80.34*	*73.901*

Source: CWC, Govt. of India

Based on the classification, it can be noted that above DL water level indicates the unprecedented situation which can extensively damage the assets and affect human lives. The verification of all 99 events is done with the DL marked events with the procured data from CWC and it is found that all 99 events have been matched with CWC data.

4.4. Data Analysis

The analysis of 99 flood events recorded during 1980-2015 is undertaken in this section. The distribution of data organized on the basis of normality of flood levels. For analysis, the average values of each variable, i.e. flood duration, severity of flood, magnitude, etc. were calculated. Other factors like Arithmetic Mean, Geometric Mean, Standard Deviation, Geometric Standard Deviation for the instances were analyzed. The Arithmetic Mean explains flood level progress in a region, whereas the Geometric Mean provides the population estimation for an event (Halgamuge and Nirmalathas 2017).

The standard deviations for 5th and 95th percentiles for arithmetic and geometric mean also analyzed to find out the variation in the population data. Further the estimate of spread of the data has been quantified through Inter-Quartile Range (IQR), the changes in the upper and lower quartile of the data, 25% and 75% respectively. The detached values, outliers, for a populated data can be known by the IQR. If the data has normal distribution then the IQR calculation found to be less effective than Standard Deviation, but the current dataset is not distributed normally. Hence, the outside values from the given range for any outliers quantify through this formula- <Q1-1.5xIQR or >Q3+1.5xIQR. Subsequently, the goodness of model fit had been applied to detect the prediction accuracy on the flood data. On 36 years data, four standard functions were applied for standard curve fitting technique: (a) linear polynomial $(y=px+q)$, (b) higher-order of polynomial functions $(y=px^3+qx^2+rx+s)$, (c) Gaussian $(y=pe^{(-((x-q)/r)2)})$, and (d) exponential $(y = pe^{(qx)})$. Among all four curve fitting techniques the linear polynomial and higher order of polynomial functions found best fitted for this data. It was found that data is noisy; therefore, the smoothing spline method was used to fit the data. Afterward, the best prediction method was selected for each event, based on the R^2 values, and the Root Mean Square Error (RMSE).

4.5. Flood Risk Assessment Indicators

The statistical analysis in this study utilized two flood indicators- flood severity and flood magnitude. These two indicators are very crucial to know the flood characteristic (Kundzewicz et al. 2013). The severity of flood is associated with the frequency of flood whereas flood magnitude is related to severity, flood duration, and affected region during the flood. Severity can be divided, on the basis of type of floods, in three categories. If the flood is large and occurs within 10-20 years, it is classified as Class-1; if the flood is larger than 20 years flood but less than 100 years then it is denominated as Class-1.5 and the extreme events when interval is more than 100 years are classified as Class-2 category in terms of flood severity. Figure 2 shows the flood severity of MGP. For the analysis of flood magnitude, equation 1 is applied:

$$\text{Magnitude (M)} = \log_{10}(\text{Severity} \times \text{Duration} \times \text{Affected Area}) \quad (1)$$

With the involvement of these two indicators, the final analysis of the flood data is completed.

Figure 2. Flood Severity of MGP

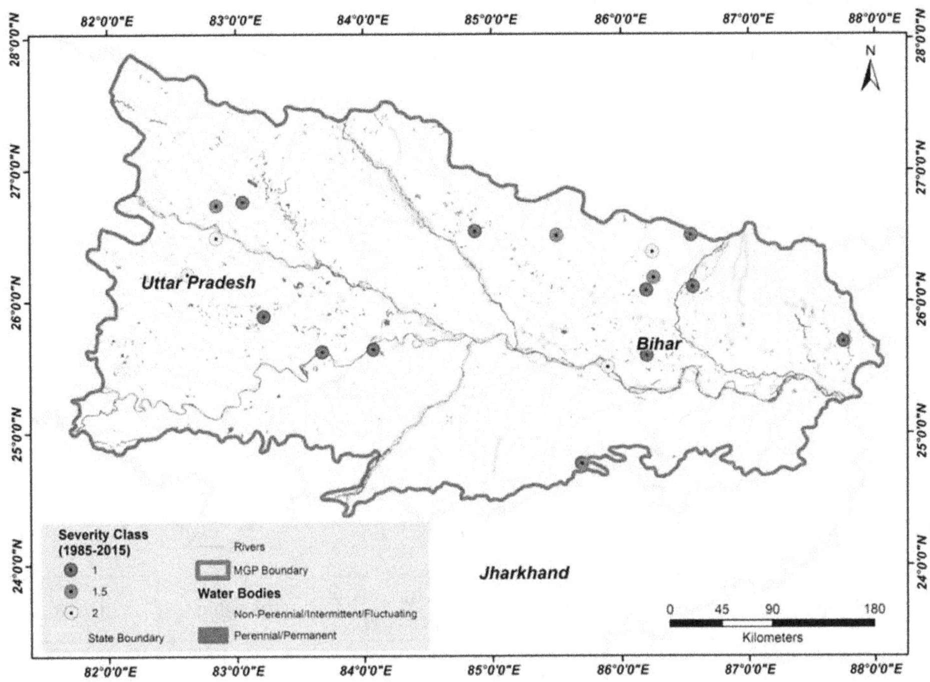

5. RESULTS

The resultant outputs of the statistical analysis are as follows:

5.1 Descriptive Analysis

The descriptive results of statistical analysis are given in table 3. The table provides the summaries of analyzed samples. Table 3 shows the mean, geometric mean, standard deviation, Geometric Standard Deviation (GSD), 25th and 75th percentile and 5th and 95th percentiles for significant variables / items of 99 flood events which are, flood severity, flood duration, affected flood area, flood magnitude, total number of deaths and total number of displaced people.

The relation can be deduced by looking at the table between each variable if the severe class value at 75th percentile is 1.5 the flood magnitude is recorded 7.07. In the same manner, at 95th percentile, the severity is 2 and the magnitude increased to 7.65. It can be noted that if zero (0) available in any class then the geometric mean also be zero (0). For better understanding of each variable, to see the normal range and outliers, the curve fitting of samples has been completed.

Table 3. Descriptive Statistical Analysis of major flood events

Item	Mean	Geometric Mean	Standard Deviation	GSD	25th&75th Percentile	5th&95th Percentile
Severity class (1,1.5 or 2)	1.18	1.14	0.71	1.63	1&1.5	1&2
Flood Duration in Days	21.29	12.40	56.07	22.22	6.75&27.5	1&80.3
Affected flood area (km^2)	248663.04	100762.88	611214.26	61.03	38330.25&313375	2587.5&866975
Flood Magnitude *(M)*=log_{10}*(Severity × Duration × Affected Area)*	6.25	6.13	2.93	1.73	5.76&7.07	3.50&7.65
Number of Death	555.21	0	1904.24	53.18	41&689	9.8&2702.8
Number of displaced people	6160921.57	0	23292663.06	0	21250&5574540	0&32940800

Table 4. Best-fitted model for each variable

Item	Best Fitted Model	Coefficients (p, q & r)	R^2 Value	RMSE
Number of Fatalities (Figure 4)	Polynomial $f(x) = px^2 + qx + r$	34.41, -53.69, 500.1	0.0067	916.9
Affected Area (Figure 5)	Linear $f(x) = px + q$	-3650, 7.483e+06	0.009314	1.736e+05
Flood Magnitude (Figure 6)	Linear $f(x) = px + q$	-0.03857, 83.61	0.1081	0.8478
Flood Duration (Figure 7)	Polynomial $f(x) = px^2 + qx + r$	-1.587, -1.833	0.02695	12.1
Total affected or displaced people (Figure 8)	Linear $f(x) = px + q$	-1.341e+05, 2.705e+08	0.09991	3.713e+06

5.2 Curve Fitting of Data and Error Estimation

The curve fitting and error estimation are carried out for the finalized 99 flood events in this section. The charts were prepared for total damage, causalities, magnitude, displacement, and total affected areas. As the magnitude is being calculated on the basis of severity class, therefore, curve fitting of severity is not done.

The goodness of model fit is found deemed suitable on the basis of R-squared and RMSE values. Model validation, perhaps, the utmost essential part for selection of prediction at a definite confidence level. Although, the higher degree of R^2 doesn't guarantee that the prediction models are best fitted for the noisy data sometimes

(Halgamuge and Nirmalathas 2017). The results found with the best-fitted model with a low R^2value in this study. For the prediction, the model selected here is plotted with the lower and upper confidence level (Cl) at 95% for all variables.

The polynomial with higher order (2) is the best-fitted model found for the variable of total number of death. The normal range found is 0-2000 for the computed variable (figure 3). Six events for the years 1992, 1993, 1994, 1998, 2004 and 2008 have taken lives of more than 2000 people in each event. The range for flood affected areas is found between 0 and 6.5 lakh sq.km for the study period. The outliers in figure 4 indicate the recorded flood event when the flood affected area was more than 7 lakh sq.km. The study finds that the years 1987, 1990, 1993, 2004 and 2005 recorded such floods.

Flood magnitude of any event defines the severity as well as the impact of an event. Figure 5 depicts such events. It is noted that almost every event was impactful and lay between 4 and 8 ranges of magnitude which impacted one or more classes during that time.

From figure 6, it is noted that the flood duration was ranging between 0-50 days for maximum years. Only five (05) such events are detected which was lasted more than 50 days for the years 1998, 2002, 2003, 2004 and 2007.

The range of total number of affected people (x10 million) in each year is displayed in figure 7. The range found in fitted model is between 0-2 crore people for the study period. Nine (09) devastating events for the years 1980, 1995, 1997, 1998, 1999, 2000 (02), 2002 and 2004 in which more than two (02) crore people, in each event, were affected or displaced during the flood. The best-fitted model for each variable / item is described in table 4.

Figure 3. Curve fitting of the total number of death (causalities)

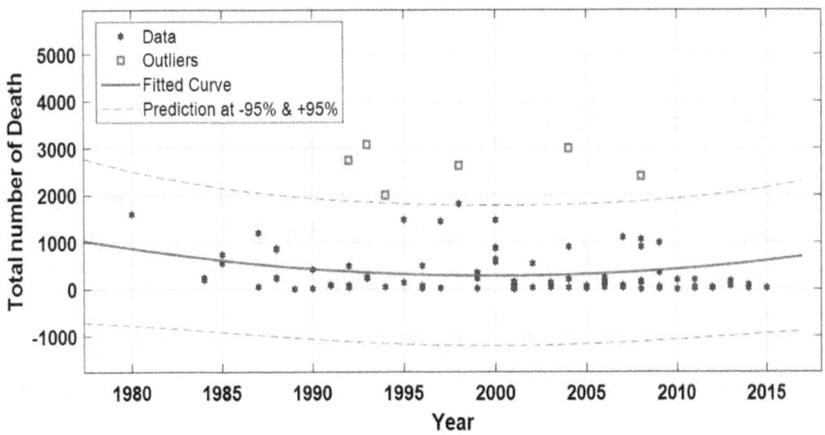

Figure 4. Curve fitting of the total affected area (sq.km.)

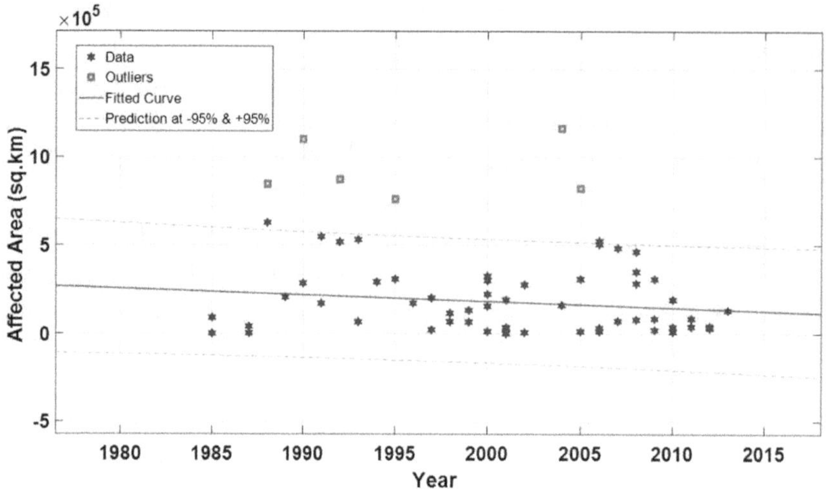

Figure 5. Curve fitting of flood magnitude

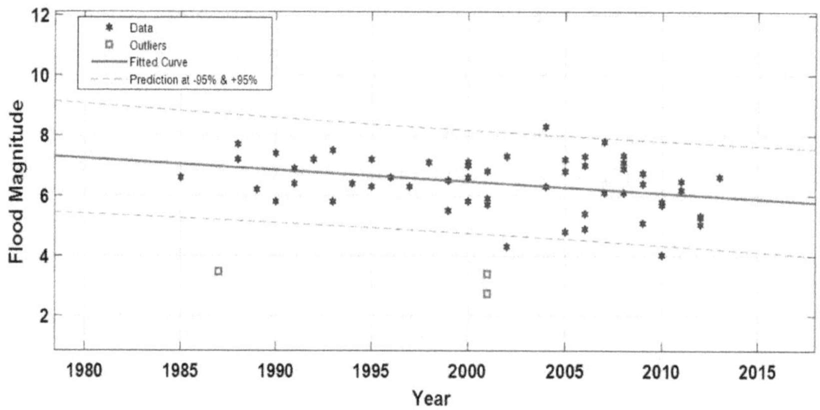

5.3 Top ten Floods of MGP

From all major 99 flood events, top ten (10) events were selected on the basis of total number of fatalities, no. of affected people and damage occurred during the flood. Table 5 lists the statistics. The maximum number of fatalities, 1771, was recorded in 1998 flood while the least (872) in 2011. The maximum severity is 1.5 for the selected events but not 2 because the class 2 severity also collected those events which cover more than study area; hence, those have been excluded from the selection process.

Figure 6. Curve fitting of flood duration (in days)

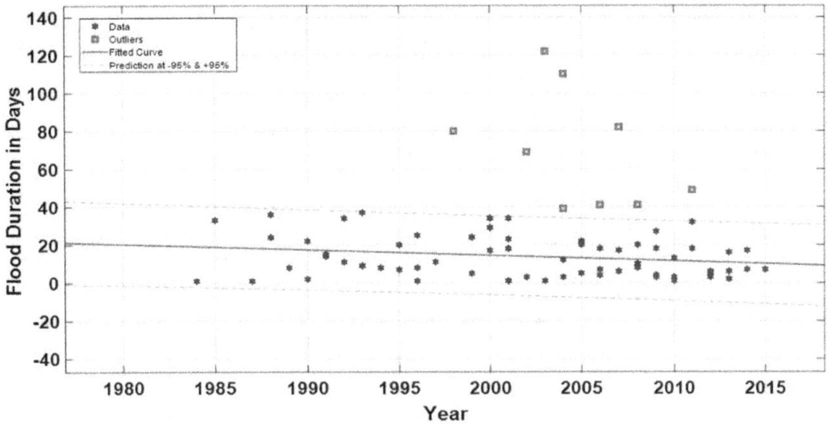

Figure 7. Curve fitting of total affected or displaced people

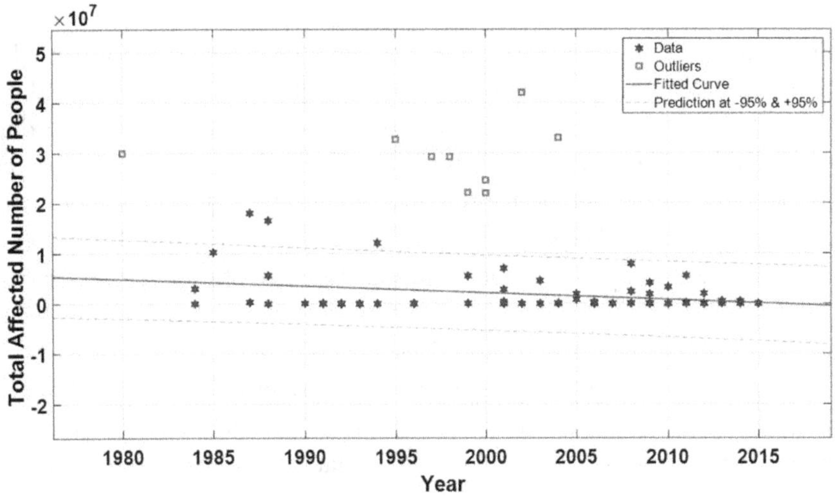

6. CONCLUSION

The systematic and historical records of flood provided the overall nature of flood and the recurrence period during the flood time in MGP. The statistical analysis of major flood events leads us to understand the average range of affected variables during a flood event. By utilizing the CWC hydrological stations data for the period of 1980-2015, the systematic records of large flood events have provided the understanding of recurrence of large flood events and how the responses of it were

Table 5. Top 10 Floods in MGP (Bihar and Uttar Pradesh)

Year	Total Death	Duration in Days	Total Affected (Nos.)	Damage (in Crore)	Severity	Affected Area (in sq km)	Magnitude (M)
1998	1771	80	2,56,61,000	1139.61	1.5	50350	6.78
1993	1542	37	90,16,000	770.92	1.5	18590	6.01
2007	1457	17	57,80,000	2242.24	1.0	18800	5.50
1980	1376	45	3,78,00,000	423.93	1	77800	6.54
2008	1308	41	87,88,000	940.98	1.5	8820	5.73
1987	1281	10	2,83,29,000	1497.86	1	47640	5.68
2003	1204	122	2,22,74,000	1794.02	1	38630	6.67
2004	885	110	2,99,85,000	2310.65	1.5	27000	6.65
1985	873	33	2,44,60,000	2437.48	1.5	48200	6.38
2011	872	49	2,88,7000	1541.89	1.5	5250	5.59

Source: Computed by the Author from different sources
(Selection of events are based on no. of deaths, affected people and damages)

recorded by the gauging stations of MGP. The years 1980, 1982, 1983, 1992, 1994, 1996, 2003 and 2013 are typical years as all five selected stations recorded water level above DL. The Gandhi Ghat stations had crossed the DL for 29 times during the 36 years of recorded period. The confluence of major rivers- Ganga, Ghaghara, Gandak, and Son caused the probability of crossing the water level above danger mark in this region which brings more probability of major floods in this region. The statistical analysis of major flood events provided a range of each selected variables such as the number of dead persons, affected lands, etc., during a devastating flood. The goodness of fit test, with 95% confidence level (upper and lower), successfully explained the maximum number of common flood events in each selected category. The outliers collected after the test gave the numbers of either very significant or very weak flood events. It further describes the nature of most extreme floods and the damages occurred during these floods in terms of loss of lives, loss of property, damage to agriculture and affected or displaced people. Nonetheless, this study is helpful for policy makers, decision makers, research scientists, and administration to compare and evaluate the nature of flood during the study period for MGP. Further, it can be useful to develop the flood mitigation measures.

REFERENCES

Arora, A., Pandey, M., Siddiqui, M. A., Hong, H., & Mishra, V. N. (2019). Spatial Flood Susceptibility Prediction in Middle Ganga Plain: Comparison of Frequency Ratio and Shannon's Entropy Models. *Geocarto International*, 1–32. doi:10.1080/10106049.2019.1687594

Becerra, O., Cavallo, E., & Noy, I. (2014). Foreign Aid in the Aftermath of Large Natural Disasters. *Review of Development Economics*, *18*(3), 445–460. doi:10.1111/rode.12095

Benito, G., Lang, M., Barriendos, M., Llasat, C., Francés, F., & Ouarda, T. (2004). Use of Systematic, Palaeoflood and Historical Data for the Improvement of Flood Risk Estimation. Review of Scientific Methods. *Natural Hazards*, *31*(3), 623–643. doi:10.1023/B:NHAZ.0000024895.48463.eb

Brakenridge, Anderson, Nghiem, Caquard, & Shabaneh. (2003). *Flood Warnings, Flood Disaster Assessments, and Flood Hazard Reduction: The Roles of Orbital Remote Sensing*. Academic Press.

Brakenridge, G. R. (2010). *Global Active Archive of Large Flood Events*. Dartmouth Flood Observatory.

Centre for Research on the Epidemiology of Disasters (CRED) EM-DAT Team. (n.d.). *EM-DAT | The International Disasters Database*. Retrieved June 7, 2019 https://www.emdat.be/

CRED first. (2011). *EM-DAT: The OFDA/CRED International Disaster Database*. Université Catholique de Louvain.

CWC. (1973). *Estimation of Design Flood Peak: A Method Based on Unit Hydrograph Approach*. Author.

CWC. (2018). *Flood Damage Statistics (Statewise and for the Country as a Whole) for the Period 1953 to 2016*. Central Water Commission (CWC), Flood Forecast Monitoring Directorate, Government of India.

CWC Handbook for Flood Protection. (2012). *Handbook for Flood Protection*. Anti Erosion & River Training Works.

EMDAT. (2019). *OFDA/CRED International Disaster Database*. Université Catholique de Louvain.

EMDAT. (2018). *Total Economic Damage: 1900-2017*. OFDA/CRED International Disaster Database, Université Catholique de Louvain.

UNDP Emergency Analyst. (2008). *Situation Report: Bihar Flood 2008*. Author.

Fallis, A. (2013). Emergency Events Database (EM-DAT) and Its Applications. *Journal of Chemical Information and Modeling*, *53*(9), 1689–1699.

Flood Forecast Monitoring Directorate. (2012). *Flood Forecasting and Warning Network Performance Appraisal Report 2011*. Author.

Halgamuge, M., & Nirmalathas, A. (2017). Analysis of Large Flood Events: Based on Flood Data During 1985–2016 in Australia and India. Academic Press.

Hoyois & Guha-sapir. (2003). *Three Decades of Floods in Europe : A Preliminary Analysis of EMDAT Data . 2 . Number of Flood Disasters by UN European Region*. Academic Press.

IPCC. (2012). *Managing the Risks of Extreme Events and Disasters to Advance Climate Change Adaptation*. IPCC.

Kundzewicz, Z. W., Pińskwar, I., & Robert Brakenridge, G. (2013). Large Floods in Europe, 1985–2009. *Hydrological Sciences Journal*, *58*(1), 1–7. doi:10.1080/02626667.2012.745082

Magilligan, F. J., & Goldstein, P. S. (2001). El Niño Floods and Culture Change: A Late Holocene Flood History for the Rio Moquegua, Southern Peru. *Geology*, *29*(5), 431–434. doi:10.1130/0091-7613(2001)029<0431:ENOFAC>2.0.CO;2

(1994). Salas, José D., Ellen E. Wohl, and Robert D. Jarrett. InRossi, G., Harmancioğlu, N., & Yevjevich, V. (Eds.), *Determination of Flood Characteristics Using Systematic, Historical and Paleoflood Data BT - Coping with Floods* (pp. 111–134). Springer Netherlands.

Schmidt-Thomé, P., Greiving, S., Kallio, H., Fleischhauer, M., & Jarva, J. (2006). Economic Risk Maps of Floods and Earthquakes for European Regions. *Quaternary International*, *150*(1), 103–112. doi:10.1016/j.quaint.2006.01.024

Singh, R. L. (1971). India: A Regional Geography. National Geographical Society of India.

Sodhi, M. S., & Tang, C. S. (2014). Buttressing Supply Chains against Floods in Asia for Humanitarian Relief and Economic Recovery. *Production and Operations Management*, *23*(6), 938–950. doi:10.1111/poms.12111

Vittal, H., Ghosh, S., Karmakar, S., Pathak, A., & Murtugudde, R. (2016). Lack of Dependence of Indian Summer Monsoon Rainfall Extremes on Temperature: An Observational Evidence. *Scientific Reports*, *6*(July), 1–12. doi:10.1038rep31039 PMID:27485661

World Bank – World Development Indicators. (2019). *Internally displaced persons, new displacement associated with disasters (number of cases), 1960-2017*. The Internal Displacement Monitoring Centre. https://Www.Internal-Displacement.Org/

Chapter 13
Inland Saline Wetlandscapes:
The Missing Links for 4th Ramsar Strategic Plan (2016-2024) in India

Laxmikant Sharma
https://orcid.org/0000-0003-2911-2893
Central University of Rajasthan, India

Rajashree Naik
Central University of Rajasthan, India

Alok Raj
Central University of Rajasthan, India

ABSTRACT

Wetland ecosystems are one of the highly productive ecosystems in the world. These ecosystems have been deteriorating at a faster rate. Ramsar Convention is putting enormous effort to protect, maintain, and restore these ecosystems. Currently, the fourth phase of Strategic Plans of Ramsar Convention is going on, in which saline wetlandscapes can play vital role to attain 19 targets of this plan. In India there are 27 Ramsar sites in all the biogeographic zones; however, research work has been carried out in the past five years in only eight Ramsar sites. Currently, four years are available for the strategic plans to encourage more wetland researches. The chapter presents a case study of Sambhar Salt Lake, a Ramsar site of India that is on the verge of extinction. Normalized Difference Water Index has been calculated for three decades in 1992, 2009, and 2019, revealing the declining phases of the lake.

DOI: 10.4018/978-1-7998-5027-4.ch013

INTRODUCTION

Wetland Ecosystem

An ecosystem is composed of both biotic and abiotic factors. Biotic factor constitutes of fauna and flora and abiotic components constitutes of water, energy, nutrient, soil and gaseous constituents. There are different types of ecosystems co-existing in nature like forest ecosystem, grassland ecosystem and wetland ecosystems. However, there are also some man-made ecosystems also like rice-fields, dams and reservoirs and aquarium. Wetland ecosystems, unique in nature are among the highly productive ecosystems. Wetlands are of different types depending on their process of origin, location, physico-chemical and biological constituents, micro-climatic factors, nature of inlet and outlet of waters, hydro-ecological cycle and seasonality of water level (Lin el al., 2019).

Under the text of the Ramsar Convention (Article 1.1), wetlands are defined as "areas of marsh, fen, peatland or water, whether natural or artificial, permanent or temporary, with water that is static or flowing, fresh, brackish or salt, including areas of marine water the depth of which at low tide does not exceed six meters (rsis. Ramsar.org, 2019)". Wetlands provide enormous amount of goods and services that have been acknowledged recently (Lin el al., 2019). The provisioning services derived are food, fresh water, fiber, fuel, salt, energy and ornamental resources, regulating services like water regulation and purification, climatic regulation, wastewater treatment, prevention of desertification, pollination and soil fertility enhancement, supporting services like nutrient cycling, wildlife habitat, soil formation, biomass production and cultural services like spiritual, recreational, educational, aesthetic, and inspirational achievement.

Wetlands ecosystems being so productive but diverse in nature have been classified by Ramsar Convention based on habitat types in 1993 known as Ramsar International Wetland Classification System. According to this classification, three major classes were categorized as marine and coastal wetlands, inland wetlands and man-made wetland ecosystems. Marine and coastal wetlands were further sub divided into twelve sub-classes, inland wetlands into fourteen sub-classes and man-made wetlands into nine sub-classes. Wetlands have also been classified by (Cowardin el al., 1979) based on location, flooding pattern, origin and ecosystem services they provide into five classes as marine, estuarine, riverine, lacustrine and palustrine wetlands. The details of these wetlands are represented in Figure 1 below.

Marine wetlands are near ocean, estuarine related to estuaries, riverine wetlands come from rivers, lacustrine deal with lakes, and palustrine belong to any inland wetland that lacks flowing water. Estuarine and marine are further sub divided into subclasses as intertidal and subtidal. Riverine wetlands are again classified into four

Figure 1. Structure of wetland ecosystem

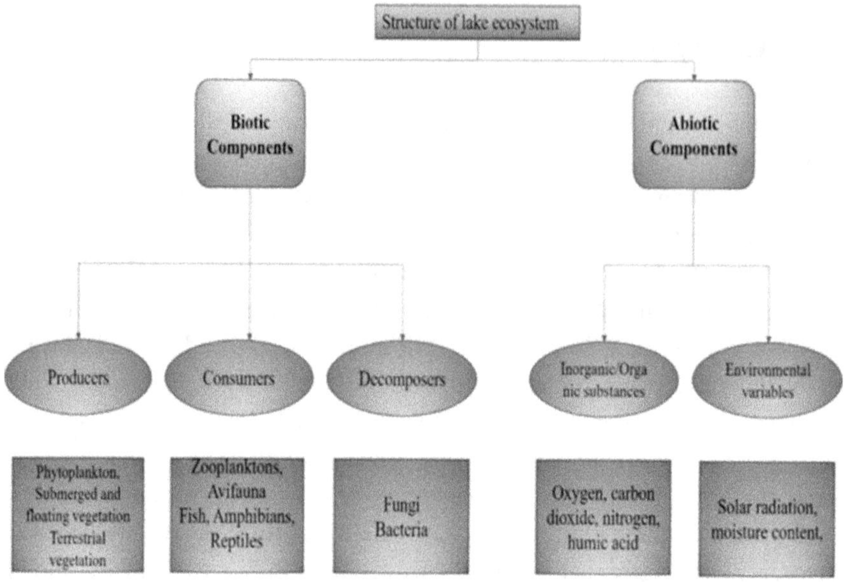

subclasses into tidal, lower perennial, upper perennial, and intermittent. Lacustrine wetlands are further classified into two sub classes into littoral and limnetic. Palustrine wetlands have no subclasses.

INLAND SALINE WETLANDSCAPE

These saline wetlands occur in all the continents along with Antarctica. Throughout the world there are about 200 Ramsar sites occupying 2,28,56,009 ha (Accessed Ramsar site link 15 January 2020). Africa and Europe have highest saline Ramsar sites out of which 38% is located in Spain (Luna el al., 2019). There are two types of saline water ecosystems of which one deals with coastal ecosystem and other epi-continental salt lakes. They are continuously subject to high to very high evaporation and low precipitation and inflow causing seasonal and inter-annual water level fluctuation (Moore, 2016). The inland salt lakes embrace a wide range of economic and non-economic goods and services. Salinity of these lakes range from 3g to 300g of salts per kilogram of Total Dissolved Solids (TDS) (Wurtsbaugh el al., 2017). These inland saline lakes provide all the types of ecosystem services. They play very critical role for flamingoes and water birds like gulls, pelicans, stilts, stints, ruffs, plovers and shovellers. Economic benefits are also derived from these lakes from both inorganic part by extracting edible like sodium chloride and non-edible salts sodium and potassium chlorides, borates, nitrates, sulphates of potassium, magnesium, calcium and sodium, calcium and sodium carbonates, and lithium chloride and also from organic biota of lake through fisheries, brine shrimps and *Artemia spp.* worth US$35–110 million (Wurtsbaugh el al., 2017). These wetlands very suitable sites for extremophiles like halophiles and halophytes, arthropods, protozoa and other invertebrates complete their partial life cycle here, amphibians and vertebrates also depend on these sites different medicinal plants as raw materials for pharmaceutical and cosmetic industries depend on these sites (Baid, 1968). However, these saline inland wetlands are under high threats of extinction due to salinization, desiccation, water channel diversion and climate change (Herbst, 2014), ground water extraction (Heintzman el al., 2017), habitat destruction (Venne el al., 2012) agriculture (Fowler el al., 2014) sedimentation (Cariveau el al., 2011), landscape transformation (Luna el al., 2019), watershed erosion (Tsai el al., 2010). The major threats are given in Table 1 below.

Table 1. Major threats of inland saline wetlandscapes

S.No.	Major Threats to the inland saline wetlandscapes
1	Tourism
2	Human intrusion
3	Hunting and poaching
4	Pollution
5	Mining and quarrying
6	Agricultural extension
7	Urban expansion
8	Domestic Sewage water disposal
9	Industrial effluents discharge
10	Micro climate modifications
11	Habitat shifting and alteration
12	Dredging
13	Land conversion
14	Vegetation pattern modification
15	Ground water extraction
16	Shipping lanes
17	Excess energy Production

WETLAND ECOSYSTEM OF INDIA

India is blessed to have total 27 Ramsar sites with 10,568.71 km^2 area which include inland, coastal and man-made wetlands as shown in Figure 2 with their reference Ramsar site numbers. These wetlands are of three types. They are inland wetlands with 3,407.19 km^2, marine wetlands with 9,029.90 km^2 and man-made wetlands with 2,645.74 km^2 (RSIS, 2019). Indian geographic condition is divided into ten biogeographic zones. They are Western Ghats zone, Trans Himalayan zone, Deccan plateau zone, Himalayan zone, Gangetic plain zone, Desert zone, Semiarid zone, North east zone and Coastal zone. Majority of the state of India share partly share their boundary with one or more zones. In the west of India, Gujarat possesses Western Ghats, coastal zone, arid zone, Maharashtra also belongs to Western Ghats, coastal zones and Deccan plateau. The Indian Himalayan region is occupied by 10 states, Jammu & Kashmir, Uttarakhand, Himachal Pradesh, Meghalaya, Sikkim, Arunachal Pradesh, Nagaland, Assam, Manipur, Mizoram, Tripura, and West Bengal (ENVIS, Assessed on 17 January 2019). Among these states, six have Ramsar sites. Jammu and Kashmir with four sites as Hokera, Surinsar-Mansar, Tsomoriri and Wular, Himachal Pradesh with three sites as Chandertal, Pong and Renuka, Manipur has only one site as Loktak, Tripura has also only one site as Rudrasagar, Assam has also one site as Deepor beel, and West Bengal has two sites as East Calcutta wetlands and Sundarbans. In short, there are twelve Ramsar sites in Indian Himalayan region. In total, desert and semi-arid zones have six Ramsar sites. They are Harike, Kanjli and Ropar Ramsar sites of Punjab, Nalsarovar of Gujarat, Keoladeo Ghana and Sambhar Salt Lake of Rajasthan. In the region of Western Ghats there are in total three Ramsar sites. Kerala has two sites as Asthamudi and Sasthankotta and Tamil Nadu with Point Calimere wetland. In Deccan plateau region, Andhra Pradesh has one site as Kolleru and Madhya Pradesh has one site Bhoj wetland. In coastal zones, Odisha has two Ramsar sites as Bhitarkanika Mangroves and Chilika lagoon. It is very clear that India has Ramsar sites in all the biogeographic zones ranging from high-altitude regions, swamps, fresh water and saline wetlands, mangroves, backwater, lagoons, mud flats, estuaries, coral reefs and inland wetlands like lakes, ponds and creeks (Prasad el al., 2002). The states with no Ramsar sites are Karnataka, Maharashtra, Telangana, Goa, Sikkim, Bihar, Jharkhand, Uttarakhand, Arunachal Pradesh, Meghalaya, Mizoram and Nagaland.

STRATEGIC PLANS OF RAMSAR CONVENTION

Ramsar Convention came into being in March 1971 as a result of continuous and long efforts of ornithologists and nature conservationist to protect deterioration of

Figure 2. Ramsar sites of India

waterbirds. 18 countries namely Belgium, Denmark, Finland, France, the Federal Republic of Germany, India, Iran, Ireland, Jordan, the Netherlands, Pakistan, South Africa, Spain, Sweden, Switzerland, Turkey, the Union of Soviet Socialist Republics and the United Kingdom joined together with inter-governmental agencies like Educational, Scientific and Cultural Organization (UNESCO), Food and Agricultural Organization (FAO) and The United Nations and non-governmental organizations like World Wide Fund (WWF) and International Union for Conservation of Nature and Natural Resources (IUCN) worked together to establish this Convention.

Ramsar Convention has been putting enormous efforts to protect, restore and conserve the wetlands designating them as Ramsar sites at global scale. The first ever site declared is Cobourg Peninsula in Australia and Sundarbans of India with site number 2370 is the most recent one. Today, there are 2370 sites out of which 1102 sites are located in Europe, 396 sites in Africa, 339 sites in Asia, 218 sites in North America, 204 sites in Latin America and 82 sites with Oceania (rsis.Ramsar.org, 2020).

In addition to designation of sites, Ramsar Convention is also cooperating with and assisting Non-Governmental Organizations (NGOs) like Birdlife International, Wetland International, World Wetland Network, Conservation International and

The Nature Conservancy; Inter-Governmental agencies like International Union for Conservation of Nature and Natural Resources (IUCN), World Wide Fund for nature (WWF) along with institutes like International Institute for Sustainable Development (IISD) and International Water Management Institute (IWMI). Ramsar has also Memorandum of Understanding (MoU) with Convention on Biological Diversity (CBD), Convention on Migratory Species (CMS), Convention on World Heritage Sites (CWHS) and United Nations Convention to Combat Desertification (UNCCD) along with Millennium Development Goals (MDGs), Sustainable Development Goals (SDGs)

Even after putting gargantuan efforts, wetland ecosystem and its landscape variables are being continuously subject to different sorts of threats. In order to reverse these factors, in March 1996 on the eve of 6th Meeting of the Conference of the Contracting Parties in Brisbane, Ramsar Convention adopted the first "Strategic Plan" for 5 years from 1997 to 2002. The foundation pillars were to bring down the concept of "wise use" to the ground along with its conservation with the assistance of Contracting Parties of the Convention, Scientific and Technical Review Panel, Standing Committee and Bureau (Convention on Wetlands, 2002). The second Strategic Plan was carried from 2003 to 2008, third was from 2009 to 2015 and the fourth is from 2016 to 2014. The foundation mission of the first phase was to increase membership for the Convention worldwide, popularize wise use concept, increase awareness about wetland ecosystems and its services at all the levels, conservation of all Ramsar sites, strengthening National and International Cooperation between Governmental, Inter-Governmental and Non-Governmental organizations for providing institutional mechanisms and resources (Convention on Wetlands, 2002).

The fourth Strategic Plan consists of visions, missions, goals and targets for conservation of wetland ecosystems. The vision is "*Wetlands are conserved, wisely used, restored and their benefits are recognized and valued by all*". There are four goals of which the first one deals with the identification of drivers of wetland degradation, second goal deals with the conservation and management of Ramsar Site Network, third goal with the wise use concept and fourth goal with the enhancement of implementation of the Convention along with 19 targets (The Fourth Ramsar Strategic Plan, 2016). These plans have been continuously prioritizing the incorporation of wetlands into local and national policies related to water, mining, energy, tourism, agriculture, infrastructure, urban development, forestry, aquaculture, industry and fisheries so that the health of the wetlands remain maintained and fit to attain the all the 17 Sustainable Development Goals (The Fourth Ramsar Strategic Plan, 2016).

ADVANCED SPATIAL INFORMATION SCIENCE FOR WETLAND MANAGEMENT

Spatial Information Science deals with the spatial information about the earth. Since 1970s, spatial information is being collected through numerous space crafts and satellites. Landsat is the largest satellite mission till now. This satellite scheme has been providing multispectral data sets with coarse to high spatial resolution data sets for assessment of natural resources from local to global scale. There are other missions also that provide both spatial. These are of immense help to wetland researchers for mapping (Jahncke el al., 2018), monitoring (Rapinel el al., 2015), classification (Ludwig el al., 2019), change detection (Davranche, 2013), soil moisture and salinity mapping (Wigmore el al., 2019), water budgeting (Sahoo el al., 2011), stages of river inundation and discharge (Cienciala, 2017), wetland health and risk assessment (Sun el al., 2016), spatial modelling (Fu el al., 2018) and ecosystem services assessment at landscape level (Fengqin, 2019).

In addition to multispectral, with the advancement in technology currently data sets with high spectral information are also available through hyperspectral sensors. This technology has very wide range of application for wetland ecosystem management like application in studies of reef (Bajjouk, 2019), saline soil mapping (Khadim el al., 2019); estuarine (Lunetta el al., 2009); interdunal lake (Timm, 2012); littoral (Karpouzli, 2004); salt marsh (Judd et al., 2007); coastal (Pengra el al., 2007); tidal marsh (Turpie el al., 2015); dissolved organic matter (Matthews el al., 2010), total dissolved solids (Thiemann and Kaufmann, 2002; Koponen el al., 2002). soil erosion (Chabrillat, 2014); soil moisture content (Yanmin, 2010), soil available nutrients (Qi el al., 2018), soil types (Vibhute, 2015), Identification along with discrimination of algal blooms and survival competition (Bi el al., 2019) and ecophysiology (Stratoulias el al., 2015).

Application of Light Detection and Ranging (LiDAR) technology are proving to be of immense help for wetland management from last few decades. This advanced technology has been applied for classification (O'Neil et al., 2019), benthic invertebrate identification (Kerfoot et al., 2019), plant community types identification (Ward et al., 2013), vegetation mapping (Geerling et al., 2007) with the data fusion of other remote sensing techniques.

WETLAND STUDIES IN INDIA

In India within this half decade from 2015 to 2020 in the tenure of 4th Strategic Plans of Ramsar Convention, only fifty research articles could be retrieved which is a very alarming indication towards the wetland ecosystems of India. Though

India has wetlands in all the ten biogeographic regions of the country, still they are neglected. However, the wetlands that have been studies cover diverse range studies related to ecology (Chatterjee et al., 2015), geology (Krishnakumar et al., 2015), biotechnology and microbiology (Kaur and Attri, 2015), degradation (Ghermandi et al., 2016), pollution (Bhattacharya et al., 2015), aquaculture (Biswas et al., 2015). The details are given in Table 1 below.

Table 2. Wetland research in India from 2015 to 2020

S.NO.	State	Wetland	Purpose	References
1	West Bengal	Sundarbans	Distribution of dissolved trace metals	Bhattacharya et al., 2015
2	Assam	Charan Beel	Fishery	Biswas et al., 2015
3	Rajasthan	Keoladeo National Park	Wetland Degradation	Chatterjee et al., 2015
4	Punjab	Harike	Molecular characterization of wetland species	Kaur and Attri, 2015
5	Kerala	Ashtamudi	Geological exploration	Krishnakumar et al., 2015
6	Rajasthan	Keoladeo National Park	Seasonal behavior of wetland plants	Middleton et al., 2015
7	Kerala	Ashtamudi	Bioaccumulation efficiency of bacteria	Chinnadurai et al., 2016
8	West Bengal	Sundarbans	Phytoremediation of mangroves	Ansari et al., 2016
9	Kerala	Akkulam-Veli	Ecosystem services evaluation	Ghermandi et al., 2016
10	Punjab	Harike	Molecular characterization of wetland species	Kaur and Joshi, 2016
11	Kerala	Kuttanad	Sustainable development planning	Kumar and Devadas, 2016
12	Andhra Pradesh	Kolleru	Geospatial analysis for changing environment	Kumar et al., 2016
13	West Bengal	Sundarbans	Trace metal assessment	Rodríguez-Iruretagoiena et al., 2016
14	West Bengal	Ramnagar Beel, Bhomra Beel, Mathura Beel and	Labour participation in wetland	Roy et al., 2016
15	West Bengal	East Kolkata Wetland Area (EKWA),	Wetland conservation risk analysis	Sarkar et al., 2016
16	West Bengal	Buxa Tiger Reserve	Limnochemical features	Chatterjee et al., 2017

continues on following page

Table 2. Continued

S.NO.	State	Wetland	Purpose	References
17	West Bengal	Sundarbans	Trace metal assessment	Chatterjee et al., 2017
18	Punjab	Ranjit Sagar Wetland	Fish diseases	Gupta and Kaur, 2017
19	Jammu	Gharana wetland reserve	Avi-fauna survey	Jamwal et al., 2017
20	Punjab	Harike	Water quality monitoring	Kaur et al., 2017
21	Andhra Pradesh	Hussain Sagar	Physical, chemical and biological characterization	Kora et al., 2017
22	Punjab	Ranjit Sagar Wetland	Molecular characterization	Gupta and Kaur, 2017
23	West Bengal	East Kolkata Wetland	Wetland shrinkage estimation	Mondal et al., 2017
24	Rajasthan	Keoladeo National Park	Wetland modelling	Patra et al., 2017
25	West Bengal	Sundarbans	Persistent organic pollutants	Pozo et al., 2017
26	Kerala	Veli-Akkulam Lake	Hydro-geochemistry analysis	Sajinkumar et al., 2017
27	Haryana	Sultanpur National Park	Land use pattern change	Singh et al., 2017
28	Uttar Pradesh	Keetham lake, Nawabganj lake and Okhla reservoir	Methane emission	Bansal et al., 2018
29	West Bengal	Sundarbans	Carbon sequestration	Gorain et al., 2018
30	Punjab	Ranjit Sagar Wetland	Molecular characterization	Gupta and Kaur, 2018
31	West Bengal	Atreyee River basin	Change detection	Pal and Saha, 2018
32	Karnataka	Bellandur wetland	Heavy metal accumulation	Ramachandra el al., 2018
33	Haryana	Bhindawas bird sanctuary	Discrimination of wetland components using hyperspectral data	Saluja el al., 2018
34	West Bengal	Punarbhaba river basin	Vulnerability assessment	Pal and Talukdar, 2018
35	Punjab	Ropar	Heavy metal assessment	Sharma el al., 2018
36	Bihar	Kanwar	Heavy metal assessment	Singh el al., 2018
37	Kerala	Vembanad	Recreational ecosystem services assessment	Sinclair el al., 2018
38	Sikkim	Tsomgo high-altitude lake	Biodiversity assessment	O'Neill, 2019
39	West Bengal	Ganga basin	Ecosystem service assessment	Everard el al., 2019
40	West Bengal	East Kolkata wetland	Vulnerability assessment	Ghosh and Das, 2019

continues on following page

Table 2. Continued

S.NO.	State	Wetland	Purpose	References
41	West Bengal	Sundarbans	Physico-chemical parameters	Gogoi el al., 2019
42	Andhra Pradesh	Godavari river basin	Mangrove management using Web-GIS	Jayakumar el al., 2019
43	Haryana	Ponds in Kurukshetra	Avian analysis	Kumar and Sharma, 2019
44	West Bengal	Nadia	Micro benthos analysis	Meena el al., 2019
45	Punjab	Ropar	Heavy metal assessment	Sharma el al., 2019
46	Bihar	Kaabar Tal	Hydrological connectivity analysis	Singh and Sinha, 2019
47	Mizoram	Pala Wetland	Trees species diversity assessment	Nohro and Jayakumar, 2020
48	West Bengal	Nararthali, Rasomati, Rasik Beel, Raichenmari-Sakobhanga, Bochamari Atiamochar-Naldubi	Foraging habit of waterbirds	Chatterjee el al., 2020
49	West Bengal	Ganga-Brahmaputra delta	Future projection of landscape	Debanshi and Pal, 2020
50	Bihar	Wetland complex of Ganga plain	Geo-morphology study	Singh and Sinha, 2020

From the above table it is observed that the researches have been conducted in thirty two wetlands viz. Vembanad, Veli-Akkulam, Tsomgo, Sundarbans, Sultanpur, Ropar, Ranjit Sagar, Ramnagar Beel, Punarbhaba, Kurukshetra ponds, Pala, Nararthali, Nadia, Kuttanad, Kolleru, Keoladeo, Keetham lake, Kanwar, Kaabar Tal, Hussain Sagar, Harike, Godavari river basin, Gharana, Ganga basin, East Kolkata wetland, Charan Beel, Buxa Tiger Reserve, Bhindawas, Bellandur, Atreyee River basin, Ashtamudi and Akkulam-Veli out of which only eight are Ramsar sites represented in Figure 3 below. The highlighted points in the figure are the Ramsar sites in which recent studies have been conducted, rest others are not designated sites. The numbers in the parenthesis indicate the number of publications for the respective wetlands.

THE MISSING LINK

There are 28 states and 9 union territories in India. India has ten biogeographic zones. As discussed earlier all the zones have wetland ecosystems, however in last half decade studies have been conducted on wetlands of thirteen states viz. Andhra Pradesh, Assam, Bihar, Haryana, Jammu, Karnataka, Kerala, Mizoram, Punjab,

Rajasthan, Sikkim, Uttar Pradesh and West Bengal over thirty-two wetlands out of which there are only eight Ramsar sites (Harike, Ropar, Keoladeo, East Kolkata Wetlands, Ganga basin, Sundarbans, Vembanada and Asthamundi). Among all the wetland sites studies, there is no single study conducted in the inland saline wetland ecosystems, neither in Sambhar Lake (Rajasthan), Nalsarovar (Gujarat), Chandertal (Himachal Pradesh) not in Tso Moriri (Ladakh). From the Figure 4, it is observed that the highest studies have been conducted in West Bengal having eighteen publications and the second highest publications in Punjab. From Figure 5 it is observed that the highest work has been conducted in Sundarbans with seven publications.

Figure 3. Representation of wetlands where research have been carried out

A CASE STUDY ON SAMBHAR SALT LAKE

This case study has been conducted at preliminary level to assess the three decadal condition of the country's largest inland saline lake "Sambhar Salt Lake" situated in semi-arid zone of Rajasthan in India. The primary aim was to unveil the deteriorating condition of the lake that has a very rich history, culture, natural resources and biodiversity. Once upon a time this lake was a haunting place for Currently, this lake is undergoing

Figure 4. States where wetland studies have been undertaken

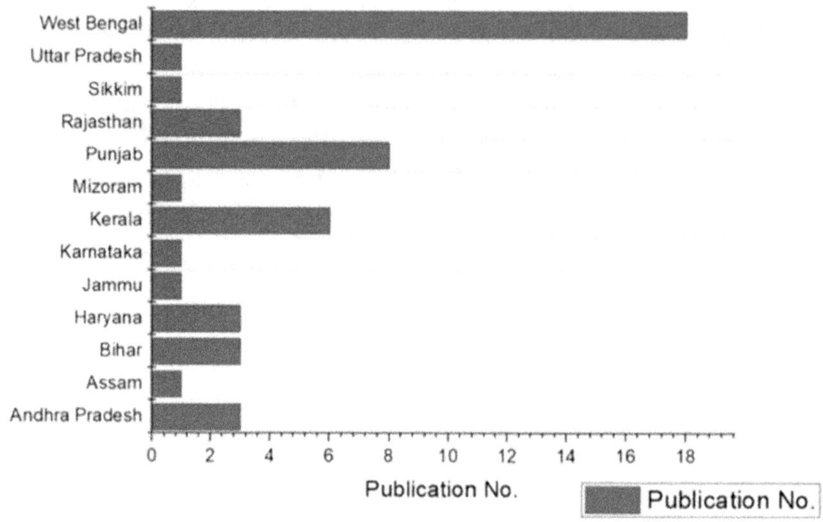

Figure 5. Wetlands that have been studies

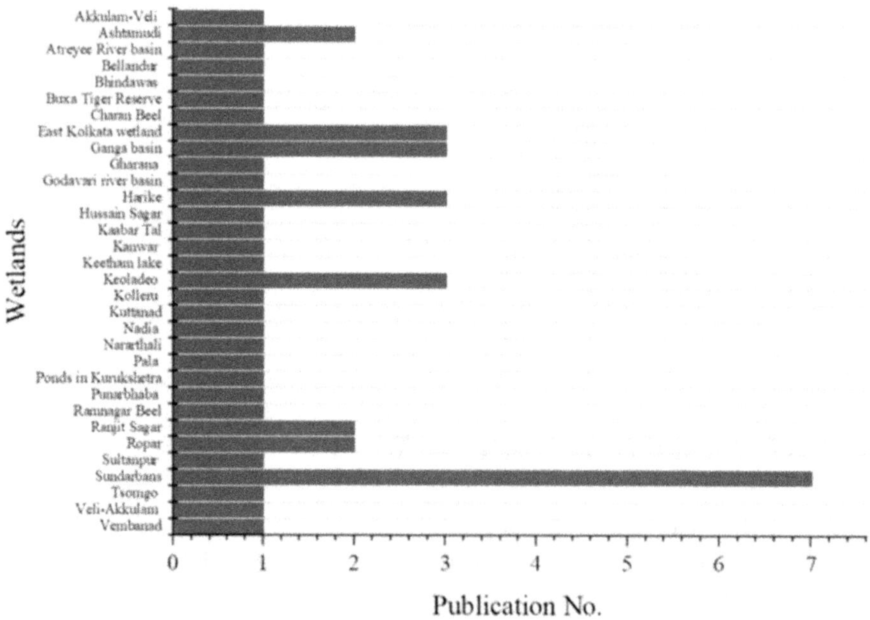

Study Area

Sambhar Salt Lake is the largest saline wetland of India situated at 26^0 52¢ to 27^0 02¢ N; 74^0 54¢ – 75^0 14¢ E as an endorheic basin (Sahay el al., 2012). It is one of the gateways to Thar desert. It has Ramsar site No. 464 and Important Bird Area No. IN073. It is surrounded by world's oldest hill range Aravalli in North, west and south-east direction in the three sides. The lake area is 230 km^2 based on the seasonal variation with length 22.5 km and width varying between 3 and 11 km (Vijay el al., 2016). The water depth fluctuates from as few as 60 cm during the dry season to about 3 m after the monsoon season. It also receives inflow from four major rivers Rupangarh, Mendha, Kharian and Khandel. The lake is divided into two unequal parts by 5.16 km long dam to which eastern part covering 77 km^2 is under public salt industry company Sambhar Salt Limited, a subsidiary of Hindustan Salt Limited and western part covering 113 km^2 is natural saline wetland (Milja el al., 2011). It used to support large number of breeding colonies of greater and lesser flamingo (*Phoenicopterus roseus*) along with other migratory birds (Kalambe el al., 2017) but illegal saltpans escalation around the lake are pernicious to habitat quality. So, the current study was conducted in 3km buffer area to the lake boundary.

METHODOLOGY

Data Used

For the current study, satellite images of Landsat mission have been used to derive the results. The data sets have been collected for the winter season as the study area is a critical visiting site for flamingoes and 28 other water birds like coots, grebes, shovellers, stilts and stints for resting, nesting and food. The details of the study area are given in Table 3 below.

Table 3. Details of satellite data used

S.No.	Satellite	Sensor	Path Row	Spatial Resolution (m)	Temporal Resolution (days)	Date of Acquisition
1	Landsat 5	Thematic Mapper (TM)	148/41	30	16 days	25-11-1992
2	Landsat 5	Enhanced Thematic Mapper (ETM)	148/41	30	16 days	08-11-2009
3	Landsat 8	Operational Land Imager (OLI)	148/41	30	16 days	20-01-2019

Methods

The Normalized Difference Water Index (NDWI) was given by McFeeters in 1996 for detecting surface waters in wetland ecosystems (McFeeters, 1996). Initially, this index was developed for Landsat Multispectral Scanner (MSS) satellite data sets, however, it is being effectively used for other datasets (Panigrahy el al., 2012; Murray el al., 2012. The NDWI is calculated using Equation.1.

Equation 1: Calculation of NDWI

NDWI = GREEN – NIR / GREEN + NIR

where Landsat TM and ETM sensors, band 2 was representing GREEN and band 4 was representing NIR but in case of Landsat 8 OLI sensor band 3 was representing as GREEN and band 6 as NIR to calculate water index. McFeeters (McFeeters, 1996) emphasized that values of NDWI greater than zero will be representing water surfaces, however, values ranging less than, or equal to zero will be considered as non-water surfaces. The values of NDWI for the study area were calculated from the Landsat images using Equation (1) in the Raster Calculator option of Spatial Analyst extension of ArcGIS 10.5. The result obtained was a single band gray-scale as floating-point grid file which was color-coded and reclassified for analysis as shown in Figure 6.

RESULTS AND DISCUSSION

The results of NDWI for Sambhar Salt Lake has been obtained for the study year 1992, 2009 and 2019 at almost ten years gap period to assess the decadal water level change detection. The result has been represented in Figure 6 below. It is very distinctly observed that surface water level in the image (a) representing 1992 is very high as major portion of the lake is covered by water. In second image (b) representing 2009 has comparatively less surface water coverage however, the eastern part of lake and the reservoir area still have surface water. In the last image (c) representing 2019, indicates an alarming situation for the lake since the surface water has drastically decreased. Decrease in water level has put adverse impact on hydro-ecological status of the lake. This may also have reverse impact on salt industry also putting pressure on local livelihood.

Figure 6. NDWI maps of Sambhar lake; (a) represents map of 1992, (b) represents map of 2009 and (c) represents map of 2019

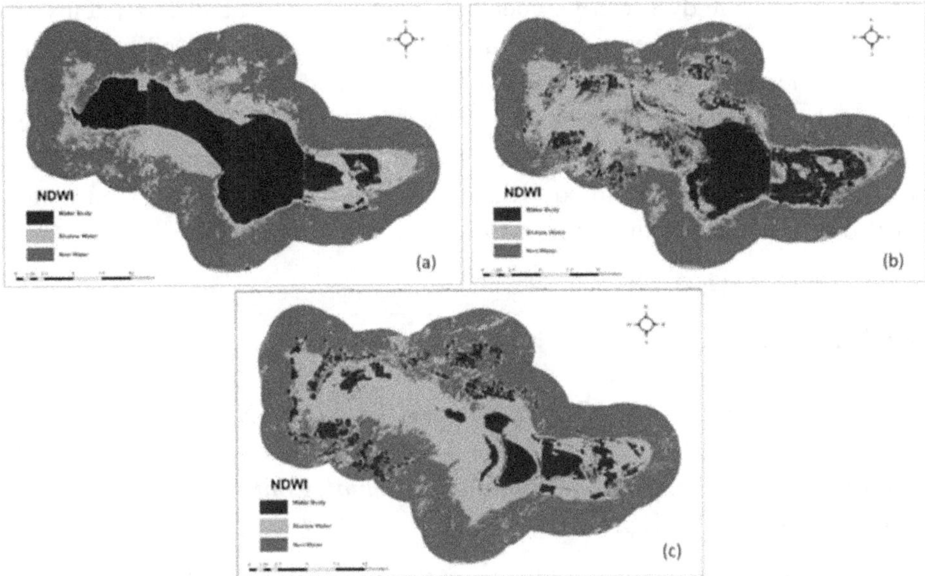

CONCLUSION

With the use of spatial datasets numerous facets of ecosystems are unveiled.

The study area receives water from four major seasonal rivers but construction of 675 illegal check dams and anicuts within 2 km buffer zone have restricted the inflow (Vijay el al., 2016) though 3km buffer zone comes under eco-sensitivity zone. The only construction in the lake area was permissible to Sambhar Salt Ltd. to dig 25 bore wells and 50 surface wells which did not harm wetland (Sangha, 2008). This indicates that increasing rate of salt pans are the responsible factors for drastic changes of this ecosystem. Mushrooming of private saltpans are prominent towards Nawa district in North indicating increase in water demand which ultimately is quenched by the extraction of water from both surface and underground using electrical pumps. Restricting the inflow of water and at the same time it's over extraction have led to acute water stress which is leading towards its conversion from a wetland to wasteland. This lake needs urgent attention for its survival lest it may be under Red List for ecosystems and in Montreux record.

FUTURE DIRECTION

From the current study it can be concluded that Indian wetland ecosystems are under least focus. Four years of the 4th Strategic Plans have passed by but only eight Ramsar sites have been studies out of 27 and inland saline wetlandscapes have been highly neglected. India is blessed to have ten biogeographic zones and each zone has Ramsar sites which need to be studies, driving factors be identified and the degraded wetland s be restored so that India can support the Ramsar Convention and also in Sustainable Development Goals.

REFERENCES

Ansari, A. A., Gill, S. S., Gill, R., Lanza, G. R., & Newman, L. (2016). Phytoremediation: Management of environmental contaminants, volume 4. *Phytoremediation: Management of Environmental Contaminants*, *4*, 1–409. doi:10.1007/978-3-319-41811-7

Baid, I. C. (1968). The Arthropod Fauna of Sambhar Salt Lake, Rajasthan, India. *Oikos Copenhagen*, *19*(2), 292–303. doi:10.2307/3565015

Bajjouk, T., Mouquet, P., Ropert, M., Quod, J. P., Hoarau, L., Bigot, L., Populus, J. (2019). Detection of changes in shallow coral reefs status: Towards a spatial approach using hyperspectral and multispectral data. *Ecological Indicators, 96*, 174–191. doi:10.1016/j.ecolind.2018.08.052

Bansal, S., Katyal, D., Saluja, R., Chakraborty, M., & Garg, J. K. (2018). Geoinformation Remotely sensed MODIS wetland components for assessing the variability of methane emissions in Indian tropical / subtropical wetlands. *Int J Appl Earth Obs Geoinformation*, *64*, 156–170. doi:10.1016/j.jag.2017.08.011

Bhattacharya, B. D., Nayak, D. C., Sarkar, S. K., Biswas, S. N., Rakshit, D., & Ahmed, M. K. (2015). Distribution of dissolved trace metals in coastal regions of Indian Sundarban mangrove wetland: A multivariate approach. *Journal of Cleaner Production*, *96*, 233–243. doi:10.1016/j.jclepro.2014.04.030

Bi, X., Lin, S., Zhu, S., Yin, H., Li, Z., & Chen, Z. (2019). Species identification and survival competition analysis of microalgae via hyperspectral microscopic images. *Optik (Stuttgart)*, *176*, 191–197. doi:10.1016/j.ijleo.2018.09.077

Biswas, P., Kohli, M. P. S., Chadha, N. K., Bhattacharjya, B. K., Debnath, D., Yengkokpam, S., & Sharma, A. P. (2015). Optimizing Stocking Density of Labeo rohita Fry in Cage Aquaculture System as a Tool for Floodplain Wetland Fisheries Management. *Proceedings of the National Academy of Sciences. India. Section B, Biological Sciences*, *85*(1), 181–190. doi:10.100740011-014-0343-6

Chabrillat, S., Milewski, R., Schmid, T., Rodriguez, M., Escribano, P., Palacios-orueta, A., & De Silvopascicultura, D. (2014). Potential of hyperspectral imagery for the spatial assessment of soil erosion stages in agricultural semi-arid Spain at different German Research Centre for Geosciences, Telegrafenberg. Potsdam, Germany: EEZA-CSIC, Departmento de Desertificación y Geoecología.

Chatterjee, A., Adhikari, S., & Mukhopadhyay, S. K. (2017). Effects of Waterbird Colonization on Limnochemical Features of a Natural Wetland on Buxa Tiger Reserve, India, During Wintering Period. *Wetlands*, *37*(1), 177–190. doi:10.100713157-016-0851-7

Chatterjee, A., Adhikari, S., Pal, S., & Mukhopadhyay, S. K. (2020). Foraging guild structure and niche characteristics of waterbirds wintering in selected sub-Himalayan wetlands of India. *Ecological Indicators*, *108*, 105693. doi:10.1016/j.ecolind.2019.105693

Chatterjee, K., Bandyopadhyay, A., Ghosh, A., & Kar, S. (2015). Assessment of environmental factors causing wetland degradation, using Fuzzy Analytic Network Process: A case study on Keoladeo National Park, India. *Ecological Modelling*, *316*, 1–13. doi:10.1016/j.ecolmodel.2015.07.029

Chinnadurai, S., Mohamed, K. S., Sharma, J., Venkatesan, V., & Kripa, V. (2016). Assessment of bio-accumulation of bacteria in oysters from shellfish growing waters in Ashtamudi Lake (Kerala, India): A RAMSAR wetland. *Regional Studies in Marine Science*, *7*, 118–122. doi:10.1016/j.rsma.2016.05.016

Chowdhury, R., Favas, P. J. C., Jonathan, M. P., Venkatachalam, P., Raja, P., & Sarkar, S. K. (2017). Bioremoval of trace metals from rhizosediment by mangrove plants in Indian Sundarban Wetland. *Marine Pollution Bulletin*, *124*(2), 1078–1088. doi:10.1016/j.marpolbul.2017.01.047 PMID:28187966

Cienciala, P., & Pasternack, G. B. (2017). Floodplain inundation response to climate, valley form, and flow regulation on a gravel-bed river in a Mediterranean-climate region. *Geomorphology*, *282*, 1–17. doi:10.1016/j.geomorph.2017.01.006

Davranche, A., Poulin, B., & Lefebvre, G. (2013). Mapping flooding regimes in Camargue wetlands using seasonal multispectral data. *Remote Sensing of Environment*, *138*, 165–171. doi:10.1016/j.rse.2013.07.015

Debanshi, S., & Pal, S. (2020). Wetland delineation simulation and prediction in deltaic landscape. *Ecological Indicators, 108*, 105757. doi:10.1016/j.ecolind.2019.105757

Everard, M., Kangabam, R., Tiwari, M. K., McInnes, R., Kumar, R., Talukdar, G. H., & Das, L. (2019). Ecosystem service assessment of selected wetlands of Kolkata and the Indian Gangetic Delta: Multi-beneficial systems under differentiated management stress. *Wetlands Ecology and Management, 27*(2–3), 405–426. doi:10.100711273-019-09668-1

Fengqin, Y., & Shuwen, Z. (2019). Ecosystem service decline in response to wetland loss in the Sanjiang Plain, Northeast China. *Ecological Engineering, 130*(January), 117–121. doi:10.1016/j.ecoleng.2019.02.009

Fowler, D. N., King, S. L., & Weindorf, D. C. (2014). *Evaluating Abiotic Influences on Soil Salinity of Inland Managed Wetlands and Agricultural Croplands in a Semi-Arid Environment*. Academic Press.

Fu, Y., Zhao, J., Peng, W., Zhu, G., Quan, Z., & Li, C. (2018). Spatial modelling of the regulating function of the Huangqihai Lake wetland ecosystem. *Journal of Hydrology (Amsterdam), 564*, 283–293. doi:10.1016/j.jhydrol.2018.07.017

Geerling, G. W., Labrador-Garcia, M., Clevers, J. G. P. W., Ragas, A. M. J., & Smits, A. J. M. (2007). Classification of floodplain vegetation by data fusion of spectral (CASI) and LiDAR data. *International Journal of Remote Sensing, 28*(19), 4263–4284. doi:10.1080/01431160701241720

Ghermandi, A., Sheela, A. M., & Justus, J. (2016). Integrating similarity analysis and ecosystem service value transfer: Results from a tropical coastal wetland in India. *Ecosystem Services, 22*(April), 73–82. doi:10.1016/j.ecoser.2016.09.014

Ghosh, S., & Das, A. (2019). Urban expansion induced vulnerability assessment of East Kolkata Wetland using Fuzzy MCDM method. *Remote Sensing Applications: Society and Environment, 13*, 191–203. doi:10.1016/j.rsase.2018.10.014

Gogoi, P., Sinha, A., Das Sarkar, S., Chanu, T. N., Yadav, A. K., Koushlesh, S. K., & Das, B. K. (2019). Seasonal influence of physicochemical parameters on phytoplankton diversity and assemblage pattern in Kailash Khal, a tropical wetland, Sundarbans, India. *Applied Water Science, 9*(7), 1–13. doi:10.100713201-019-1034-5

Gorain, P. C., Sengupta, S., Satpati, G. G., Paul, I., Tripathi, S., & Pal, R. (2018). Carbon sequestration in macroalgal mats of brackish-water habitats in Indian Sunderbans: Potential as renewable organic resource. *The Science of the Total Environment, 626*, 689–702. doi:10.1016/j.scitotenv.2018.01.106 PMID:29898555

Gupta, A., & Kaur, H. (2017). A new pathogen, *Myxobolus holzerae (Myxosporea: Myxozoa)* causing severe gill disease in an Indian major carp *Labeo rohita* in a cold-water wetland, Punjab (India). *Microbial Pathogenesis, 111*, 244–251. doi:10.1016/j.micpath.2017.08.044 PMID:28870697

Gupta, A., & Kaur, H. (2017). Morphological, histopathological and molecular characterization of *Thelohanellus theinensis n. sp. (Cnidaria: Myxosporea)* infecting an Indian major carp, Labeo bata in a cold-water wetland in Punjab (India). *Journal of Parasitic Diseases: Official Organ of the Indian Society for Parasitology, 41*(3), 629–638. doi:10.100712639-016-0856-4 PMID:28848251

Gupta, A., & Kaur, H. (2018). *Myxobolus okamurae sp. nov. (Myxosporea: Myxozoa)* causing severe gill myxoboliosis in the cyprinid Labeo bata in a coldwater wetland, Punjab (India). *Microbial Pathogenesis, 115*, 86–92. doi:10.1016/j.micpath.2017.12.041 PMID:29253599

Heintzman, L. J., Starr, S. M., Mulligan, K. R., Barbato, L. S., & McIntyre, N. E. (2017). Using Satellite Imagery to Examine the Relationship between Surface-Water Dynamics of the Salt Lakes of Western Texas and Ogallala Aquifer Depletion. *Wetlands, 37*(6), 1055–1065. doi:10.100713157-017-0940-2

Herbst, D. (2014). From mountains to the saline lakes of the Great Basin: ecosystems at risk. In *From Mountains to the Saline Lakes of the Great Basin: Ecosystems at Risk*. Ecological Society of America. https://www.Ramsar.org/wetland/india

Jahncke, R., Leblon, B., Bush, P., & Larocque, A. (2018). Mapping wetlands in Nova Scotia with multi-beam RADARSAT-2 Polarimetric SAR, optical satellite imagery, and Lidar data. *Int J Appl Earth Obs Geoinformation, 68*, 139–156. doi:10.1016/j.jag.2018.01.012

Jamwal, P. S., Chandan, P., Rattan, R., Anand, A., Kannan, P. M., & Parsons, M. H. (2017). Survey of avifauna of the Gharana wetland reserve: Implications for conservation in a semi-arid agricultural setting on the Indo-Pakistan border. *BMC Zoology, 2*(1), 1–9. doi:10.118640850-017-0016-z

Jayakumar, K., Selvam, V., & Prabavathy, V. R. (2019). A WebGIS-Based Study for Managing Mangroves of Godavari Wetland, Andhra Pradesh, India. *Photonirvachak (Dehra Dun), 47*(12), 2027–2045. doi:10.100712524-019-01046-x

Judd, C., Steinberg, S., Shaughnessy, F., & Crawford, G. (2007). Mapping salt marsh vegetation using aerial hyperspectral imagery and linear unmixing in Humboldt Bay, California mapping salt marsh vegetation using aerial hyperspectral. *BioOne, 27*(4), 1144–1152.

Kalambe, G. N., & Upasani, V. N. (2017). *Isolation and Identification of Haloalkaliphilic Archaeal Isolates from a Soda Lake in India*. doi:10.15680/IJIRSET.2017.0606048

Kaur, H., & Attri, R. (2015). Morphological and molecular characterization of Henneguya bicaudi n. sp. (Myxosporea: Myxobolidae) infecting gills of Cirrhinus mrigala (Ham.) in Harike Wetland, Punjab (India). *Parasitology Research*, *114*(11), 4161–4167. doi:10.100700436-015-4647-0 PMID:26239799

Kaur, H., Attri, R., & Joshi, J. (2016). Molecular identification of a new myxozoan, *Myxobolus dermiscalis n. sp.* (Myxosporea) infecting scales of Labeo rohita Hamilton in Harike Wetland, Punjab (India). *International Journal for Parasitology. Parasites and Wildlife*, *5*(2), 139–144. doi:10.1016/j.ijppaw.2015.10.003 PMID:27330981

Kaur, J., Walia, H., Mabwoga, S. O., & Arora, S. (2017). Water quality monitoring of an international wetland at Harike, Punjab and its impact on biological systems. *Applied Water Science*, *7*(3), 1107–1115. doi:10.100713201-015-0347-2

Kerfoot, W. C., Hobmeier, M. M., Regis, R., Raman, V. K., Brooks, C. N., Shuchman, R., … Reif, M. (2019). Lidar (light detection and ranging) and benthic invertebrate investigations: Migrating tailings threaten Buffalo Reef in Lake Superior. *Journal of Great Lakes Research*. doi:10.1016/j.jglr.2019.07.009

Khadim, F. K., Su, H., Xu, L., & Tian, J. (2019). Soil Salinity Mapping in Everglades National Park Using Remote Sensing Techniques and Vegetation Salt Tolerance. *Physics and Chemistry of the Earth*, *110*, 31–50. Advance online publication. doi:10.1016/j.pce.2019.01.004

Kora, A. J., Rastogi, L., Kumar, S. J., & Jagatap, B. N. (2017). Physico-chemical and bacteriological screening of Hussain Sagar lake: An urban wetland. *Water Science*, *31*(1), 24–33. doi:10.1016/j.wsj.2017.03.003

Krishnakumar, A., Saranya, P., Prasad, P. G., & Rakhi, C. (2015). Holocene Records of Human Driven Geological Impacts in a Ramsar Wetland of India. *Aquatic Procedia*, *4*, 373–380. doi:10.1016/j.aqpro.2015.02.050

Kumar, K. C. V. N., Demudu, G., Malini, B. H., Rao, K. N., & Kubo, S. (2016). Geospatial Analysis of the Changing Environment of Kolleru Lake, the Largest Freshwater Wetland in India. *Wetlands*, *36*(4), 745–758. doi:10.100713157-016-0787-y

Kumar, P., & Sharma, A. (2019). Wetland Birds Assemblages in Man-Made Sacred Ponds of Kurukshetra, India. *Proceedings of the Zoological Society*, *72*(1), 61–73. doi:10.100712595-018-0259-x

Kumar, S., & Devadas, V. (2016). Integrated Planning for Sustainable Development of Kuttanad Wetland Region, Kerala State. *Procedia Technology*, *24*, 1660–1667. doi:10.1016/j.protcy.2016.05.185

Ludwig, C., Walli, A., Schleicher, C., Weichselbaum, J., & Riffler, M. (2019). A highly automated algorithm for wetland detection using multi-temporal optical satellite data. *Remote Sensing of Environment, 224*, 333–351. doi:10.1016/j.rse.2019.01.017

Luna, E., Jouany, C., & Castañeda, C. (2018, February). 2019. Soil composition and plant nutrients at the interface between crops and saline wetlands in arid environments in NE Spain. *Catena, 173*, 384–393. doi:10.1016/j.catena.2018.10.032

Matthews, M. W., Bernard, S., & Winter, K. (2010). Remote sensing of cyanobacteria-dominant algal blooms and water quality parameters in Zeekoevlei, a small hypertrophic lake, using MERIS. *Remote Sensing of Environment, 114*(9), 2070–2087. doi:10.1016/j.rse.2010.04.013

McFeeters, S. K. (1996). The use of the Normalized Difference Water Index (NDWI) in the delineation of open water features. *International Journal of Remote Sensing*, *17*(7), 1425–1432. doi:10.1080/01431169608948714

Meena, D.K., Lianthuamluaia, L., Mishal, P., Swain, H. S., Naskar, B. K., Saha, S., Das, B. K. (2019). Assemblage patterns and community structure of macro-zoobenthos and temporal dynamics of eco-physiological indices of two wetlands, in lower gangetic plains under varying ecological regimes: A tool for wetland management. *Ecological Engineering, 130*, 1–10. doi:10.1016/j.ecoleng.2019.02.002

Middleton, B. A., Van der Valk, A. G., & Davis, C. B. (2015). Responses to water depth and clipping of twenty-three plant species in an Indian monsoonal wetland. *Aquatic Botany*, *126*, 38–47. doi:10.1016/j.aquabot.2015.06.004

Milja, T. E., Prathish, K. P., & Rao, T. P. (2011). Synthesis of surface imprinted nanospheres for selective removal of uranium from simulants of Sambhar Salt Lake and ground water. *Journal of Hazardous Materials*, *188*(1-3), 384–390. doi:10.1016/j.jhazmat.2011.01.121 PMID:21345587

Mondal, B., Dolui, G., Pramanik, M., Maity, S., & Sarathi, S. (2017). Urban expansion and wetland shrinkage estimation using a GIS-based model in the East Kolkata Wetland, India. *Ecological Indicators*, *83*(July), 62–73. doi:10.1016/j.ecolind.2017.07.037

Moore, J. N. (2016). Recent desiccation of Western Great Basin saline lakes: Lessons from Lake Abert, Oregon, USA. *The Science of the Total Environment*, *554*, 142–154. doi:10.1016/j.scitotenv.2016.02.161 PMID:26950628

Murray, N. J., Phinn, S. R., Clemens, R. S., Roelfsema, C. M., & Fuller, R. A. (2012). Continental scale mapping of tidal flats across East Asia using the Landsat archive. *Remote Sensing*, *4*(11), 3417–3426. doi:10.3390/rs4113417

Nohro, S., & Jayakumar, S. (2020). Tree species diversity and composition of the Pala Wetland Reserve Forest, Mizoram, Indo-Burma hotspot, India. *Biocatalysis and Agricultural Biotechnology, 23*, 101474. doi:10.1016/j.bcab.2019.101474

O'Neil, G. L., Saby, L., Band, L. E., & Goodall, J. L. (2019). Effects of LiDAR DEM Smoothing and Conditioning Techniques on a Topography-Based Wetland Identification Model. *Water Resources Research*, *55*(5), 4343–4363. doi:10.1029/2019WR024784

O'Neill, A. R. (2019). Evaluating high-altitude Ramsar wetlands in the Eastern Himalayas. *Global Ecology and Conservation*, *20*, e00715. doi:10.1016/j.gecco.2019.e00715

Pal, S., & Saha, T. K. (2018). Identifying dam-induced wetland changes using an inundation frequency approach: The case of the Atreyee River basin of Indo-Bangladesh. *Ecohydrology & Hydrobiology*, *18*(1), 66–81. doi:10.1016/j.ecohyd.2017.11.001

Pal, S., & Talukdar, S. (2018). Drivers of vulnerability to wetlands in Punarbhaba river basin of India- Bangladesh. *Ecological Indicators*, *93*(April), 612–626. doi:10.1016/j.ecolind.2018.05.043

Panigrahy, S., Murthy, T. V. R., Patel, J. G., & Singh, T. S. (2012). Wetlands of India: Inventory and assessment at 1:50,000 scale using geospatial techniques. *Curr. Sci., 102*, 852–856.

Patra, A., Tushar, J., & Dubey, B. (2017). Modeling and simulation of a wetland park: An application to Keoladeo National Park, India. *Mathematics and Computers in Simulation, 134*, 54–78. doi:10.1016/j.matcom.2016.10.001

Pengra, B. W., Johnston, C. A., & Loveland, T. R. (2007). Mapping an invasive plant, *Phragmites australis*, in coastal wetlands using the EO-1 Hyperion hyperspectral sensor. *Remote Sensing of Environment*, *108*(1), 74–81. doi:10.1016/j.rse.2006.11.002

Pozo, K., Sarkar, S. K., Estellano, V. H., Mitra, S., Audi, O., Kukucka, P., & Corsolini, S. (2017). Passive air sampling of persistent organic pollutants (POPs) and emerging compounds in Kolkata megacity and rural mangrove wetland Sundarban in India: An approach to regional monitoring. *Chemosphere*, *168*, 1430–1438. doi:10.1016/j.chemosphere.2016.09.055 PMID:27912912

Qi, H., Paz-Kagan, T., Karnieli, A., Jin, X., & Li, S. (2018). Evaluating calibration methods for predicting soil available nutrients using hyperspectral VNIR data. *Soil and Tillage Research, 175*, 267–275. doi:10.1016/j.still.2017.09.006

Ramachandra, T. V., Sudarshan, P. B., Mahesh, M. K., & Vinay, S. (2018). Spatial patterns of heavy metal accumulation in sediments and macrophytes of Bellandur wetland, Bangalore. *Journal of Environmental Management, 206*, 1204–1210. doi:10.1016/j.jenvman.2017.10.014 PMID:29157887

Rapinel, S., Hubert-moy, L., & Clément, B. (2015). Combined use of LiDAR data and multispectral earth observation imagery for wetland habitat mapping. *International Journal of Applied Earth Observation and Geoinformation, 37*, 56–64. doi:10.1016/j.jag.2014.09.002

Rodríguez-Iruretagoiena, A., Chowdhury, R., Gredilla, A., deVallejuelo, S. F. O., de Diego, A., Sarkar, S. K., & Venkatachalam, P. (2016). Uptake and Distribution of Trace Elements in Dominant Mangrove Plants of the Indian Sundarban Wetland. *Bulletin of Environmental Contamination and Toxicology, 97*(5), 721–727. doi:10.100700128-016-1919-z PMID:27650375

Roy, M. B., Roy, P. K., Samal, N. R., & Mazumdar, A. (2016). A Comparative Study of Labour Participation in Three Wetland Regions of West Bengal, India. *Journal of The Institution of Engineers (India): Series A, 97*(2), 81–91. doi:10.100740030-016-0150-6

Sahay, H., Mahfooz, S., & Singh, A. K. (2012). *Exploration and characterization of agriculturally and industrially important haloalkaliphilic bacteria from environmental samples of hypersaline Sambhar lake*. doi:10.100711274-012-1131-1

Sahoo, A. K., Pan, M., Troy, T. J., Vinukollu, R. K., Sheffield, J., & Wood, E. F. (2011). Reconciling the global terrestrial water budget using satellite remote sensing. *Remote Sensing of Environment, 115*(8), 1850–1865. doi:10.1016/j.rse.2011.03.009

Sajinkumar, K. S., Revathy, A., & Rani, V. R. (2017). Hydrogeochemistry and spatio-temporal changes of a tropical coastal wetland system: Veli-Akkulam Lake, Thiruvananthapuram, India. *Applied Water Science, 7*(3), 1521–1534. doi:10.100713201-015-0333-8

Saluja, R., Prasad, S., & Garg, J. K. (2018). Field spectroradiometry for discrimination of wetland components: A case study of a tropical inland wetland in India. *Wetlands Ecology and Management, 26*(5), 915–930. doi:10.100711273-018-9620-0

Sarkar, S., Parihar, S. M., & Dutta, A. (2016). Fuzzy risk assessment modelling of East Kolkata Wetland Area: A remote sensing and GIS based approach. *Environmental Modelling & Software, 75*, 105–118. doi:10.1016/j.envsoft.2015.10.003

Sharma, S., Kaur, I., & Nagpal, A. K. (2018). Estimation of arsenic, manganese and iron in mustard seeds, maize grains, groundwater and associated human health risks in Ropar wetland, Punjab, India, and its adjoining areas. *Environmental Monitoring and Assessment, 190*(7), 385. Advance online publication. doi:10.100710661-018-6763-7 PMID:29884925

Sharma, S., Nagpal, A. K., & Kaur, I. (2019). Appraisal of heavy metal contents in groundwater and associated health hazards posed to human population of Ropar wetland, Punjab, India and its environs. *Chemosphere, 227*, 179–190. doi:10.1016/j.chemosphere.2019.04.009 PMID:30986600

Sinclair, M., Ghermandi, A., & Sheela, A. M. (2018). A crowdsourced valuation of recreational ecosystem services using social media data: An application to a tropical wetland in India. *The Science of the Total Environment, 642*, 356–365. doi:10.1016/j.scitotenv.2018.06.056 PMID:29906727

Singh, A. K., Sathya, M., Verma, S., & Jayakumar, S. (2018). Health risk assessment of heavy metals in crop grains grown on open soils of Kanwar wetland, India. *Euro-Mediterranean Journal for Environmental Integration, 3*(1), 29. Advance online publication. doi:10.100741207-018-0073-x

Singh, M., & Sinha, R. (2019). Evaluating dynamic hydrological connectivity of a floodplain wetland in North Bihar, India using geostatistical methods. *The Science of the Total Environment, 651*, 2473–2488. doi:10.1016/j.scitotenv.2018.10.139 PMID:30336437

Singh, M., & Sinha, R. (2020). Distribution, diversity, and geomorphic evolution of floodplain wetlands and wetland complexes in the Ganga plains of north Bihar, India. *Geomorphology, 351*, 106960. doi:10.1016/j.geomorph.2019.106960

Singh, P., Javed, S., Shashtri, S., Singh, R. P., Vishwakarma, C. A., & Mukherjee, S. (2017). Influence of changes in watershed landuse pattern on the wetland of Sultanpur National Park, Haryana using remote sensing techniques and hydrochemical analysis. *Remote Sensing Applications: Society and Environment, 7*, 84–92. doi:10.1016/j.rsase.2017.07.002

Stratoulias, D., Balzter, H., Zlinszky, A., & Tóth, V. R. (2015). Assessment of ecophysiology of lake shore reed vegetation based on chlorophyll fluorescence, field spectroscopy and hyperspectral airborne imagery. *Remote Sensing of Environment, 157*, 72–84. doi:10.1016/j.rse.2014.05.021

Sun, T., Lin, W., Chen, G., Guo, P., & Zeng, Y. (2016). Wetland ecosystem health assessment through integrating remote sensing and inventory data with an assessment model for the Hangzhou Bay, China. *The Science of the Total Environment, 566–567*, 627–640. doi:10.1016/j.scitotenv.2016.05.028 PMID:27236628

Thiemann, S., & Kaufmann, H. (2002). Lake water quality monitoring using hyperspectral airborne data – a semi empirical multisensory and multi-temporal approach for the Ecklenburg Lake District, Germany. *Remote Sensing of Environment, 81*(2-3), 228–237. doi:10.1016/S0034-4257(01)00345-5

Timm, B. C., & McGarigal, K. (2012). Fine-scale remotely-sensed cover mapping of coastal dune and salt marsh ecosystems at Cape Cod National Seashore using Random Forests. *Remote Sensing of Environment, 127*, 106–117. doi:10.1016/j.rse.2012.08.033

Tsai, J. S., Venne, L. S., McMurry, S. T., & Smith, L. M. (2010). Vegetation and land use impact on water loss rate in playas of the Southern High Plains, USA. *Wetlands, 30*(6), 1107–1116. doi:10.100713157-010-0117-8

Turpie, K. R., Klemas, V. V., Byrd, K., Kelly, M., & Jo, Y. H. (2015). Prospective HyspIRI global observations of tidal wetlands. *Remote Sensing of Environment, 167*, 206–217. doi:10.1016/j.rse.2015.05.008

Venne, L. S., Tsai, J. S., Cox, S. B., Smith, L. M., & McMurry, S. T. (2012). Amphibian community richness in cropland and grassland playas in the Southern High Plains, USA. *Wetlands, 32*(4), 619–629. doi:10.100713157-012-0305-9

Vibhute, A. D., Kale, K. V., Dhumal, R. K., & Mehrotra, S. (2015). Soil Type Classification and Mapping using Hyperspectral Remote Sensing Data. *International Conference on Man and Machine Interfacing (MAMI), 1*. 10.1109/MAMI.2015.7456607

Vijay, R., Pinto, S.M., Kushwaha, V.K., Pal, S., & Nandy, N. (2016). A multi-temporal analysis for change assessment and estimation of algal bloom in Sambhar Lake, Rajasthan. *Environmental Monitoring and Assessment, 188*(510), 1–10. doi:10.100710661-016-5509-7

Ward, R. D., Burnside, N. G., Joyce, C. B., & Sepp, K. (2013). The use of medium point density LiDAR elevation data to determine plant community types in Baltic coastal wetlands. *Ecological Indicators, 33*, 96–104. doi:10.1016/j.ecolind.2012.08.016

Wigmore, O., Mark, B., Mckenzie, J., Baraer, M., Lautz, L., & Blanca, C. (2019). Sub-metre mapping of surface soil moisture in proglacial valleys of the tropical Andes using a multispectral unmanned aerial vehicle. *Remote Sensing of Environment, 222*, 104–118. doi:10.1016/j.rse.2018.12.024

Wurtsbaugh, W. A., Miller, C., Null, S. E., Derose, R. J., Wilcock, P., Hahnenberger, M., & Moore, J. (2017). Decline of the world' s saline lakes. *Nature Geoscience*, (October), 1–8. doi:10.1038/NGEO3052

Yanmin, Y., Youqi, C., Yingbin, H. P., & Na, W. (2010). Soil Moisture Monitoring Using Hyper-Spectral Remote Sensing Technology. *International Conference on Geoscience and Remote Sensing*, 373–376. 10.1109/IITA-GRS.2010.5604219

Chapter 14
Geomorphic Changes Related to Anthropogenic Interference Along the Ganga River From Rishikesh to Haridwar, Uttarakhand, India:
Geomorphic Response of the Ganga to Tehri Dam

Manish Pandey
https://orcid.org/0000-0001-8291-2043
University Center for Research and Development (UCRD), Chandigarh University, Mohali, India

Aman Arora
Department of Geography, Faculty of Natural Sciences, Jamia Millia Islamia, New Delhi, India

Rajesh Kumar
Center for the Study of Regional Development, Jawaharlal Nehru University, India

Vijendra Kumar Pandey
Department of Geography, Kirori Mal College, University of Delhi, New Delhi, India

Akshay Kumar
https://orcid.org/0000-0002-4217-4944
Department of Remote Sensing, Birla Institute of Technology, Mesra, India

DOI: 10.4018/978-1-7998-5027-4.ch014

ABSTRACT

The authors, in this study, have quantified the changes in the geomorphic activity of the Ganga River between Rishikesh and Haridwar by using Survey of India (SOI) 1:5000 scale topographic sheets, satellite data, digital elevation model (DEM), and hydrological data from published resources. They have also correlated the change in settlement area in Haridwar and Rishikesh and connected it to the changing geomorphic features in the downstream reach of the Tehri dam site. The study finds that the change in geomorphology of the river in the studied reach of the Ganga River is attributable to reduced water and sediment supply. The study area being in the active tectonic zone, bound between MCT in the north and HFT in the south, may have experienced offsetting of the reduction in sediment supply by dam in the post dam era (after 2006) due to massive earthquakes delivering sediment directly delivered to streams through enhanced landslide and other mass wasting processes.

INTRODUCTION

Topography and landforms act as the memory of factors that sculpted them. Understanding and quantifying the linkage between landforms and their control factors is and has always been of utmost importance to geomorphologists. Climate, tectonics, base level change (local as well as local), anthropogenic changes and biotic controls are the chief factors of denudation responsible for constructing, modifying, and destructing landforms (Pandey, 2014; Aman *et al.*, 2018; Arora *et al.*, 2019; Pandey, *et al.*, 2020). Changes in climate, tectonics and base level and their effects on topographic forms and the processes thereof have got comprehensive research exposure to earth science community since long but the biotic controls and their role in construction/modification landform by changing the rate of water and sediment flux rate including hydrogeomorphic response have started a little late. Biotic controls include both plants and animals. From among the animal kingdom, humans have lately been identified as one of the intensely active controlling factors of topography at different scales (Robock and Liu, 1994; Gedney *et al.*, 2006; Braje and Erlandson, 2013; Anthony *et al.*, 2014; BalaSundareshwaran *et al.*, 2019; Gomez *et al.*, 2019; Wilkinson and McElroy, 2007).

The construction of dams across the river courses result in change of the water and sediment supply regimes downstream of these structures (Ranasinghe *et al.*, 2019; Stähly *et al.*, 2019; Warrick *et al.*, 2019; Binh *et al.*, 2020; Brenna *et al.*, 2020; De Rego *et al.*, 2020). Dams also change the ecological balance in the upstream and downstream reaches (Kingsford, 2000; BEDNAREK, 2001; Bednarek and Hart, 2005; Richter and Thomas, 2007; Maavara *et al.*, 2020). Overall, the imbalance between demand and supply of water and sediment alter the ecosystem downstream of the dam by the changing the trend and pattern of aggradation and degradation. Wohl, (2011), based on rivers in a variety of topographic and environmental settings, suggest the metrics and 'form and process' combinations that differentiate rivers pre-and-post extensive human intervention stage from the start of the industrial era. Three metrics advised by Schmidt and Wilcock, (2008) chiefly represent 'the drivers of channel change downstream from dams': 1) ratio between sediment supply and transport capacity; 2) a Shields number; 3) a ratio of post-dam to pre-dam flood discharge. Merritt and Cooper, (2000), based on their study's results, suggest changes in the channel planform and vegetation composition and canopy and compared them with the conditions before river regulation process started on the studies river reaches. There are other geometric indices measuring channel morphometric properties e.g. active channel belt, floodplain width, sinuosity, braiding indices, etc. are suggested by fluvial scientists as the measure of river regulation with dam construction.

Land use such as croplands, plantations, urban, pastures have expanded at an alarming rate in the last century because of growing population demand (Foley, 2005). Expansion of agricultural activity and timber extraction, at the expense of forested lands, has led to loss of ~7-11 km^2 during the last 300 years (Foley, 2005). These changes in different land use types in varying direction have complex response when it comes to the water and sediment transport and yield at watershed scale (Liu *et al.*, 2008).

Land use/ land cover (LULC hereafter) changes alter water and sediment flow pathways and quantity in many ways (Daniels, 2008; Schilling *et al.*, 2010). Increase in impervious surfaces such as concrete structures and buildings for different purposes lead to increase in the peak discharge of streams in the downstream reaches (Houben *et al.*, 2006; Nasta *et al.*, 2017). As urbanization expands, it leads to alteration in the processes like water infiltration and evapotranspiration, ultimately causing changes in the peak water discharge and sedimentation as well as soil erosion (Olley and Wasson, 2003; Ragab *et al.*, 2003). Redfern *et al.*, (2016) report that the common urban surfaces communicate with water flow in complex manner resulting into greater rates of infiltration than often reported and assumed by previous studies till date, and contradict "a commonly adopted, but simplified classification of the hydrological properties of urban surfaces". Giri *et al.*, (2019) suggests, based on coupled climate-land use change modelling exercise, that the expansion of urbanization would cause lower streamñow.

The Bhagirathi River, a tributary which, after its confluence with the Alaknanda River at Deoprayag in Uttarakhand, becomes the Ganga River, was regulated by a dam at Tehri. The dam construction, which started in 1978 and was opened for operational activities in 2006, is one of the highest dams in Asia (height: 260.5 m). No study till date has assessed the impact of regulation of the flow due to Tehri dam on the geomorphic and hydrological characteristics of the Bhagirathi River downstream of the dam. In this work, we have attempted to fill this research gap by assessing the effects of Tehri dam on the channel morphology, peak discharge, sinuosity, and braiding index of the Ganga river channel between Rishikesh and Haridwar between 1964-67 and 2019. Global dam database generated and used by Lehner et al., (2011b, 2011a); survey of India (SOI) topographic sheets (1964-67), Landsat TM (30 m resolution) have been utilized for extracting the geomorphic features of the pre-dam and post-dam periods and for calculation of different indices to show the activity of neotectonic activity SRTM DEM of 1-arc second has been used. Active structural features of the Himalayan Orogen used by Taylor and Yin, (2009) and other studies have been used in this study to locate and infer the tectonic/neotectonic activity in the study area. The cross-sectional passing by of the HFT and the anticline noted in the structural features which are considered to be still active structures in the Himalaya are influencing the sediment flux in the study area.

Inset A pinpoints the location of the study area expanded in B with reference to the surrounding state boundaries. Sections C, D, and E are composite showing the studied reach of the Ganga river between Rishikesh and Haridwar overlaid over the Sentinel 2 imageries in natural color composite, False color infrared FCC and Atmospheric penetration FCC respectively.

The main objective of this study is to investigate how changes in human interference in terms of dam construction and human settlements encroachments in the upper reaches have brought about changing river behaviour in the geomorphic contexts in the Ganga river valley between Rishikesh and Haridwar.

Table 1. Detailed data specifications used in this study

Data type	Spatial reference for all	Year	Remarks
SOI Toposheets	Projection: UTM Datum: WGS1984 Spheroid: WGS1984 UTM Zone: 44N	1964-67	Used for Geomorphological mapping
Landsat satellite imagery		2011	Used for Geomorphological mapping
SRTM DEM		2000	For visualization
Global Human Settlement (GHS)		1975-2014	Settlement area change detection
Global Reservoir and Dam (GRanD) database		2011	For acquiring dam related information about Tehri dam

Figure 1. Location map of the study area

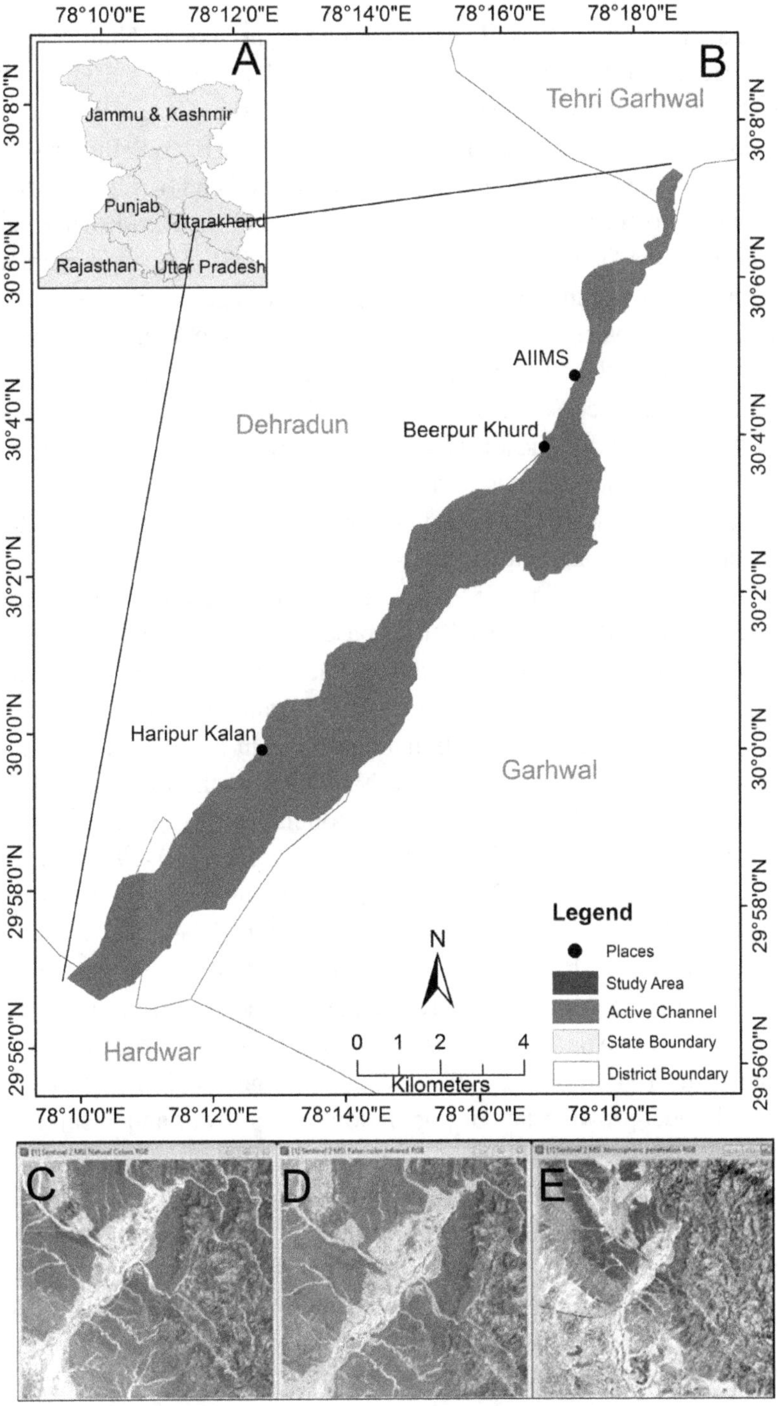

Study Area Description

The studied reach of the Ganga River is located in the state of Uttarakhand, India. It extends from latitude 29^0 56'N to 30^0 18'N and longitude 78^0 10'E to 78^0 18'E. Starting from Rishikesh downstream to Haridwar, this Ganga River channel is bound between main boundary thrust (MBT) in the north and Himalayan Frontal Thrust (HFT) in the south (Figure 1). This channel stretch forms part of the Ganga River Valley zero (GRV-0; stretching from Gaumukh to Haridwar) zone as per classification of the Ganga River by Singh *et al.*, (2007). The main characteristics of the Ganga river channel in this zone are: the average river slope of 1280 cm/km, 5-6 levels of unpaired or rarely paired terraces, most of them occurring around 50m to 250m above the river bed level, and much of the sediment supply comes from mass wasting processes. Sinha *et al.*, (2017) have put this stretch of the Ganga between Rishikesh and Haridwar Class-II category wherein the river experiences transition from bedrock reaches to alluvial reaches. The regional landscape falls under the Sub-Himalayan range zone with sandstone, shale, boulder-conglomerate, and siliciclastic piedmont facies composites.

The first phase of 260.5 m (855 ft.) high and 575 m (1,886 ft.) long, with 20 m (66 ft.) crest width and 1,128 m (3,701 ft.) base width, rock Tehri dam construction was finished in 2006 and it was opened the same year (Adhikari, 1970; Fink, 2000; Sheet, 2003; Prasad *et al.*, 2007). It caused LULC alteration in the area leading to changes in the changes in the gradational behaviour in the downstream reaches.

The population in the towns like Rishikesh and Haridwar have changed substantially during the period of study. For instance, the population of Haridwar, as per census of India 2011, was recorded to be 1890422. The decade 1901-1911 has registered a population growth rate of 12.05% for Haridwar municipality. It further felt a declining trend during 1951-61 with 4.57% growth rate. Due to the development of Bharat Heavy Electrical Limited (BHEL) during the 1964-65, the growth rate hiked to 45.71% during that decade 1961-71. Rishikesh, under the Hardwar Development Authority (HDA), also had the similar experience of population growth rate during the decades after 1941. It was 26.29% during 1941 and hiked to 49.30% during the 1951. The decadal growth rates of population for Haridwar and Rishikesh for the decade ending 2001 were 16.0 and 33.83 percentage respectively (Census of India, 2001). In the 2011, Rishikesh has population of 102469 (Census of India, 2011).

The growing population is being accommodated in the settlement areas causing impermeable surfaces which ultimately lead to increased surface flow and decreased infiltration of water in the groundwater reservoirs. The cumulative area of settlement in the cities of Rishikesh and Haridwar has increased approximately threefold since 1975-2014 (Figure 2, Table 2)

Figure 2. Changes in Geomorphic features along the studied reach of the Ganga River between Rishikesh and Haridwar

Part A and B are geomorphological features of the two times periods 1964-67 and 2011. Inset in A and B are zoomed in for greater clarity and visibility in C and D for the respective period geomorphic maps.

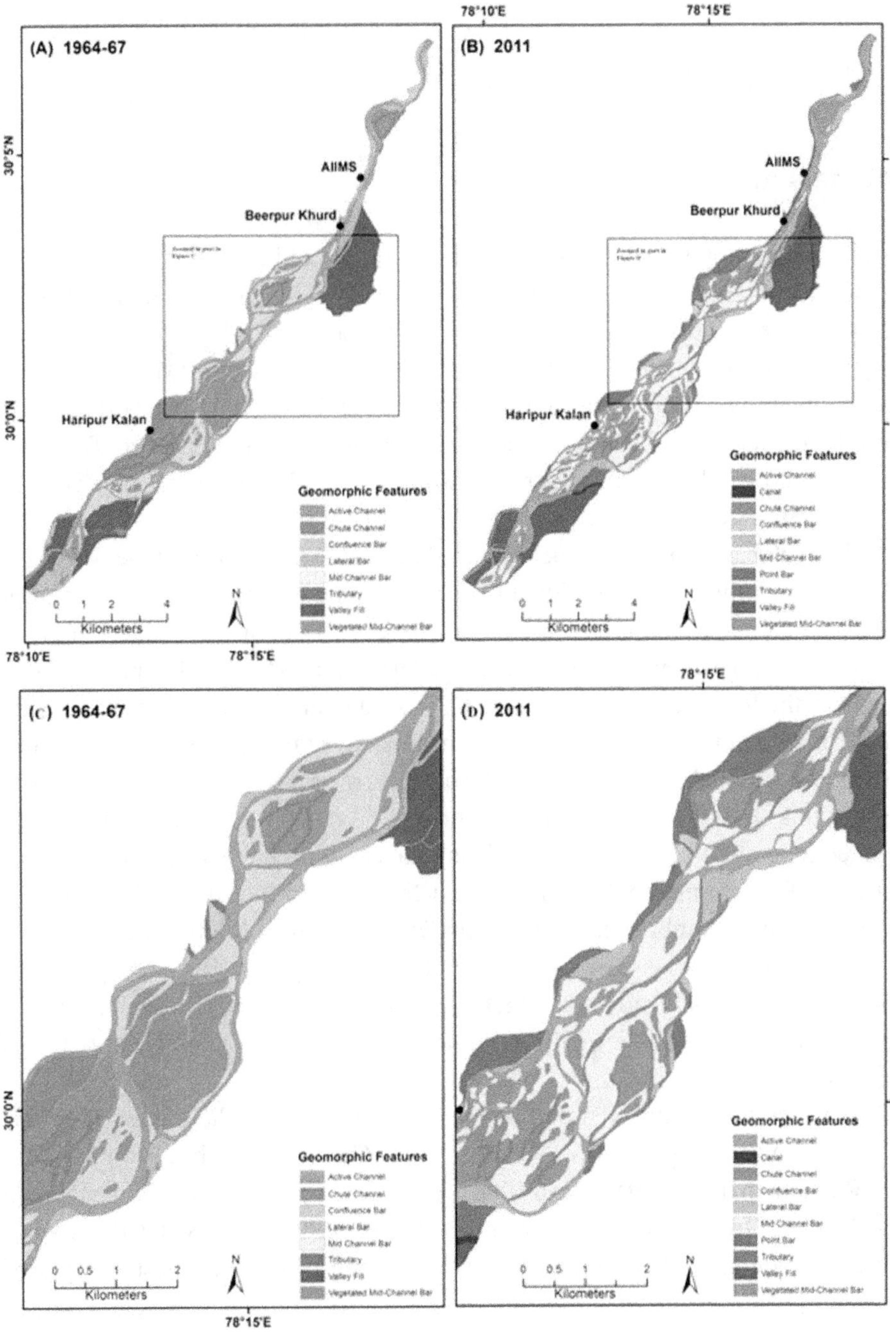

Table 2. Growth of settlement area for the period 1975 to 2014

<table>
<tr><th>Year</th><th>Cells</th><th>Area (km²)</th><th>Change (km²)</th></tr>
<tr><td>1975</td><td>26387</td><td>38.54</td><td rowspan="2">81.78</td></tr>
<tr><td>2014</td><td>82374</td><td>120.32</td></tr>
<tr><td colspan="4">Growth of settlement coverage from 1975 to 2014: 212.17%</td></tr>
</table>

Materials and Methods

Survey of India topographical sheets 53J/4, 53J/G8, and 53K/1 at 1: 50,000 have been used for extraction of geomorphic features of the Ganga channel for the period 1964-67. For extraction of geomorphology for the period 2011, Google Earth Pro® imagery has been onscreen digitized at 1:5000. SRTM 30m DEM data has been used in this study. Dam data layer downloaded from the Global Reservoir and Dam (GRanD) database has been used for extracting information related to Tehri dam and its reservoir. The detailed characteristics of the GRanD database is discussed in the technical documentation by Lehner *et al.*, (2011b).

GHSL (global human settlement layer) dataset downloaded from the web portal of European Commission for Global Human Settlement (https://ghsl.jrc.ec.europa.eu/) which has been computed using supervised classification method from Landsat satellite series, has been used in this study. The Global Human Settlement (GHS) team from the Joint Research Centre (JRC), Institute for the Protection and Security of the Citizen, European Commission aims to analyze the spatial extent of human settlements at global level to monitor the changes over period of time. The datasets used for this study have a spatial resolution of 38 meters. The population grid is prepared by combining the GHSL built-up dataset with the census data collected by Center for International Earth Science Information Network (CIESIN), The Earth Institute, Columbia University (https://www.ciesin.columbia.edu/data/hrsl/). Further, the accuracy assessment is also completed by the JRC agency, and the kappa coefficient analyzed 0.84, which shows the data is quite usable for different application purposes. Table 1 enlists detailed characteristics of all the datasets used in this study.

The water level data recorded at Haridwar gauge and discharge (G/D) and Garhmukteshwar (acquired from Central Water Commission) has been compiled from Raghunath, (2006, 2009) and Kamal *et al.*, (2017) and computed deviation of peak discharge from mean discharge during the period has been used in this study to investigate the changing hydrological behavior during pre-dam and post-dam period.

Figure 3. Changes in the settlement areas during the period 1975-2014

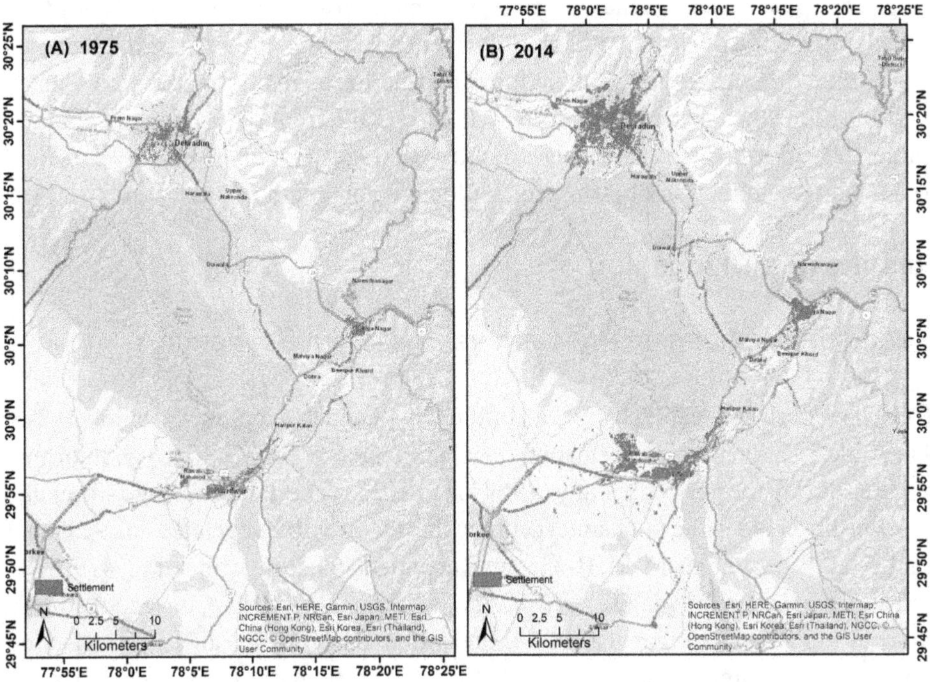

Table 3. Geomorphic changes during the period of study

S.No.	Geomorphic Features	Area in Hectare (1964-67)	Area in Hectare (2011)	Changes (%)
1	Active Channel	676.06	788.99	14.31
2	Vegetated Mid-Channel Bar	733.45	485.19	-51.17
3	Lateral Bar	381.54	252.52	-51.10
4	Chute Channel	248.66	151.35	-64.29
5	Mid-Channel Bar	651.56	777.35	16.18
6	Confluence Bar	29.72	12.07	-146.26
7	Valley Fill	804.56	756.64	-6.33
8	Canal	0.00	33.15	100.00
9	Point Bar	0.00	394.45	100.00
10	Tributary	8.03	1.52	-429.25

The following steps have been taken to gain an understanding about changes in the geomorphic features of the Ganga river channel between Rishikesh and Haridwar due to obstruction of flow of Bhagirathi River at Tehri and growing settlement area to accommodate the growing population: 1) extraction of geomorphic features, listed in Table 3, for the period between 1964-67 and 2011 from the datasets detailed in Table 1; 2) built-up areas have been quantified using Landsat 5TM time series data to compare the direction of change of settlement and geomorphic features.

Results and Discussion

Geomorphic Changes: Pre-Dam vs Post-Dam Scenario

There have been noticeable changes in different in-channel (feature serial numbers 2-6, 9 in Table 3) and out-channel (feature serial numbers 7,8,10 in Table 3) geomorphic features during the period of study (1964-67 to 2011). The areal coverage shrinkage has been observed in the in-channel geomorphic features like vegetated-mid-channel bar (-51.17%), lateral bar (-51.10%), chute channel (-64.29%), tributary (-429.25%) and confluence bar (-146.26%) whereas active channel (14.31%), unvegetated-mid channel bar (16.18%), canal (100%), and point bar (100%), all have registered expansion in the area as seen in Figure 2 (Table 3). The massive reduction in different channel bars like mid-channel bar, confluence bar, tributary, chute channel and lateral bars etc. have been caused by massive reduction in supply of sediment curbed by construction of Tehri dam as the general river morphology behavioral change is regulated by several factors including sediment supply, water discharge, alteration of surface in the river valley by LULC metamorphosis, tectonic perturbation, sea-level variations, etc. (Galay, 1983; Hickin, 1984; Surian and Rinaldi, 2003; Keesstra *et al.*, 2005). In the studied reach of the Ganga River, there has been construction of canals after the Operational opening of the Tehri barrage in 2006 as one of the primary purposes of the dam was to construct network of canals for irrigation to enhance agricultural productivity. And hence, there is 100% increase in canal area. The 100% increase in point bars is attributable to channel morphosis of some of the reaches due to decreased water and sediment supply caused by flow regulation.

The channel planform geometry remain to be of braided channel style even after the construction of Tehri dam, causing flow regulation and sediment supply reduction, but the length of braid channels and their braiding index have changed (Figure 2 A, B, C & D).

Since the beginning of construction of Tehri dam in 1978, the hydrological response of the Ganga has changed, as evident in the peak flood discharge deviation from mean at Haridwar and Garhmukteshwar G/D sites (Shukla, 2014; Kamal *et al.*, 2017). Figure 2 in the paper by Kamal *et al.*, (2017) portrays the drastic downward trend in the annual peak discharge value at Haridwar between 1880-1943 and at Garhmukteshwar between 1970-2010. The peak flood discharge at Haridwar site has experienced a plateau top during the period 1943-1975 with only some years like 1963-1966 deviating from the general stable general trend (Figure 4).

Figure 4. Deviation of peak discharge values from long term mean discharge at Haridwar (uppaer and lower left) and Garhmukteshwar (lower right) gauging sites

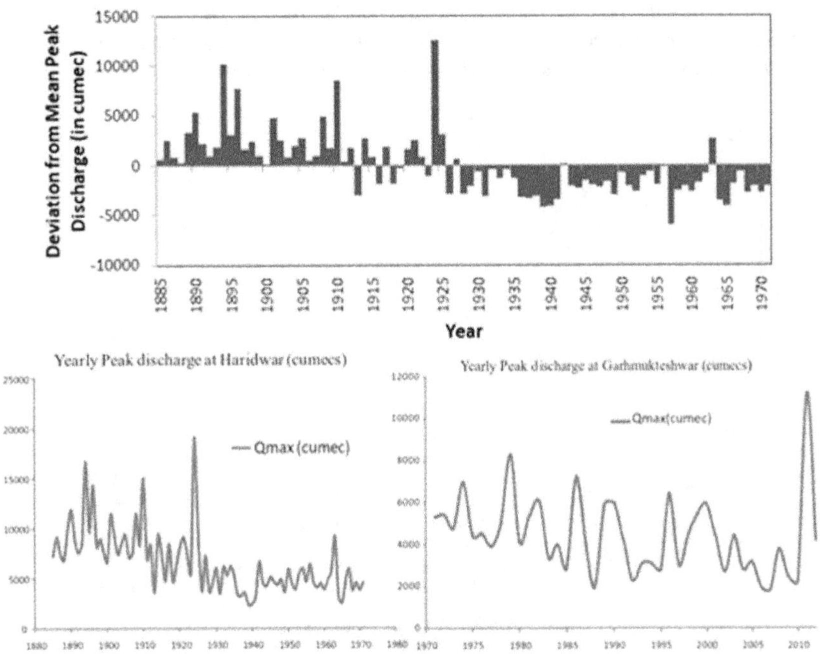

The location of Himalayan Frontal Thrust (HFT) fault system (Figure 5A) at the southern end of the studied channel, Main Central Thrust (MCT) north of the Tehri dam site, dense cluster of earthquake epicentres in the entire area (Figure 5C) and still continuing plate movement of the Indian plate in north eastern direction facilitates the sediment delivery and hence complicates the attempts to signal out effects of dam and settlement encroachment on the sediment supplied to the downstream of the dam site. Studies suggest that, in tectonically active areas, sediment delivery to streams increases by a factor of three to four in the epicentral area even after 5-6 years later of large seismic events (Dadson *et al.*, 2004; Hovius *et al.*, 2011). Following the logic of these two studies by Dadson *et al.*, (2004) and Hovius *et al.*, (2011), the reduction in sediment supply to the studied reach of the Ganga River might be even higher than being evident in the form reduced area of geomorphic features because the increment in sediment supply due to past large earthquakes might have attenuated the lowing of sediment supplied.

Figure 5. Tehri dam vis-à-vis location of Himalayan Frontal Thrust (HFT) fault (A). B, C, and D show distribution of structures, earthquakes epicenters, and plate movement direction of the Indian Plate respectively (after Taylor and Yin, 2009) with study area confined in the yellow box marked with 1 in each.
Part E shows Google Earth Pro® perspective images of the studied reach by a) looking towards the Gangetic plains from mountainous areas (towards southwest), b) zoomed-in part showing valley south of Rishikesh, and c) shows cross-sectional view of the reach (the image is 90⁰ left rotated).

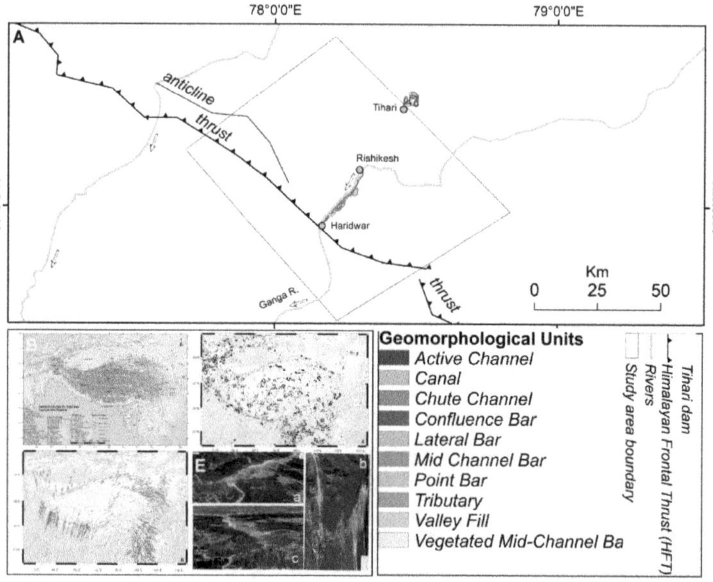

The settlement area changes computed using GHSL dataset reveal the changing surface character of the area. Though other surface alterations might also have affected the sediment quantities delivered to the channel, this study has attempted to analyze the settlement area changes only and compared it with the trend in water discharge at Hardwar and Rishikesh sites. Figure 3, there are three settlement areas viz — Rishikesh, Haridwar, and Dehradun which prominently appear as the main population clusters. The two cities along the Ganga, Rishikesh and Haridwar, have undergone 212.17% (81.78 km^2) expansion during 1975-2014 (Table 2). The lack of settlement area during some of the years just before the beginning of the Tehri dam and just after opening of the dam is needed to establish clear cut relationship between changing peak discharge and increasing settlement areas. Extensive study of metamorphosis of different land uses and their transitions and transformation from one to another and their effect on hydrological and geomorphological behavior of the river both downstream and upstream is entirely a new dimension and needs an independent study over this topic.

CONCLUSION

The results indicate that the changed water and sediment flow regimes downstream of the dam site, when compared to the pre-dam period, have altered the aggradation and degradation trend and pattern; increased area of bars of different types; change in the braiding index, active floodplain area, channel belt area, and channel width. The neotectonic activity and seismic events appear to attenuate the effects of reduced sediment and water supply in the post-dam period by delivering extra sediment quantity to the streams through mass wasting processes. The hydrological response of the dam construction (as evident in the peak discharge variability seen in pre-dam and post-dam scenarios) and change in human settlement area over the study period, along with tectonoclimatic activities, have been some of the major causes of the moderate channel metamorphosis in the form of areal-change in geomorphic features.

ACKNOWLEDGMENT

The authors are grateful to the Department of Civil Engineering, Chandigarh University for providing lab access for carrying out analyses using geospatial software package ArcMap.

Declaration: The authors declare no conflict of interest about this research work.

REFERENCES

Adhikari, B. R. (1970). Tehri Dam: An Engineering Marvel. *Hydro Nepal: Journal of Water. Energy & Environment*. Advance online publication. doi:10.3126/hn.v5i0.2481

Aman, A., Pandey, M., Singh, A., & Siddiqui, M.A. (2018). Study of Landscape Evolution in North Koel River Basin, Jharkhand, India: Tectonic and Structural Implications Based on Hypsometric Analysis. *Forum Geografic, 17*(2), 111–117. Doi:10.5775/fg.2018.073.d

Anthony, E. J., Marriner, N., & Morhange, C. (2014). Human influence and the changing geomorphology of Mediterranean deltas and coasts over the last 6000 years: From progradation to destruction phase? *Earth-Science Reviews*, *139*, 336–361. Advance online publication. doi:10.1016/j.earscirev.2014.10.003

Arora, A., Pandey, M., Siddiqui, M. A., Hong, H., & Mishra, V. N. (2019). Spatial flood susceptibility prediction in Middle Ganga Plain: Comparison of frequency ratio and Shannon's entropy models. *Geocarto International*, 1–32. doi:10.1080/10106049.2019.1687594

BalaSundareshwaran, A., Abdul Rahaman, S., Balasubramani, K., Kumaraswamy, K., & Ramkumar, M. (2019). Habitat Risk Assessment Along Coastal Tamil Nadu, India—An Integrated Methodology for Mitigating Coastal Hazards. Coastal Zone Management. Doi:10.1016/b978-0-12-814350-6.00023-9

Bednarek, A. T. (2001). Undamming Rivers: A Review of the Ecological Impacts of Dam Removal. *Environmental Management*, *27*(6), 803–814. doi:10.1007002670010189 PMID:11393315

Bednarek, A. T., & Hart, D. D. (2005). Modifying dam operations to restore rivers: Ecological responses to tennessee river dam mitigation. *Ecological Applications*, *15*(3), 997–1008. doi:10.1890/04-0586

Braje, T. J., & Erlandson, J. M. (2013). Looking forward, looking back: Humans, anthropogenic change, and the Anthropocene. *Anthropocene*, *4*, 116–121. Advance online publication. doi:10.1016/j.ancene.2014.05.002

Brenna, A., Surian, N., & Mao, L. (2020). Response of A Gravel - Bed River To Dam Closure: Insights From Sediment Transport Processes And Channel Morphodynamics. *Earth Surface Processes and Landforms*, *45*(3), 756–770. doi:10.1002/esp.4750

Dadson, S. J., Hovius, N., Chen, H., Dade, W. B., Lin, J.-C., Hsu, M.-L., Lin, C.-W., Horng, M.-J., Chen, T.-C., Milliman, J., & Stark, C. P. (2004). Earthquake-triggered increase in sediment delivery from an active mountain belt. *Geology*, *32*(8), 733. doi:10.1130/G20639.1

Daniels, J. M. (2008). Distinguishing allogenic from autogenic causes of bed elevation change in late Quaternary alluvial stratigraphic records. *Geomorphology*, *101*(1–2), 159–171. doi:10.1016/j.geomorph.2008.05.022

De Rego, K., Lauer, J. W., Eaton, B., & Hassan, M. (2020). A decadal-scale numerical model for wandering, cobble-bedded rivers subject to disturbance. *Earth Surface Processes and Landforms*, *45*(4), 912–927. doi:10.1002/esp.4784

Fink, A. K. (2000). *Tehri hydro power complex on the Bhagirathi river in India.* Hydrotechnical Construction., doi:10.1023/A:1004187208788

Foley, J. A. (2005). Global Consequences of Land Use. *Science*, *309*(5734), 570–574. doi:10.1126cience.1111772 PMID:16040698

Galay, V. J. (1983). Causes of river bed degradation. *Water Resources Research*, *19*(5), 1057–1090. doi:10.1029/WR019i005p01057

Gedney, N., Cox, P. M., Betts, R. A., Boucher, O., Huntingford, C., & Stott, P. A. (2006). Detection of a direct carbon dioxide effect in continental river runoff records. *Nature*, *439*(7078), 835–838. doi:10.1038/nature04504 PMID:16482155

Giri, S., Arbab, N. N., & Lathrop, R. G. (2019). Assessing the potential impacts of climate and land use change on water fluxes and sediment transport in a loosely coupled system. *Journal of Hydrology (Amsterdam)*, *577*, 123955. doi:10.1016/j.jhydrol.2019.123955

Gomez, C., Hart, D. E., Wassmer, P., Kenta, I., Matsui, H., & Shimizu, M. (2019). *Coastal Evolution, Geomorphic Processes and Sedimentary Records in the Anthropocene*. Forum Geografi. doi:10.23917/forgeo.v33i1.7551

Hickin, E. J. (1984). Vegetation and river channel dynamics. *The Canadian Geographer. Geographe Canadien*, *28*(2), 111–126. doi:10.1111/j.1541-0064.1984.tb00779.x

Houben, P., Hoffmann, T., Zimmermann, A., & Dikau, R. (2006). Land use and climatic impacts on the Rhine system (RheinLUCIFS): Quantifying sediment fluxes and human impact with available data. *Catena*, *66*(1–2), 42–52. doi:10.1016/j.catena.2005.07.009

Hovius, N., Meunier, P., Lin, C.-W., Chen, H., Chen, Y.-G., Dadson, S., Horng, M.-J., & Lines, M. (2011). Prolonged seismically induced erosion and the mass balance of a large earthquake. *Earth and Planetary Science Letters*, *304*(3–4), 347–355. doi:10.1016/j.epsl.2011.02.005

Kamal, V., Mukherjee, S., Singh, P., Sen, R., Vishwakarma, C. A., Sajadi, P., Asthana, H., & Rena, V. (2017). Flood frequency analysis of Ganga river at Haridwar and Garhmukteshwar. *Applied Water Science*, *7*(4), 1979–1986. doi:10.100713201-016-0378-3

Keesstra, S. D., van Huissteden, J., Vandenberghe, J., Van Dam, O., de Gier, J., & Pleizier, I. D. (2005). Evolution of the morphology of the river Dragonja (SW Slovenia) due to land-use changes. *Geomorphology*, *69*(1–4), 191–207. doi:10.1016/j.geomorph.2005.01.004

Kingsford, R. T. (2000). Ecological impacts of dams, water diversions and river management on floodplain wetlands in Australia. *Austral Ecology*, *25*(2), 109–127. doi:10.1046/j.1442-9993.2000.01036.x

Lehner, B., Liermann, C. R., Revenga, C., Vörösmarty, C., Fekete, B., Crouzet, P., Döll, P., Endejan, M., Frenken, K., & Magome, J. (2011b). Global reservoir and dam (GRanD) database. *Technical Documentation. Version*, *1*, 1–14.

Lehner, B., Liermann, C. R., Revenga, C., Vörösmarty, C., Fekete, B., Crouzet, P., Döll, P., Endejan, M., Frenken, K., Magome, J., Nilsson, C., Robertson, J. C., Rödel, R., Sindorf, N., & Wisser, D. (2011a). High-resolution mapping of the world's reservoirs and dams for sustainable river-flow management. *Frontiers in Ecology and the Environment*, *9*(9), 494–502. doi:10.1890/100125

Lehner, B., Reidy Liermann, C., Revenga, C., Vorosmarty, C., Fekete, B., Crouzet, P., Doll, P., Endejan, M., Frenken, K., & Magome, J. (2011c). Global Reservoir and Dam Database, Version 1 (GRanDv1): Dams. *Revision*, *01*. Advance online publication. doi:10.7927/H4N877QK

Liu, M., Tian, H., Chen, G., Ren, W., Zhang, C., & Liu, J. (2008). Effects of Land-Use and Land-Cover Change on Evapotranspiration and Water Yield in China During 1900-2000 1. *JAWRA Journal of the American Water Resources Association*, *44*(5), 1193–1207. doi:10.1111/j.1752-1688.2008.00243.x

Maavara, T., Chen, Q., Van Meter, K., Brown, L. E., Zhang, J., Ni, J., & Zarfl, C. (2020). River dam impacts on biogeochemical cycling. *Nature Reviews Earth & Environment*, *1*(2), 103–116. doi:10.103843017-019-0019-0

Merritt, D. M., & Cooper, D. J. (2000). Riparian vegetation and channel change in response to river regulation: A comparative study of regulated and unregulated streams in the Green River Basin, USA. *Regulated Rivers: Research and Management*, *16*(6), 543–564. doi:10.1002/1099-1646(200011/12)16:6<543::AID-RRR590>3.0.CO;2-N

Nasta, P., Palladino, M., Ursino, N., Saracino, A., Sommella, A., & Romano, N. (2017). Assessing long-term impact of land-use change on hydrological ecosystem functions in a Mediterranean upland agro-forestry catchment. *The Science of the Total Environment*, *605–606*, 1070–1082. doi:10.1016/j.scitotenv.2017.06.008 PMID:28715855

Olley, J. M., & Wasson, R. J. (2003). Changes in the flux of sediment in the Upper Murrumbidgee catchment, Southeastern Australia, since European settlement. *Hydrological Processes*, *17*(16), 3307–3320. doi:10.1002/hyp.1388

Pandey, M., Ray, Y., & Arora, A. (2020). Structural control on the landscape Evolution of Son Alluvial Fan System in Ganga foreland basin. In Advances in Remote Sensing for Natural Resource Monitoring. Wiley and Sons.

Pandey, M. K. (2014). *Confluence Dynamics of Lower Son River with River Ganga: Tectonic and Climatic Perspective* (Unbpublished Ph.D. Thesis). Banaras Hindu University.

Prasad, G. M., Narang, J. L., & Singhal, S. (2007). *Design aspects of the dam*. Water and Energy International.

Ragab, R., Bromley, J., Rosier, P., Cooper, J. D., & Gash, J. H. C. (2003). Experimental study of water fluxes in a residential area: 1. Rainfall, roof runoff and evaporation: the effect of slope and aspect. *Hydrological Processes*, *17*(12), 2409–2422. doi:10.1002/hyp.1250

Raghunath, H. M. (2006). *Hydrology: principles, analysis and design*. New Age International.

Raghunath, H. M. (2009). *Hydrology: principle analysis and design* (revised edition). New age international publishers.

Ranasinghe, R., Wu, C. S., Conallin, J., Duong, T. M., & Anthony, E. J. (2019). Disentangling the relative impacts of climate change and human activities on fluvial sediment supply to the coast by the world's large rivers: Pearl River Basin, China. *Scientific Reports*, *9*(1), 9236. doi:10.103841598-019-45442-2 PMID:31239484

Redfern, T. W., Macdonald, N., Kjeldsen, T. R., Miller, J. D., & Reynard, N. (2016). Current understanding of hydrological processes on common urban surfaces. *Progress in Physical Geography: Earth and Environment*, *40*(5), 699–713. doi:10.1177/0309133316652819

Richter, B. D., & Thomas, G. A. (2007). Restoring Environmental Flows by Modifying Dam Operations. *Ecology and Society*, *12*(1), art12. doi:10.5751/ES-02014-120112

Robock, A., & Liu, Y. (1994). The volcanic signal in Goddard Institute for Space Studies and 3d model simulations. *Journal of Climate*, *7*, 44–55. doi:10.1175/1520-0442(1994)007<0044:TVSIGI>2.0.CO;2

Schilling, K. E., Chan, K.-S., Liu, H., & Zhang, Y.-K. (2010). Quantifying the effect of land use land cover change on increasing discharge in the Upper Mississippi River. *Journal of Hydrology (Amsterdam)*, *387*(3–4), 343–345. doi:10.1016/j.jhydrol.2010.04.019

Schmidt, J. C., & Wilcock, P. R. (2008). Metrics for assessing the downstream effects of dams. *Water Resources Research*, *44*(4). Advance online publication. doi:10.1029/2006WR005092

Sheet, I. R. N. F. (2003). Tehri dam. *Environment.*

Shukla, P. (2014). *Cumulative Impact Assessment on River Ganga from Hardiwar to Garhmukteshwar with Focus on Point Source Discharges*. GRIN Verlag.

Singh, M., Singh, I. B., & Müller, G. (2007). Sediment characteristics and transportation dynamics of the Ganga River. *Geomorphology*, *86*(1–2), 144–175. doi:10.1016/j.geomorph.2006.08.011

Sinha, R., Mohanta, H., Jain, V., & Tandon, S. K. (2017). Geomorphic diversity as a river management tool and its application to the Ganga River, India. *River Research and Applications*, *33*(7), 1156–1176. doi:10.1002/rra.3154

Stähly, S., Franca, M. J., Robinson, C. T., & Schleiss, A. J. (2019). Sediment replenishment combined with an artificial flood improves river habitats downstream of a dam. *Scientific Reports*, *9*(1), 5176. doi:10.103841598-019-41575-6 PMID:30914717

Surian, N., & Rinaldi, M. (2003). Morphological response to river engineering and management in alluvial channels in Italy. *Geomorphology*, *50*(4), 307–326. doi:10.1016/S0169-555X(02)00219-2

Taylor, M., & Yin, A. (2009). Active structures of the Himalayan-Tibetan orogen and their relationships to earthquake distribution, contemporary strain field, and Cenozoic volcanism. *Geosphere*, *5*(3), 199–214. doi:10.1130/GES00217.1

Van Binh, D., Kantoush, S., & Sumi, T. (2020). Changes to long-term discharge and sediment loads in the Vietnamese Mekong Delta caused by upstream dams. *Geomorphology*, *353*, 107011. doi:10.1016/j.geomorph.2019.107011

Warrick, J. A., Stevens, A. W., Miller, I. M., Harrison, S. R., Ritchie, A. C., & Gelfenbaum, G. (2019). World's largest dam removal reverses coastal erosion. *Scientific Reports*, *9*(1), 13968. doi:10.103841598-019-50387-7 PMID:31562373

Wilkinson, B. H., & McElroy, B. J. (2007). The impact of humans on continental erosion and sedimentation. *Geological Society of America Bulletin*, *119*(1–2), 140–156. doi:10.1130/B25899.1

Wohl, E. (2011). What should these rivers look like? Historical range of variability and human impacts in the Colorado Front Range, USA. *Earth Surface Processes and Landforms*, *36*(10), 1378–1390. doi:10.1002/esp.2180

Compilation of References

(1994). Salas, José D., Ellen E. Wohl, and Robert D. Jarrett. InRossi, G., Harmancioğlu, N., & Yevjevich, V. (Eds.), *Determination of Flood Characteristics Using Systematic, Historical and Paleoflood Data BT - Coping with Floods* (pp. 111–134). Springer Netherlands.

Adam, E., Mutanga, O., & Rugege, D. (2010). Multispectral and hyperspectral remote sensing for identification and mapping of wetland vegetation: A review. *Wetlands Ecology and Management*, *18*(3), 281–296. doi:10.100711273-009-9169-z

Adger, W. N., Barnett, J., Brown, K., Marshall, N., & O'brien, K. (2013). Cultural dimensions of climate change impacts and adaptation. *Nature Climate Change*, *3*(2), 112–117. doi:10.1038/nclimate1666

Adhikari, B. R. (1970). Tehri Dam: An Engineering Marvel. *Hydro Nepal: Journal of Water. Energy & Environment*. Advance online publication. doi:10.3126/hn.v5i0.2481

Afshari, S., Tavakoly, A. A., & Rajib, M. A. (2018a). Comparison of new generation low-complexity flood inundation mapping tools with a hydrodynamic model. *Journal of Hydrology (Amsterdam)*, *556*, 539–556. doi:10.1016/j.jhydrol.2017.11.036

Agunbiade, M. E., Rajabifard, A., & Bennett, R. (2016). What scope for integrating land management policies, land administration processes and data infrastructures for housing production in Nigeria? *Journal of Housing and the Built Environment*, *31*(1), 51–68. doi:10.100710901-015-9443-8

Ahmed, Shahid, bin Harun, Ismail, Nawaz, & Shamsudin (2014). Assessment of groundwater potential zones in an arid regionbased on catastrophe theory. *Earth Science Informatics*. Advance online publication. doi:10.100712145-014-0173-3

Ahmed, R., Siqueira, P., & Hensley, S. (2013). A study of forest biomass estimates from lidar in the northern temperate forests of New England. *Remote Sensing of Environment*, *130*, 121–135. doi:10.1016/j.rse.2012.11.015

Ajadi, O., Meyer, F., & Webley, P. (2016). Change Detection in Synthetic Aperture Radar Images Using a Multiscale-Driven Approach. *Remote Sensing*, *8*(6), 482. doi:10.3390/rs8060482

Akinyemi, F. O. (2017). Land change in the central Albertine rift: Insights from analysis and mapping of land use-land cover change in north-western Rwanda. *Applied Geography (Sevenoaks, England)*, *87*, 127–138. doi:10.1016/j.apgeog.2017.07.016

Akpodiogaga-a, P., & Odjugo, O. (2010). General overview of climate change impacts in Nigeria. *Journal of Human Ecology (Delhi, India)*, *29*(1), 47–55. doi:10.1080/09709274.2010.11906248

Alao, J. S., & Shuaibu, R. B. (2013). Agroforestry practices and concepts in sustainable land use systems in Nigeria. *Journal of Horticulture and Forestry*, *5*(10), 156–159.

Alsdorf, D. E., Rodríguez, E., & Lettenmaier, D. P. (2007a). Measuring surface water from space. *Reviews of Geophysics*, *45*(2), RG2002. Advance online publication. doi:10.1029/2006RG000197

Alves, D., Soares, J. V., Amaral, S., Mello, E., Almeida, S., da Silva, O. F., & Silveira, A. (1997). Biomass of primary and secondary vegetation in Rondônia, Western Brazilian Amazon. *Global Change Biology*, *3*(5), 451–461. doi:10.1046/j.1365-2486.1997.00081.x

Amalo, L. F., Hidayat, R., & Haris. (2017). Comparison between remote-sensing based drought indices in East Java. *Earth and Environmental Science*, *54*(1), 012009. Advance online publication. doi:10.1088/1755-1315/54/1/012009

Aman, A., Pandey, M., Singh, A., & Siddiqui, M.A. (2018). Study of Landscape Evolution in North Koel River Basin, Jharkhand, India: Tectonic and Structural Implications Based on Hypsometric Analysis. *Forum Geografic, 17*(2), 111–117. Doi:10.5775/fg.2018.073.d

Ammar, A., Michel, R., Ouessar, M., & Ritsema, C. (2016). Identification of suitable sites for rainwater harvesting structures in arid and semi-arid regions: A review. *ISWCR*, *4*, 108–120.

Anand, A., Singh, S. K., & Kanga, S. (2018). Estimating the change in Forest Cover Density and Predicting NDVI for West Singhbhum using Linear Regression. *ESSENCE Int. J. Env. Rehab. Conserv.*, *IX*(1), 193–203.

Anderson, J. E., Plourde, L. C., Martin, M. E., Braswell, B. H., Smith, M. L., Dubayah, R. O., Hofton, M., & Blair, J. (2008). Integrating waveform lidar with hyperspectral imagery for inventory of a northern temperate forest. *Remote Sensing of Environment: Remote Sensing Data Assimilation Special Issue*, *112*(4), 1856–1870. doi:10.1016/j.rse.2007.09.009

Andrew, M. E., & Ustin, S. L. (2008). The role of environmental context in mapping invasive plants with hyperspectral image data. *Remote Sensing of Environment*, *112*(12), 4301–4317. doi:10.1016/j.rse.2008.07.016

Andrew, M. E., & Ustin, S. L. (2009). Habitat suitability modelling of an invasive plant with advanced remote sensing data. *Diversity & Distributions*, *15*(4), 627–640. doi:10.1111/j.1472-4642.2009.00568.x

Ansari, A. A., Gill, S. S., Gill, R., Lanza, G. R., & Newman, L. (2016). Phytoremediation: Management of environmental contaminants, volume 4. *Phytoremediation: Management of Environmental Contaminants*, *4*, 1–409. doi:10.1007/978-3-319-41811-7

Anselin, A., Meire, P. M., & Anselin, L. (1989). Multicriteria techniques in ecological evaluation: An example using the analytical hierarchy process. *Biological Conservation, 49*, 215–229.

Anthony, E. J., Marriner, N., & Morhange, C. (2014). Human influence and the changing geomorphology of Mediterranean deltas and coasts over the last 6000 years: From progradation to destruction phase? *Earth-Science Reviews, 139*, 336–361. Advance online publication. doi:10.1016/j.earscirev.2014.10.003

Areola, O. (1987). The political reality of conservation in Nigeria. *Conservation in Africa: people, policies and practice*, 277-292.

Areola, O., & Ofomata, G. E. K. (1978). Land use control and natural resource management in Nigeria. *GeoJournal, 2*(5), 429–436. doi:10.1007/BF00156220

Ariza-Porras, C., Bravo, G., Villamizar, M., Moreno, A., Castro, H., Galindo, G., Cabera, E., Valbuena, S., & Lozano, P. (2017, September). CDCol: A geoscience data cube that meets Colombian needs. In *Advances in Computing. CCC: Colombian Conference on Computing. Communications in Computer and Information Science*, vol. 735 (pp. 87-99). Springer. 10.1007/978-3-319-66562-7_7

Arora, A., Pandey, M., Siddiqui, M. A., Hong, H., & Mishra, V. N. (2019). Spatial Flood Susceptibility Prediction in Middle Ganga Plain: Comparison of Frequency Ratio and Shannon's Entropy Models. *Geocarto International*, 1–32. doi:10.1080/10106049.2019.1687594

Arowolo, A. O., Deng, X., Olatunji, O. A., & Obayelu, A. E. (2018). Assessing changes in the value of ecosystem services in response to land-use/land-cover dynamics in Nigeria. *The Science of the Total Environment, 636*, 597–609. doi:10.1016/j.scitotenv.2018.04.277 PMID:29723833

Asadu, C. L., Ezeaku, P. I., & Nnaji, G. U. (2004). Land use and soil management situations in Nigeria: An analytical review of changes. *Outlook on Agriculture, 33*(1), 27–37. doi:10.5367/000000004322877755

Asner, G. P., Knapp, D. E., Kennedy-Bowdoin, T., Jones, M. O., Martin, R. E., Boardman, J., & Hughes, R. F. (2008). Invasive species detection in Hawaiian rainforests using airborne imaging spectroscopy and LiDAR. *Remote Sensing of Environment, 112*(5), 1942–1955. doi:10.1016/j.rse.2007.11.016

Attema, E., Davidson, M., Floury, N., Levrini, G., Rosich, B., Rommen, B., & Snoeij, P. (2008). *Sentinel-1 ESA's new European radar observatory.* Paper presented at the 7th European Conference on Synthetic Aperture Radar.

Atzberger, C., Darvishzadeh, R., Immitzer, M., Schlerf, M., Skidmore, A., & Maire, G. (2015). Comparative analysis of different retrieval methods for mapping grassland leaf area index using airborne imaging spectroscopy. *Int. J. Appl. L. Earth Obs., 43*, 19–31. doi:10.1016/j.jag.2015.01.009

Avitabile, V., Baccini, A., Friedl, M. A., & Schmullius, C. (2012). Capabilities and limitations of Landsat and land cover data for aboveground woody biomass estimation of Uganda. *Remote Sensing of Environment, 117*, 366–380. doi:10.1016/j.rse.2011.10.012

Babalola, S. O., & Uyi, I. O. (2019). A web-based land information system, tool for online land administration in Akure Nigeria. International Archives of the Photogrammetry, Remote Sensing & Spatial Information Sciences. XLII-4/W16, 137-142. doi:10.5194/isprs-archives-XLII-4-W16-137-2019

Babalola, K. H., & Hull, S. A. (2019). Using a domain model of social tenure to record land rights: A Case Study of Itaji-Ekiti, Ekiti State, Nigeria. *South African Journal of Geomatics*, *8*(2), 221–237.

Babalola, S. O., Rahman, A. A., Choon, L. T., & Van Oosterom, P. J. M. (2015). Possibilities of land administration domain model (LADM) implementation in Nigeria. *ISPRS Annals of The Photogrammetry. Remote Sensing and Spatial Information Sciences*, *2*(2), 155.

Baban, S. J., & wan-Yusof, K. (2003). Modelling optimum sites for locating reservoirs in tropical environments. *Water Resources Management*, *17*, 1–17.

Baghdadi, N., Gaultier, S., & King, C. (2002). Retrieving surface roughness and soil moisture from synthetic aperture radar (SAR) data using neural networks. *Canadian Journal of Remote Sensing*, *28*(5), 701–711. doi:10.5589/m02-066

Baghdadi, N., Holah, N., & Zribi, M. (2006). Soil moisture estimation using multi-incidence and multi-polarization ASAR data. *International Journal of Remote Sensing*, *27*(10), 1907–1920. doi:10.1080/01431160500239032

Baid, I. C. (1968). The Arthropod Fauna of Sambhar Salt Lake, Rajasthan, India. *Oikos Copenhagen*, *19*(2), 292–303. doi:10.2307/3565015

Baig, M. H. A., Zhang, L., Shuai, T., & Tong, Q. (2014). Derivation of a tasselled cap transformation based on Landsat 8 at-satellite reflectance. *Remote Sensing Letters*, *5*(5), 423–431. doi:10.1080/2150704X.2014.915434

Bajjouk, T., Mouquet, P., Ropert, M., Quod, J. P., Hoarau, L., Bigot, L., Populus, J. (2019). Detection of changes in shallow coral reefs status: Towards a spatial approach using hyperspectral and multispectral data. *Ecological Indicators, 96*, 174–191. doi:10.1016/j.ecolind.2018.08.052

BalaSundareshwaran, A., Abdul Rahaman, S., Balasubramani, K., Kumaraswamy, K., & Ramkumar, M. (2019). Habitat Risk Assessment Along Coastal Tamil Nadu, India—An Integrated Methodology for Mitigating Coastal Hazards. Coastal Zone Management. Doi:10.1016/b978-0-12-814350-6.00023-9

Ballanti, L., Blesius, L., Hines, E., & Kruse, B. (2016). Tree Species Classification Using Hyperspectral Imagery: A Comparison of Two Classifiers. *Remote Sensing*, *8*(6), 445. doi:10.3390/rs8060445

Ballantyne, C. K. (2002). Paraglacial geomorphology. *Quaternary Science Reviews*, *21*(18), 1935–2017. doi:10.1016/S0277-3791(02)00005-7

Banai-Kashani, R. (1989). A new method for site suitability analysis: The analytic hierarchy process. *Environmental Management*, *13*, 685–693.

Bansal, S., Katyal, D., Saluja, R., Chakraborty, M., & Garg, J. K. (2018). Geoinformation Remotely sensed MODIS wetland components for assessing the variability of methane emissions in Indian tropical / subtropical wetlands. *Int J Appl Earth Obs Geoinformation*, *64*, 156–170. doi:10.1016/j.jag.2017.08.011

Barik, R. K. (2019). CloudGanga: cloud computing based SDI model for ganga river basin management in India. In *Geospatial Intelligence: Concepts, Methodologies, Tools, and Applications* (pp. 278–297). IGI Global. doi:10.4018/978-1-5225-8054-6.ch013

Barnsley, M. J., Lewis, P., O'Dwyer, S., Disney, M. I., Hobson, P., Cutter, M., & Lobb, D. (2000). On the potential of CHRIS/PROBA for estimating vegetation canopy properties from space. *Remote Sensing Reviews*, *19*(1-4), 171–189. doi:10.1080/02757250009532417

Becerra, O., Cavallo, E., & Noy, I. (2014). Foreign Aid in the Aftermath of Large Natural Disasters. *Review of Development Economics*, *18*(3), 445–460. doi:10.1111/rode.12095

Bednarek, A. T. (2001). Undamming Rivers: A Review of the Ecological Impacts of Dam Removal. *Environmental Management*, *27*(6), 803–814. doi:10.1007002670010189 PMID:11393315

Bednarek, A. T., & Hart, D. D. (2005). Modifying dam operations to restore rivers: Ecological responses to tennessee river dam mitigation. *Ecological Applications*, *15*(3), 997–1008. doi:10.1890/04-0586

Behera, M. D., Kushwaha, S. P. S., & Roy, P. S. (2005). Rapid assessment of biological richness in a part of Eastern Himalaya: An integrated three-tier approach. *Forest Ecology and Management*, *207*(3), 363–384. doi:10.1016/j.foreco.2004.10.070

Benito, G., Lang, M., Barriendos, M., Llasat, C., Francés, F., & Ouarda, T. (2004). Use of Systematic, Palaeoflood and Historical Data for the Improvement of Flood Risk Estimation. Review of Scientific Methods. *Natural Hazards*, *31*(3), 623–643. doi:10.1023/B:NHAZ.0000024895.48463.eb

Bhanja, S. N., Mukherjee, A., Rodell, M., Wada, Y., Chattopadhyay, S., Velicogna, I., Pangaluru, K., & Famiglietti, J. S. (2017). Groundwater rejuvenation in parts of India influenced by water-policy change implementation. *Scientific Reports*, *7*(1), 7453. doi:10.103841598-017-07058-2 PMID:28785088

Bhatt, A., Bisht, A. K., Rawal, R., & Dhar, U. (2007). Assessment of status and biomass of Swertia angustifolia: a high value Himalayan medicinal plant. *Afr. J. Plant Sci, 1*(1), 1-6.

Bhattacharya, B. D., Nayak, D. C., Sarkar, S. K., Biswas, S. N., Rakshit, D., & Ahmed, M. K. (2015). Distribution of dissolved trace metals in coastal regions of Indian Sundarban mangrove wetland: A multivariate approach. *Journal of Cleaner Production*, *96*, 233–243. doi:10.1016/j.jclepro.2014.04.030

Bhattacharyya, A. (1943). On a measure of divergence between two statistical populations defined by their probability distributions. *Bulletin of the Calcutta Mathematical Society*, *35*, 99–109.

Bisnath, S. B., & Langley, R. B. (2002). High-precision, kinematic positioning with a single GPS receiver. *Navigation*, *49*(3), 161–169. doi:10.1002/j.2161-4296.2002.tb00265.x

Biswas, P., Kohli, M. P. S., Chadha, N. K., Bhattacharjya, B. K., Debnath, D., Yengkokpam, S., & Sharma, A. P. (2015). Optimizing Stocking Density of Labeo rohita Fry in Cage Aquaculture System as a Tool for Floodplain Wetland Fisheries Management. *Proceedings of the National Academy of Sciences. India. Section B, Biological Sciences*, *85*(1), 181–190. doi:10.100740011-014-0343-6

Bi, X., Lin, S., Zhu, S., Yin, H., Li, Z., & Chen, Z. (2019). Species identification and survival competition analysis of microalgae via hyperspectral microscopic images. *Optik (Stuttgart)*, *176*, 191–197. doi:10.1016/j.ijleo.2018.09.077

Boken, V. K., Hoogenboom, G., Kogan, F. N., Hook, J. E., Thomas, D. L., & Harrison, K. A. (2004). Potential of using NOAA-AVHRR data for estimating irrigated area to help solve an inter-state water dispute. *International Journal of Remote Sensing*, *25*(12), 2277–2286. doi:10.1080/01431160310001618077

Bookhagen, B., & Burbank, D. W. (2006). Topography, relief, and TRMM-derived rainfall variations along the Himalaya. *Geophysical Research Letters*, *33*(8), L08405. doi:10.1029/2006GL026037

Borgatti, L., & Soldati, M. (2010). Landslides as a geomorphological proxy for climate change: A record from the Dolomites (northern Italy). *Geomorphology*, *120*(1), 56–64. doi:10.1016/j.geomorph.2009.09.015

Boroushaki, S., & Malczewski, J. (2008). Implementing an extension of the analytical hierarchy process using ordered weighted averaging operators with fuzzy quantifiers in ArcGIS. *Computers & Geosciences*, *34*, 399–410.

Bouhennache, R., & Bouden, T. (2014). Change Detection In Urban Land Cover Using Landsat Images Satellites, A Case Study In Algiers Town. *IEEExplore*, *14*, 4799–562.

Bowler, D. E., Buyung-Ali, L. M., Healey, J. R., Jones, J. P., Knight, T. M., & Pullin, A. S. (2012). Does community forest management provide global environmental benefits and improve local welfare? *Frontiers in Ecology and the Environment*, *10*(1), 29–36. doi:10.1890/110040

Bozdag, C. E., Kahraman, C., & Ruan, D. (2003). Fuzzy group decision making for selection among computer integrated manufacturing systems. *Computers in Industry*, *51*, 13–29.

Bradford, J. B., Jensen, N. R., Domke, G. M., & D'Amato, A. W. (2013). Potential increases in natural disturbance rates could offset forest management impacts on ecosystem carbon stocks. *Forest Ecology and Management*, *308*, 178–187. doi:10.1016/j.foreco.2013.07.042

Braje, T. J., & Erlandson, J. M. (2013). Looking forward, looking back: Humans, anthropogenic change, and the Anthropocene. *Anthropocene*, *4*, 116–121. Advance online publication. doi:10.1016/j.ancene.2014.05.002

Brakenridge, Anderson, Nghiem, Caquard, & Shabaneh. (2003). *Flood Warnings, Flood Disaster Assessments, and Flood Hazard Reduction: The Roles of Orbital Remote Sensing*. Academic Press.

Brakenridge, G. R. (2010). *Global Active Archive of Large Flood Events*. Dartmouth Flood Observatory.

Brenna, A., Surian, N., & Mao, L. (2020). Response of A Gravel - Bed River To Dam Closure: Insights From Sediment Transport Processes And Channel Morphodynamics. *Earth Surface Processes and Landforms*, *45*(3), 756–770. doi:10.1002/esp.4750

Brisco, B., Brown, R. J., Hirose, T., McNairn, H., & Staenz, K. (1998). Precision agriculture and the role of remote sensing: A review. *Canadian Journal of Remote Sensing*, *24*(3), 315–327. do i:10.1080/07038992.1998.10855254

Brivio, P. A., Colombo, R., Maggi, M., & Tomasoni, R. (2002). Integration of remote sensing data and GIS for accurate mapping of flooded areas. *International Journal of Remote Sensing*, *23*, 429–441. doi:10.1080/01431160010014729

Brovkina, O., Novotny, J., Cienciala, E., Zemek, F., & Russ, R. (2017). Mapping forest aboveground biomass using airborne hyperspectral and LiDAR data in the mountainous conditions of Central Europe. *Ecological Engineering*, *100*, 219–230. doi:10.1016/j.ecoleng.2016.12.004

Brown, S., & Iverson, L. R. (1992). Biomass estimates for tropical forests. *World Resource Review, 4*(3), 366-384.

Brown, S. (1996). *Tropical forests and the global carbon cycle: estimating state and change in biomass density. In Forest ecosystems, forest management and the global carbon cycle*. Springer.

Brown, S. (1997). *Estimating biomass and biomass change of tropical forests: a primer* (Vol. 134). Food & Agriculture Org.

Brown, S. (2002). Measuring carbon in forests: Current status and future challenges. *Environmental Pollution*, *116*(3), 363–372. doi:10.1016/S0269-7491(01)00212-3 PMID:11822714

Brown, S. L., Schroeder, P., & Kern, J. S. (1999). Spatial distribution of biomass in forests of the eastern USA. *Forest Ecology and Management*, *123*(1), 81–90. doi:10.1016/S0378-1127(99)00017-1

Buckley, J. J. (1985). Fuzzy hierarchical analysis. *Fuzzy Sets and Systems*, *17*, 233–247.

Bulcock, H. H., & Jewitt, G. P. W. (2010). Spatial mapping of leaf area index using hyperspectral remote sensing for hydrological applications with a particular focus on canopy interception. *Hydrology and Earth System Sciences*, *14*(2), 383–392. doi:10.5194/hess-14-383-2010

Bull, W. B. (2009). *Tectonically active landscapes* (1st ed.). John Wiley and Sons. doi:10.1002/9781444312003

Bull, W. B., & McFadden, L. D. (1977). Tectonic geomorphology north and south of the Garlock fault, California. Geomorphology in arid regions. *Proceedings of the eighth annual geomorphology symposium*, 115-138.

Buyukozkam, G., & Feyzioglu, O. (2004). A fuzzy-logic-based decision-making approach for new product development. *International Journal of Production Economics*, *90*, 27–45.

Campbell, J. B., & Wynne, R. H. (2011). *Introduction to remote sensing*. Guilford Press.

Carlson, K. M., Asner, G. P., Hughes, R. F., Ostertag, R., & Martin, R. E. (2007). Hyperspectral remote sensing of canopy biodiversity in Hawaiian lowland rainforests. *Ecosystems (New York, N.Y.), 10*(4), 536–549. doi:10.100710021-007-9041-z

Casas, A., Riaño, D., Ustin, S. L., Dennison, P., & Salas, J. (2014). Estimation of water-related biochemical and biophysical vegetation properties using multitemporal airborne hyperspectral data and its comparison to MODIS spectral response. *Remote Sensing of Environment, 148*, 28–41. doi:10.1016/j.rse.2014.03.011

Castro-Esau, K. L., Sanchez-Azofeifa, G. A., & Caelli, T. (2004). Discrimination of lianas and trees with leaf level hyperspectral data. *Remote Sensing of Environment, 90*(3), 353–372. doi:10.1016/j.rse.2004.01.013

Caughlin, T. T., Graves, S. J., Asner, G. P., van Breugel, M., Hall, J. S., Martin, R. E., Ashton, M. S., & Bohlman, S. A. (2016). A hyperspectral image can predict tropical tree growth rates in single-species stands. *Ecological Applications, 26*(8), 2367–2373. doi:10.1002/eap.1436

Cebeci, U. (2009). Fuzzy AHP-based decision support system for selecting ERP systems in textile industry by using balanced scorecard. *Expert Systems with Applications, 36*, 8900–8909.

Centre for Research on the Epidemiology of Disasters (CRED) EM-DAT Team. (n.d.). *EM-DAT | The International Disasters Database*. Retrieved June 7, 2019 https://www.emdat.be/

Chabrillat, S., Milewski, R., Schmid, T., Rodriguez, M., Escribano, P., Palacios-orueta, A., & De Silvopascicultura, D. (2014). Potential of hyperspectral imagery for the spatial assessment of soil erosion stages in agricultural semi-arid Spain at different German Research Centre for Geosciences, Telegrafenberg. Potsdam, Germany: EEZA-CSIC, Departmento de Desertificación y Geoecología.

Chabukdhara, M., & Singh, O. P. (2016). Coal mining in northeast India: An overview of environmental issues and treatment approaches. *International Journal of Coal Science & Technology, 3*(2), 87–96. doi:10.100740789-016-0126-1

Chakraborty, K., Saikom, V., Borah, S. B., Kalita, M., Gupta, C., Meitei, L. R., ... Raju, P. L. N. (2019). Forest biometric parameter extraction using unmanned aerial vehicle to aid in forest inventory data collection. *Current Science, 117*(7).

Chakraborty, K., Chabukdhara, M., Mondal, P. P., Borah, S. B., Sarma, K. K., Sudhakar, S., & Raju, P. L. N. (2017). Status of forest fire – A case study from northeast India. *Asian Conference on Remote Sensing*.

Chakraborty, K., Joshi, P. K., & Sarma, K. K. (2009). Land use/land cover dynamics in Umngot watershed of Meghalaya using geospatial tools. *Journal of the Indian Society of Remote Sensing, 37*(1), 99–106. doi:10.100712524-009-0015-9

Chakraborty, K., Mondal, P. P., Chabukdhara, M., & Sudhakar, S. (2014b). Forest fire scenario and challenges of mitigation during fire season in North East India. *The International Archives of the Photogrammetry, Remote Sensing and Spatial Information Sciences*, *40*(8), 27–33. doi:10.5194/isprsarchives-XL-8-27-2014

Chakraborty, K., & Sudhakar, S. (2014a). The expansion and impact of cement manufacturing units and mining areas in Lumshnong, Jaintia hills, Meghalaya. *Current Science*, 997–1000.

Chakraborty, K., Sudhakar, S., Sarma, K. K., Raju, P. L. N., & Das, A. K. (2018). Recognizing the rapid expansion of rubber plantation–a threat to native forest in parts of northeast India. *Current Science*, *114*(1), 207. doi:10.18520/cs/v114/i01/207-213

Champion, S. H., & Seth, S. K. (1968). *A revised survey of the forest types of India*. A revised survey of the forest types of India.

Chandramouli, C. (2011). Census of India 2011: rural urban distribution of population. *Registrar General and Census Commissioner, Ministry of Home Affairs*. http://censusindia. gov. in/2011-prov-results/paper2/data_files/india/Rural_Urban_2011. pdf

Chang, C. I. (2007). *Hyperspectral Data Exploitation: Theory and Applications*. John Wiley & Sons. doi:10.1002/0470124628

Chang, D. Y. (1996). Application of the extent analysis method on fuzzy AHP. *European Journal of Operational Research*,, 649–655.

Chang, T.-H., & Wang, T.-C. (2009). Using the fuzzy multi-criteria decision making approach for measuring the possibility of successful knowledge management. *Information Sciences*, *179*, 355–370.

Charbonneau, F., Trudel, M., & Fernandes, R. (2005). Use of Dual Polarization and Multi-Incidence SAR for soil permeability mapping. *Advanced Synthetic Aperture Radar (ASAR) 2005*.

Chatterjee, A., Adhikari, S., Pal, S., & Mukhopadhyay, S. K. (2020). Foraging guild structure and niche characteristics of waterbirds wintering in selected sub-Himalayan wetlands of India. *Ecological Indicators*, *108*, 105693. doi:10.1016/j.ecolind.2019.105693

Chatterjee, A., Adhikari, S., & Mukhopadhyay, S. K. (2017). Effects of Waterbird Colonization on Limnochemical Features of a Natural Wetland on Buxa Tiger Reserve, India, During Wintering Period. *Wetlands*, *37*(1), 177–190. doi:10.100713157-016-0851-7

Chatterjee, K., Bandyopadhyay, A., Ghosh, A., & Kar, S. (2015). Assessment of environmental factors causing wetland degradation, using Fuzzy Analytic Network Process: A case study on Keoladeo National Park, India. *Ecological Modelling*, *316*, 1–13. doi:10.1016/j.ecolmodel.2015.07.029

Chatterjee, S., Saikia, A., Dutta, P., Ghosh, D., Pangging, G., & Goswami, A. K. (2006). *Background paper on biodiversity significance of North East India, for the study on natural resources, water and environment nexus for development and growth in North Eastern India*. Forest Conservation Programme.

Chavan, B., & Rasal, G. (2010). Sequestered standing carbon stock in selective tree species grown in University campus at Aurangabad, Maharashtra, India. *International Journal of Engineering Science and Technology*, *2*(7), 3003–3007.

Chen, C., Park, T., Wang, X., Piao, S., Xu, B., Chaturvedi, R. K., ... Tømmervik, H. (2019). China and India lead in greening of the world through land-use management. *Nature Sustainability*, *2*(2), 122.

Chen, B., Pang, Y., Li, Z., North, P., Rosette, J., Sun, G., Suárez, J., Bye, I., & Lu, H. (2019). Potential of Forest Parameter Estimation Using Metrics from Photon Counting LiDAR Data in Howland Research Forest. *Remote Sensing*, *11*(7), 856. doi:10.3390/rs11070856

Chen, Q. (2010). Retrieving vegetation height of forests and woodlands over mountainous areas in the Pacific Coast region using satellite laser altimetry. *Remote Sensing of Environment*, *114*(7), 1610–1627. doi:10.1016/j.rse.2010.02.016

Chidebelu, A. N., Ames, G. W., & Brown, E. E. (1980). Land use in Nigeria. *Food Policy*, *5*(4), 306–310. doi:10.1016/0306-9192(80)90059-7

Chigbu, U. E. (2019). Concept and Approach to Land Management Interventions for Rural Development in Africa. In *Geospatial Technologies for Effective Land Governance* (pp. 1–14). IGI Global. doi:10.4018/978-1-5225-5939-9.ch001

Chinnadurai, S., Mohamed, K. S., Sharma, J., Venkatesan, V., & Kripa, V. (2016). Assessment of bio-accumulation of bacteria in oysters from shellfish growing waters in Ashtamudi Lake (Kerala, India): A RAMSAR wetland. *Regional Studies in Marine Science*, *7*, 118–122. doi:10.1016/j.rsma.2016.05.016

Chowdhury, R., Favas, P. J. C., Jonathan, M. P., Venkatachalam, P., Raja, P., & Sarkar, S. K. (2017). Bioremoval of trace metals from rhizosediment by mangrove plants in Indian Sundarban Wetland. *Marine Pollution Bulletin*, *124*(2), 1078–1088. doi:10.1016/j.marpolbul.2017.01.047 PMID:28187966

Chu, H.-J., Wang, C.-K., Kong, S.-J., & Chen, K.-C. (2016). Integration of full-waveform LiDAR and hyperspectral data to enhance tea and areca classification. *Journal GIScience & Remote Sensing*, *53*(4), 542–559. doi:10.1080/15481603.2016.1177249

Chutima, P., & Suwanfuji, P. (1998). Fuzzy analytical hierarchy process part routing in FMS. *Thammasat International Journal of Science and Technology*, *3*(2), 29–47.

Cienciala, P., & Pasternack, G. B. (2017). Floodplain inundation response to climate, valley form, and flow regulation on a gravel-bed river in a Mediterranean-climate region. *Geomorphology*, *282*, 1–17. doi:10.1016/j.geomorph.2017.01.006

Clark, M. L., Roberts, D. A., Ewel, J. J., & Clark, D. B. (2011). Estimation of tropical rain forest aboveground biomass with small-footprint lidar and hyperspectral sensors. *Remote Sensing of Environment*, *xxx*(11), xxx–xxx. doi:10.1016/j.rse.2010.08.029

Collier, P. (2010). The political economy of natural resources. *Social Research*, *77*(4), 1105–1132.

Collins, J. B., & Woodcock, C. E. (1996). An assessment of several linear change detection techniques for mapping forest mortality using multitemporal landsat TM data. *Remote Sensing of Environment*, *56*(1), 66–77. doi:10.1016/0034-4257(95)00233-2

Congalton, R. G., & Green, K. (2002). *Assessing the accuracy of remotely sensed data: principles and practices*. CRC Press.

Cong, J., & Ding, Y. (1994). FlowMap: An optimal technology mapping algorithm for delay optimization in lookup-table based FPGA designs. *IEEE Transactions on Computer-Aided Design of Integrated Circuits and Systems*, *13*(1), 1–12. doi:10.1109/43.273754

Correa-Berger, B. P. (2006). *Developing a habitat suitability model for the spotted turtle using a hybrid-deductive approach* (MSc dissertation). New York: Rochester Institute of Technology.

Costache, R. (2019). Flood Susceptibility Assessment by Using Bivariate Statistics and Machine Learning Models - A Useful Tool for Flood Risk Management. *Water Resources Management*, *33*, 3239–3256. doi:10.100711269-019-02301-z

CRED first. (2011). *EM-DAT: The OFDA/CRED International Disaster Database*. Université Catholique de Louvain.

Crist, E. P., & Cicone, R. C., 1984b. A physically-based transformation of thematic mapper data---The TM Tasseled Cap. *Geoscience and Remote Sensing, IEEE Transactions on, 22*(3), 256-263.

Crist, E. P. (1985). A TM Tasseled Cap equivalent transformation for reflectance factor data. *Remote Sensing of Environment*, *17*(3), 301–306. doi:10.1016/0034-4257(85)90102-6

Crist, E. P., & Cicone, R. (1984a). Application of the tasseled cap concept to simulated Thematic Mapper data. *Photogrammetric Engineering and Remote Sensing*, *50*(3), 343–352.

Crist, E. P., & Kauth, R. J. (1986). The tasseled cap de-mystified. *Photogrammetric Engineering and Remote Sensing*, *52*(1), 81–86.

Cruz-Barahona, C. (2019). *Metodología para calcular costos y beneficios ambientales del Centro de Transferencia y Transformación de Materiales*.

CWC Handbook for Flood Protection. (2012). *Handbook for Flood Protection*. Anti Erosion & River Training Works.

CWC. (1973). *Estimation of Design Flood Peak: A Method Based on Unit Hydrograph Approach*. Author.

CWC. (2018). *Flood Damage Statistics (Statewise and for the Country as a Whole) for the Period 1953 to 2016*. Central Water Commission (CWC), Flood Forecast Monitoring Directorate, Government of India.

DAC. (2009). *Manual for Drought Management*. Department of Agriculture & Cooperation, Ministry of Agriculture, Government of India.

Dadhwal, V., Sehgal, V., Singh, R., & Rajak, D. (2003). Wheat yield modelling using satellite remote sensing with weather data: Recent Indian experience. *Mausam (New Delhi), 54*(1), 253–262.

Dadson, S. J., Hovius, N., Chen, H., Dade, W. B., Lin, J.-C., Hsu, M.-L., Lin, C.-W., Horng, M.-J., Chen, T.-C., Milliman, J., & Stark, C. P. (2004). Earthquake-triggered increase in sediment delivery from an active mountain belt. *Geology, 32*(8), 733. doi:10.1130/G20639.1

Dagdeviren, M., Yavuz, S., & Kilinc¸, N. (2009). Weapon selection using the AHP and TOPSIS methods under fuzzy environment. *Expert Systems with Applications, 36*, 8143–8151.

Dagdeviren, M., & Yuksel, I. (2008). Developing a fuzzy analytic hierarchy process (AHP) model for behavior-based safety management. *Information Sciences, 178*, 1717–1733.

Dalezios, N. R., Blanta, A., Spyropoulos, N. V., & Tarquis, A. M. (2014). Risk identification of agricultural drought for sustainable Agroecosystems. *Natural Hazards and Earth System Sciences, 14*(9), 2435–2448. doi:10.5194/nhess-14-2435-2014

Dalponte, M., Bruzzone, L., & Gianelle, D. (2008). Fusion of Hyperspectral and LIDAR Remote Sensing Data for Classification of Complex Forest Areas. *IEEE Transactions on Geoscience and Remote Sensing, 46*(5), 1416–1426. doi:10.1109/TGRS.2008.916480

Daniels, J. M. (2008). Distinguishing allogenic from autogenic causes of bed elevation change in late Quaternary alluvial stratigraphic records. *Geomorphology, 101*(1–2), 159–171. doi:10.1016/j.geomorph.2008.05.022

Darvishzadeh, R., Atzberger, C., Skidmore, A., & Schlerf, M. (2011). Mapping grassland leaf area index with airborne hyperspectral imagery: A comparison study of statistical approaches and inversion of radiative transfer models. *ISPRS Journal of Photogrammetry and Remote Sensing, 66*(6), 894–906. doi:10.1016/j.isprsjprs.2011.09.013

Das, D. J. (2012). *A study on impact of shifting cultivation on Landuse/landcoverin Mizoram, using Geospatial tools* (Ph.D. thesis). FRI (deemed) University.

Das, P., Palchaudhuri, M., Biswas, S., & Deka, C. (2017). *Analysis of Agricultural Drought Severity Using MODIS Satellite Images*. Academic Press.

Davranche, A., Poulin, B., & Lefebvre, G. (2013). Mapping flooding regimes in Camargue wetlands using seasonal multispectral data. *Remote Sensing of Environment, 138*, 165–171. doi:10.1016/j.rse.2013.07.015

De Jong, S. M., Pebesma, E. J., & Lacaze, B. (2003). Above-ground biomass assessment of Mediterranean forests using airborne imaging spectrometry: The DAIS Peyne experiment. *International Journal of Remote Sensing, 24*(7), 1505–1520. doi:10.1080/01431160210145560

De Rego, K., Lauer, J. W., Eaton, B., & Hassan, M. (2020). A decadal-scale numerical model for wandering, cobble-bedded rivers subject to disturbance. *Earth Surface Processes and Landforms, 45*(4), 912–927. doi:10.1002/esp.4784

Debanshi, S., & Pal, S. (2020). Wetland delineation simulation and prediction in deltaic landscape. *Ecological Indicators, 108*, 105757. doi:10.1016/j.ecolind.2019.105757

Deering, D. W. (1978). *Rangeland reflectance characteristics measured by aircraft and spacecraft sensors* (Ph.D. Dissertation). Texas A&M University, College Station, TX.

Deka, R. L., Mahanta, C., Pathak, H., Nath, K. K., & Das, S. (2013). Trends and fluctuations of rainfall regime in the Brahmaputra and Barak basins of Assam, India. *Theoretical and Applied Climatology, 114*(1-2), 61–71. doi:10.100700704-012-0820-x

Deng, H. (1999). Multicriteria analysis with fuzzy pairwise comparisons. *International Journal of Approximate Reasoning, 21*, 215–231.

Dhanda, P., Nandy, S., Kushwaha, S.P.S., Ghosh, S., Murthy, Y.V.N. K., & Dadhwal, V.K. (2017). Optimizing spaceborne LiDAR and very high resolution optical sensor parameters for biomass estimation at ICESat/GLAS footprint level using regression algorithms. *Prog Phys Geog.*, 1-21. Doi:10.1177/0309133317693443

Dheeravath, V., Thenkabail, P. S., Chandrakantha, G., Noojipady, P., Reddy, G. P. O., Biradar, C. M., Gumma, M. K., & Velpuri, M. (2010). Irrigated areas of India derived using MODIS 500 m time series for the years 2001–2003. *ISPRS Journal of Photogrammetry and Remote Sensing, 65*(1), 42–59. doi:10.1016/j.isprsjprs.2009.08.004

Dikshit, K. R., & Dikshit, J. K. (2014). Natural Vegetation: Forests and Grasslands of North-East India. In *North-East India: Land, People and Economy* (pp. 213–255). Springer. doi:10.1007/978-94-007-7055-3_9

District Census 2011. (2011). Census2011.co.in

Dong, J., Metternicht, G., Hostert, P., Fensholt, R., & Chowdhury, R. R. (2019). Remote sensing and geospatial technologies in support of a normative land system science: Status and prospects. *Current Opinion in Environmental Sustainability, 38*, 44–52. doi:10.1016/j.cosust.2019.05.003

Dostálová, A., Naeimi, V., Wagner, W., Elefante, S., Cao, S., & Persson, H. (2016). *Geocoding uncertainty analysis for the automated processing of Sentinel-1 data using Sentinel-1 Toolbox software.* Paper presented at the Image and Signal Processing for Remote Sensing XXII.

Dowdy, S. H., Wearden, S., & Chilko, D. (2004). *Statistics for Research* (3rd ed.). John Wiley and Sons Inc.

Dubayah, R. O., & Drake, J. B. (2000). Lidar remote sensing for forestry. *Journal of Forestry, 98*, 44–46.

Duncanson, L. I., Niemann, K. O., & Wulder, M. A. (2010). Estimating forest canopy height and terrain relief from GLAS waveform metrics. *Remote Sensing of Environment, 114*(1), 138–154. doi:10.1016/j.rse.2009.08.018

Du, T. L. T., Bui, D. D., Nguyen, M. D., & Lee, H. (2018). Satellite-Based, Multi-Indices for Evaluation of Agricultural Droughts in a Highly Dynamic Tropical Catchment, Central Vietnam. *Water (Basel), 10*(5), 659. doi:10.3390/w10050659

Dutta, D., Kundu, A., Patel, N. R., Saha, S. K., & Siddique, A. R. (2015). Assessment of agricultural drought in Rajasthan (India) using remote sensing derived Vegetation Condition Index (VCI) and Standardized Precipitation Index (SPI). *The Egyptian Journal of Remote Sensing and Space Sciences, 18*(1), 53–63. doi:10.1016/j.ejrs.2015.03.006

Eastman, J. R., Kyem, P., & Toledano, J. (1993). A procedure for multi-objective decision making in GIS under conditions of conflicting objectives. EGIS '93.

Eastman, J. R., Toledano, J., Jin, W., & Kyem, P. A. K. (1992). Participatory procedures for multi-criteria evaluations in GIS. In GIS'92, Buffalo, NY.

Ebhuoma, E. E., & Simatele, D. M. (2019). 'We know our Terrain': Indigenous knowledge preferred to scientific systems of weather forecasting in the Delta State of Nigeria. *Climate and Development, 11*(2), 112–123. doi:10.1080/17565529.2017.1374239

Eggleston, S., Buendia, L., Miwa, K., Ngara, T., & Tanabe, K. (2006). *2006 IPCC guidelines for national greenhouse gas inventories* (Vol. 5). Institute for Global Environmental Strategies Hayama.

Ehiorobo, A. O. (2018). Efficient Resource Allocation and Utilization: The Missing Link in Nigeria's Quest for Sustainable Development. *Economics and Business, 32*(1), 264–275. doi:10.2478/eb-2018-0020

El Hajj, M., Baghdadi, N., Zribi, M., & Angelliaume, S. (2016). Analysis of Sentinel-1 radiometric stability and quality for land surface applications. *Remote Sensing, 8*(5), 406. doi:10.3390/rs8050406

El-Din, H. K., Abd El Munim, H. E., & Mahdi, H. Decision-making in fuzzy environment: a survey. In *Application of Decision Science in Business and Management*. IntechOpen.

El-Rabbany, A. (2002). *Introduction to GPS: the global positioning system*. Artech house.

EMDAT. (2018). *Total Economic Damage: 1900-2017*. OFDA/CRED International Disaster Database, Université Catholique de Louvain.

EMDAT. (2019). *OFDA/CRED International Disaster Database*. Université Catholique de Louvain.

Enaruvbe, G. O., & Atafo, O. P. (2016). Analysis of deforestation pattern in the Niger Delta region of Nigeria. *Journal of Land Use Science, 11*(1), 113–130. doi:10.1080/1747423X.2014.965279

Eneh, O. C. (2011). Managing Nigeria's environment: The unresolved issues. *Journal of Environmental Science and Technology, 4*(3), 250–263. doi:10.3923/jest.2011.250.263

Erensal, Y. C., Oncan, T., & Demircan, M. L. (2006). Determining key capabilities in technology management using fuzzy analytic hierarchy process: A case study of Turkey. *Information Sciences, 176*, 2755–2770.

Essien, E., & Cyrus, S. (2019). Detection of urban development in Uyo (Nigeria) using remote sensing. *Land (Basel)*, *8*(6), 102. doi:10.3390/land8060102

Eugster, P., Thiede, R. C., Scherler, D., Stübner, K., Sobel, E. R., & Strecker, M. R. (2018). Segmentation of the Main Himalayan Thrust Revealed by Low-Temperature Thermochronometry in the Western Indian Himalaya. *Tectonics*, *37*(8), 2710–2726. doi:10.1029/2017TC004752

Everard, M., Kangabam, R., Tiwari, M. K., McInnes, R., Kumar, R., Talukdar, G. H., & Das, L. (2019). Ecosystem service assessment of selected wetlands of Kolkata and the Indian Gangetic Delta: Multi-beneficial systems under differentiated management stress. *Wetlands Ecology and Management*, *27*(2–3), 405–426. doi:10.100711273-019-09668-1

Ewak, J. O. (2010). Expanding mandate and corporate social responsibility in the management of National Parks and protected areas in Nigeria. *Journal of Agriculture and Environment for International Development*, *104*(1/2), 25–38.

Eze, E. B., & Neba, N. E. (2006). A conceptual framework for ecologically and socially sustainable land management and agricultural development in Africa. *International Journal of Environment and Sustainable Development*, *5*(3), 275–286. doi:10.1504/IJESD.2006.010897

Fallis, A. (2013). Emergency Events Database (EM-DAT) and Its Applications. *Journal of Chemical Information and Modeling*, *53*(9), 1689–1699.

Famoriyo, S. (1982). Impinging upon land resources management in Nigeria. In Efficient Use of Nigerian Land Resources (pp. 289-300). Academic Press.

Famoriyo, S. (1984). Administration of land allocation in Nigeria. *Land Use Policy*, *1*(3), 217–224. doi:10.1016/0264-8377(84)90065-6

Fayad, I., Baghdadi, N., Bailly, J.-S., Barbier, N., Gond, V., Hérault, B., Hajj, M. E., Fabre, F., & Perrin, J. (2016). Regional Scale Rain-Forest Height Mapping Using Regression-Kriging of Spaceborne and Airborne LiDAR Data: Application on French Guiana. *Remote Sensing*, *8*(3), 240. doi:10.3390/rs8030240

Fengqin, Y., & Shuwen, Z. (2019). Ecosystem service decline in response to wetland loss in the Sanjiang Plain, Northeast China. *Ecological Engineering*, *130*(January), 117–121. doi:10.1016/j.ecoleng.2019.02.009

Fink, A. K. (2000). *Tehri hydro power complex on the Bhagirathi river in India*. Hydrotechnical Construction., doi:10.1023/A:1004187208788

Flint, J. J. (1974). Stream gradient as a function of order, magnitude, and discharge. *Water Resources Research*, *10*(5), 969–973. doi:10.1029/WR010i005p00969

Flood Forecast Monitoring Directorate. (2012). *Flood Forecasting and Warning Network Performance Appraisal Report 2011*. Author.

Foley, J. A. (2005). Global Consequences of Land Use. *Science*, *309*(5734), 570–574. doi:10.1126cience.1111772 PMID:16040698

Fontanelli, G., Paloscia, S., Zribi, M., & Chahbi, A. (2013). Sensitivity analysis of X-band SAR to wheat and barley leaf area index in the Merguellil Basin. *Remote Sensing Letters*, *4*(11), 1107–1116. doi:10.1080/2150704X.2013.842285

Foody, G. M., Curran, P. J., Groom, G. B., & Munro, D. C. (1989). Multi-temporal airborne synthetic aperture radar data for crop classification. *Geocarto International*, *4*(3), 19–29. doi:10.1080/10106048909354218

Foody, G. M., Cutler, M. E., McMorrow, J., Pelz, D., Tangki, H., Boyd, D. S., & Douglas, I. (2001). Mapping the biomass of Bornean tropical rain forest from remotely sensed data. *Global Ecology and Biogeography*, *10*(4), 379–387. doi:10.1046/j.1466-822X.2001.00248.x

Foody, G. M., & Mathur, A. (2004). Toward intelligent training of supervised image classifications: Directing training data acquisition for SVM classification. *Remote Sensing of Environment*, *93*(1-2), 107–117. doi:10.1016/j.rse.2004.06.017

Fowler, D. N., King, S. L., & Weindorf, D. C. (2014). *Evaluating Abiotic Influences on Soil Salinity of Inland Managed Wetlands and Agricultural Croplands in a Semi-Arid Environment*. Academic Press.

Francis, R. J., Rose, J., & Vranesic, Z. (1991). *Technology mapping of lookup table-based FPGAs for performance*. Paper presented at the 1991 IEEE International Conference on Computer-Aided Design Digest of Technical Papers. 10.1109/ICCAD.1991.185334

Franklin, S. E. (2001). *Remote sensing for sustainable forest management*. CRC press. doi:10.1201/9781420032857

Freeman, A., & Saatchi, S. S. (2004). On the detection of Faraday rotation in linearly polarized L-band SAR backscatter signatures. *IEEE Transactions on Geoscience and Remote Sensing*, *42*(8), 1607–1616. doi:10.1109/TGRS.2004.830163

Freeman, A., Villasenor, J., Klein, J. D., Hoogeboom, P., & Groot, J. (1994). On the use of multi-frequency and polarimetric radar backscatter features for classification of agricultural crops. *International Journal of Remote Sensing*, *15*(9), 1799–1812. doi:10.1080/01431169408954210

Friedl, M. A., & Brodley, C. E. (1997). Decision tree classification of land cover from remotely sensed data. *Remote Sensing of Environment*, *61*(3), 399–409. doi:10.1016/S0034-4257(97)00049-7

Fuchs, G. R. (1982). *Explanations of the geologic-tectonic map of the Himalaya; Geol.* Surv.

Fung, T. (1990). An assessment of TM imagery for land-cover change detection. *IEEE Transactions on Geoscience and Remote Sensing*, *28*(4), 681–684. doi:10.1109/TGRS.1990.572980

Furey, T. S., Cristianini, N., Duffy, N., Bednarski, D. W., Schummer, M., & Haussler, D. (2000). Support vector machine classification and validation of cancer tissue samples using microarray expression data. *Bioinformatics (Oxford, England)*, *16*(10), 906–914. doi:10.1093/bioinformatics/16.10.906 PMID:11120680

Fu, Y., Zhao, J., Peng, W., Zhu, G., Quan, Z., & Li, C. (2018). Spatial modelling of the regulating function of the Huangqihai Lake wetland ecosystem. *Journal of Hydrology (Amsterdam), 564*, 283–293. doi:10.1016/j.jhydrol.2018.07.017

Galay, V. J. (1983). Causes of river bed degradation. *Water Resources Research, 19*(5), 1057–1090. doi:10.1029/WR019i005p01057

Gan, T. Y., Zunic, F., Kuo, C.-C., & Strobl, T. (2012). Flood mapping of Danube River at Romania using single and multi-date ERS2-SAR images. *International Journal of Applied Earth Observation and Geoinformation, 18*, 69–81. doi:10.1016/j.jag.2012.01.012

Gao, Q., Zribi, M., Escorihuela, M., Baghdadi, N., & Segui, P. (2018). Irrigation mapping using Sentinel-1 time series at field scale. *Remote Sensing, 10*(9), 1495. doi:10.3390/rs10091495

Gao, X., Huete, A., Ni, W., & Miura, T. (2000). Optical-biophysical relationships of vegetation spectra without background contamination. *Remote Sensing of Environment, 74*(3), 609–620. doi:10.1016/S0034-4257(00)00150-4

García-Sopo, M. Á., Cuartero, A., Rodríguez, P. G., & Plaza, A. 2015. Hyperspectral And Lidar Data Integration And Classification. *Proceedings of the International Geoscience and Remote Sensing Symposium*, 57–60. 10.1109/IGARSS.2015.7325696

Gedney, N., Cox, P. M., Betts, R. A., Boucher, O., Huntingford, C., & Stott, P. A. (2006). Detection of a direct carbon dioxide effect in continental river runoff records. *Nature, 439*(7078), 835–838. doi:10.1038/nature04504 PMID:16482155

Geerling, G. W., Labrador-Garcia, M., Clevers, J. G. P. W., Ragas, A. M. J., & Smits, A. J. M. (2007). Classification of floodplain vegetation by data fusion of spectral (CASI) and LiDAR data. *International Journal of Remote Sensing, 28*(19), 4263–4284. doi:10.1080/01431160701241720

Gefu, J. O., & Gilles, J. L. (1990). Pastoralists, ranchers and the state in Nigeria and North America: A comparative analysis. *Nomadic Peoples*, 34–50.

Ghermandi, A., Sheela, A. M., & Justus, J. (2016). Integrating similarity analysis and ecosystem service value transfer: Results from a tropical coastal wetland in India. *Ecosystem Services, 22*(April), 73–82. doi:10.1016/j.ecoser.2016.09.014

Ghosh, S., & Das, A. (2019). Urban expansion induced vulnerability assessment of East Kolkata Wetland using Fuzzy MCDM method. *Remote Sensing Applications: Society and Environment, 13*, 191–203. doi:10.1016/j.rsase.2018.10.014

Gibril, M. B. A., Bakar, S. A., Yao, K., Idrees, M. O., & Pradhan, B. (2017). Fusion of RADARSAT-2 and multispectral optical remote sensing data for LULC extraction in a tropical agricultural area. *Geocarto International, 32*(7), 735–748. doi:10.1080/10106049.2016.1170893

Gilbert, N. (2012). India's forest area in doubt. *NATNews, 489*(7414), 14. PMID:22962698

Gillard, R., Gouldson, A., Paavola, J., & Van Alstine, J. (2016). Transformational responses to climate change: Beyond a systems perspective of social change in mitigation and adaptation. *Wiley Interdisciplinary Reviews: Climate Change*, *7*(2), 251–265. doi:10.1002/wcc.384

Giri, S., Arbab, N. N., & Lathrop, R. G. (2019). Assessing the potential impacts of climate and land use change on water fluxes and sediment transport in a loosely coupled system. *Journal of Hydrology (Amsterdam)*, *577*, 123955. doi:10.1016/j.jhydrol.2019.123955

Gitelson, A. A., Kogan, F., Zakarin, E., Spivak, L., & Lebed, L. (1998). Using AVHRR data for quantitative estimation of vegetation conditions: Calibration and validation. *Advances in Space Research*, *22*(5), 673–676. doi:10.1016/S0273-1177(97)01129-0

Giuliani, G., Chatenoux, B., De Bono, A., Rodila, D., Richard, J. P., Allenbach, K., Dao, H., & Peduzzi, P. (2017). Building an earth observations data cube: Lessons learned from the swiss data cube (SDC) on generating analysis ready data (ARD). *Big Earth Data*, *1*(1-2), 100–117. do i:10.1080/20964471.2017.1398903

Glenn, N. F., Mundt, J. T., Weber, K. T., Prather, T. S., Lass, L. W., & Pettingill, J. (2005). Hyperspectral data processing for repeat detection of small infestations of Leafy spurge. *Remote Sensing of Environment*, *95*(3), 399–412. doi:10.1016/j.rse.2005.01.003

GlobalFRA. (GFRA). (2015). How are the world's forests changing? (2nd ed.). Food and Agriculture Organization of the United Nations.

Goetz, A. F. H., Vane, G., Solomon, J. E., & Rock, B. N. (1985). Imaging spectrometry for earth remote-sensing. *Science*, *228*(4704), 1147–1153. doi:10.1126cience.228.4704.1147

Gogoi, P., Sinha, A., Das Sarkar, S., Chanu, T. N., Yadav, A. K., Koushlesh, S. K., & Das, B. K. (2019). Seasonal influence of physicochemical parameters on phytoplankton diversity and assemblage pattern in Kailash Khal, a tropical wetland, Sundarbans, India. *Applied Water Science*, *9*(7), 1–13. doi:10.100713201-019-1034-5

GoI. (1999). *Economic survey*. Economic Division, Ministry of Finance, Government of India.

Goklany, I. M. (2012). Is climate change the number one threat to humanity? *Wiley Interdisciplinary Reviews: Climate Change*, *3*(6), 489–508. doi:10.1002/wcc.194

Gomez, C., Hart, D. E., Wassmer, P., Kenta, I., Matsui, H., & Shimizu, M. (2019). *Coastal Evolution, Geomorphic Processes and Sedimentary Records in the Anthropocene*. Forum Geografi. doi:10.23917/forgeo.v33i1.7551

Gong, P., Pu, R., Biging, G. S., & Larrieu, M. R. (2003). Estimation of Forest Leaf Area Index Using Vegetation Indices Derived from Hyperion Hyperspectral Data. *IEEE Transactions on Geoscience and Remote Sensing*, *41*(6), 1–8. doi:10.1109/TGRS.2003.812910

Gong, P., Pu, R., & Miller, J. R. (1995). Coniferous Forest Leaf Area Index Estimation along the 0regon Transect Uslng Compact Airborne Spectrogtaphic lmager Data. *Photogrammetric Engineering and Remote Sensing*, *61*(9).

Goodman, N. R. (1963). Statistical analysis based on a certain multivariate complex Gaussian distribution (an introduction). *Annals of Mathematical Statistics*, *34*(1), 152–177. doi:10.1214/aoms/1177704250

Goodwin, N., Turner, R., & Merton, R. (2005). Classifying eucalyptus forests with high spatial and spectral resolution imagery. An investigation of individual species and vegetation communities. *Australian Journal of Botany*, *53*(4), 337–345. doi:10.1071/BT04085

Gorain, P. C., Sengupta, S., Satpati, G. G., Paul, I., Tripathi, S., & Pal, R. (2018). Carbon sequestration in macroalgal mats of brackish-water habitats in Indian Sunderbans: Potential as renewable organic resource. *The Science of the Total Environment*, *626*, 689–702. doi:10.1016/j.scitotenv.2018.01.106 PMID:29898555

Gorte, R. W. (2009). *US tree planting for carbon sequestration*. Academic Press.

Goswami, J., Tajo, L., & Sarma, K. K. (2010). Bamboo resources mapping using satellite technology. *Current Science*, 650–653.

Govender, M., Chetty, K., Naiken, V., & Bulcock, H. (2008). A comparison of satellite hyperspectral and muyltispectral remote sensing imagery for improved classification and mapping of vegetation. *Water S.A.*, *34*(2), 147–154. doi:10.4314/wsa.v34i2.183634

Gregoire, T. G., Valentine, H. T., & Furnival, G. M. (1995). Sampling methods to estimate foliage and other characteristics of individual trees. *Ecology*, *76*(4), 1181–1194. doi:10.2307/1940925

Grimaldi, S., Li, Y., Pauwels, V. R. N., & Walker, J. P. (2016). Remote Sensing-Derived Water Extent and Level to Constrain Hydraulic Flood Forecasting Models: Opportunities and Challenges. *Surveys in Geophysics*, *37*(5), 977–1034. doi:10.100710712-016-9378-y

Gungor, Z., Serhadlioglu, G., & Kesen, E. (2009). A fuzzy AHP approach to personnel selection problem. *Applied Soft Computing*, *9*, 641–646.

Gupta, A., & Kaur, H. (2017). A new pathogen, *Myxobolus holzerae (Myxosporea: Myxozoa)* causing severe gill disease in an Indian major carp *Labeo rohita* in a cold-water wetland, Punjab (India). *Microbial Pathogenesis*, *111*, 244–251. doi:10.1016/j.micpath.2017.08.044 PMID:28870697

Gupta, A., & Kaur, H. (2017). Morphological, histopathological and molecular characterization of *Thelohanellus theinensis n. sp. (Cnidaria: Myxosporea)* infecting an Indian major carp, Labeo bata in a cold-water wetland in Punjab (India). *Journal of Parasitic Diseases: Official Organ of the Indian Society for Parasitology*, *41*(3), 629–638. doi:10.100712639-016-0856-4 PMID:28848251

Gupta, A., & Kaur, H. (2018). *Myxobolus okamurae sp. nov. (Myxosporea: Myxozoa)* causing severe gill myxoboliosis in the cyprinid Labeo bata in a coldwater wetland, Punjab (India). *Microbial Pathogenesis*, *115*, 86–92. doi:10.1016/j.micpath.2017.12.041 PMID:29253599

Gupta, V., Jamir, I., Kumar, V., & Devi, M. (2017). Geomechanical characterization of slopes for assessing rockfall hazards in the Upper Yamuna Valley, Northwest Higher Himalaya, India. *Himalayan Geology*, *38*(2), 156–170.

Gupta, V., Nautiyal, H., Kumar, V., Jamir, I., & Tandon, R. (2016). Landslide hazards around Uttarkashi township, Garhwal Himalaya, after the tragic flash flood in June 2013. *Natural Hazards*, *80*(3), 1689–1707. doi:10.100711069-015-2048-4

Guzzetti, F., Cardinali, M., & Reichenbach, P. (1996). The influence of structural setting and lithology on landslide type and pattern. *Environmental & Engineering Geoscience*, *2*(4), 531–555. doi:10.2113/gseegeosci.II.4.531

Hairiah, K., Sitompul, S., van Noordwijk, M., & Palm, C. (2001). *Methods for sampling carbon stocks above and below ground*. ICRAF Bogoi.

Halgamuge, M., & Nirmalathas, A. (2017). Analysis of Large Flood Events: Based on Flood Data During 1985–2016 in Australia and India. Academic Press.

Hall, C. A., & Uhlig, J. (1991). Refining estimates of carbon released from tropical land-use change. *Canadian Journal of Forest Research*, *21*(1), 118–131. doi:10.1139/x91-016

Halme, E., Pellikka, P., & Mõttus, M. (2019). Utility of hyperspectral compared to multispectral remote sensing data in estimating forest biomass and structure variables in Finnish boreal forest. *Int J Appl Earth Obs Geoinformation*, *83*, 101942. doi:10.1016/j.jag.2019.101942

Han, W. J., & Tsay, W. D. (1998). *Formulation of quality strategy using analytic hierarchy process*. In 27th annual meeting of the Western Decision Science Institute, University of Northern Colorado.

Hansen, H. S. (2005). GIS-based multi-criteria analysis of wind farm development. In H. Hauska, & H. Tveite (Ed.), *Proceedings of the 10th Scandinavian research conference on geographical information science (ScanGIS'2005)* (pp. 75–87). Stockholm, Sweden: Royal Institute of Technology.

Han, T., Wulder, M., White, J., Coops, N., Alvarez, M., & Butson, C. (2007). An Efficient Protocol to Process Landsat Images for Change Detection With Tasselled Cap Transformation. *Geoscience and Remote Sensing Letters, IEEE*, *4*(1), 147–151. doi:10.1109/LGRS.2006.887066

Hawbaker, T. J., Keuler, N. S., Lesak, A. A., Gobakken, T., Contrucci, K., & Radeloff, V. C. (2009). Improved estimates of forest vegetation structure and biomass with a LiDAR-optimized sampling design. *Journal of Geophysical Research. Biogeosciences*, *114*(G2).

Hayashi, M., Saigusa, N., Oguma, H., & Yamagata, Y. (2013). Forest canopy height estimation using ICESat/GLAS data and error factor analysis in Hokkaido, Japan. *ISPRS Journal of Photogrammetry and Remote Sensing*, *81*, 12–18. doi:10.1016/j.isprsjprs.2013.04.004

Haygarth, P. M., & Ritz, K. (2009). The future of soils and land use in the UK: Soil systems for the provision of land-based ecosystem services. *Land Use Policy*, *26*, S187–S197. doi:10.1016/j.landusepol.2009.09.016

Heintzman, L. J., Starr, S. M., Mulligan, K. R., Barbato, L. S., & McIntyre, N. E. (2017). Using Satellite Imagery to Examine the Relationship between Surface-Water Dynamics of the Salt Lakes of Western Texas and Ogallala Aquifer Depletion. *Wetlands*, *37*(6), 1055–1065. doi:10.100713157-017-0940-2

Henderson, F. M., & Lewis, A. J. (1998). Principles and applications of imaging radar. *Manual of Remote Sensing, 2.*

Herbst, D. (2014). From mountains to the saline lakes of the Great Basin: ecosystems at risk. In *From Mountains to the Saline Lakes of the Great Basin: Ecosystems at Risk.* Ecological Society of America. https://www.Ramsar.org/wetland/india

Herrick, J. E., Bestelmeyer, B. T., Archer, S., Tugel, A. J., & Brown, J. R. (2006). An integrated framework for science-based arid land management. *Journal of Arid Environments, 65*(2), 319–335. doi:10.1016/j.jaridenv.2005.09.003

Hickin, E. J. (1984). Vegetation and river channel dynamics. *The Canadian Geographer. Geographe Canadien, 28*(2), 111–126. doi:10.1111/j.1541-0064.1984.tb00779.x

Hoekman, B., & Djankov, S. (1997). Determinants of the export structure of countries in Central and Eastern Europe. *The World Bank Economic Review, 11*(3), 471–487. doi:10.1093/wber/11.3.471

Holben, B. (1986). Characteristics of maximum-value composite images from temporal AVHRR data. *International Journal of Remote Sensing, 7*(11), 1417–1434. doi:10.1080/01431168608948945

Houben, P., Hoffmann, T., Zimmermann, A., & Dikau, R. (2006). Land use and climatic impacts on the Rhine system (RheinLUCIFS): Quantifying sediment fluxes and human impact with available data. *Catena, 66*(1–2), 42–52. doi:10.1016/j.catena.2005.07.009

Hovius, N., Meunier, P., Lin, C.-W., Chen, H., Chen, Y.-G., Dadson, S., Horng, M.-J., & Lines, M. (2011). Prolonged seismically induced erosion and the mass balance of a large earthquake. *Earth and Planetary Science Letters, 304*(3–4), 347–355. doi:10.1016/j.epsl.2011.02.005

Hovius, N., Stark, C. P., & Allen, P. A. (1997). Sediment flux from a mountain belt derived by landslide mapping. *Geology, 25*(3), 231–234. doi:10.1130/0091-7613(1997)025<0231:SFFAMB>2.3.CO;2

Howard, A. D., & Kerby, G. (1983). Channel changes in badlands. *Geological Society of America Bulletin, 94*(6), 739–752. doi:10.1130/0016-7606(1983)94<739:CCIB>2.0.CO;2

Hoyois & Guha-sapir. (2003). *Three Decades of Floods in Europe : A Preliminary Analysis of EMDAT Data . 2 . Number of Flood Disasters by UN European Region.* Academic Press.

Hua, L., Wang, H., Sui, H., Wardlow, B., Hayes, M. J., & Wang, J. (2019). Mapping the Spatial-Temporal Dynamics of Vegetation Response Lag to Drought in a Semi-Arid Region. *Remote Sensing, 11*(16), 1873. doi:10.3390/rs11161873

Huang, C., Wylie, B., Zhang, L., Homer, C., & Zylstra, G. (2005). *tasseled.pdf (application/pdf Object).* Retrieved September 11, 2009, from http://landcover.usgs.gov/pdf/tasseled.pdf

Huang, R., & Zhu, J. (2013). Integration of Hyperspectral Image and Lidar Data through Tri-training for Land Cover Semi-supervised Classification. In F. Bian, Y. Xie, X. Cui, & Y. Zeng (Eds.), *Geo-Informatics in Resource Management and Sustainable Ecosystem. Communications in Computer and Information Science* (Vol. 398). Springer. doi:10.1007/978-3-642-45025-9_57

Huang, X., Ziniti, B., Torbick, N., & Ducey, M. J. (2018). Assessment of forest above ground biomass estimation using multi-temporal C-band sentinel-1 and polarimetric L-band PALSAR-2 data. *Remote Sensing, 10*(9), 1424. doi:10.3390/rs10091424

Huang, Y., Chen, Z. X., Tao, Y. U., Huang, X. Z., & Gu, X. F. (2018). Agricultural remote sensing big data: Management and applications. *Journal of Integrative Agriculture, 17*(9), 1915–1931. doi:10.1016/S2095-3119(17)61859-8

Hudak, A. T., Lefsky, M. A., Cohen, W. B., & Berterretche, M. (2002). Integration of LiDARand Landsat ETM+ data for estimating and mapping forest canopy height. *Remote Sensing of Environment, 82*(2-3), 397–416. doi:10.1016/S0034-4257(02)00056-1

Huesca, M., García, M., Roth, K. L., Casas, A., & Ustin, S. L. (2016). Canopy structural attributes derived from AVIRIS imaging spectroscopy data in a mixed broadleaf/conifer forest. *Remote Sensing of Environment, 182*, 208–226. doi:10.1016/j.rse.2016.04.020

Husch, B., Beers, T. W., & Kershaw, J. A. Jr. (2002). *Forest mensuration.* John Wiley & Sons.

Hycza, T., Stereńczak, K., & Bałazy, R. (2018). Potential use of hyperspectral data to classify forest tree species. *New Zealand Journal of Forestry Science, 48*(18), 1–13. doi:10.118640490-018-0123-9

Ian, H., Cornelius, S., & Carver, S. (2010). *An introduction to GIS.* Pearson Education Limited.

ICAR. Meghalaya. (2007). *Vision 2025.* Retrieved from: http://www.kiran.nic.in/pdf/reports/Vision_2025_ICARNEH.pdf

Iizuka, K., Yonehara, T., Itoh, M., & Kosugi, Y. (2018). Estimating Tree Height and Diameter at Breast Height (DBH) from Digital surface models and orthophotos obtained with an unmanned aerial system for a Japanese Cypress (Chamaecyparis obtusa) Forest. *Remote Sensing, 10*(1), 13.

Imhoff, M. L. (1993). Radar backscatter/biomass saturation: Observations and implications for global biomass assessment. *Proceedings of IGARSS'93-IEEE International Geoscience and Remote Sensing Symposium.*

Indian State Forest Report (ISFR). (2009). *Forest Survey of India.* Retrieved from http://fsi.nic.in/forest-report-2009

Indian State Forest Report (ISFR). (2015). *Forest Survey of India.* Retrieved from http://fsi.nic.in/forest-report-2015

Indian State Forest Report (ISFR). (2017). *Forest Survey of India.* Retrieved from http://fsi.nic.in/forest-report-2017

Indian State Forest Report (ISFR). (2019). *Forest Survey of India.* Retrieved from http://fsi.nic.in/forest-report-2019

IPCC. (2012). *Managing the Risks of Extreme Events and Disasters to Advance Climate Change Adaptation.* IPCC.

Irvin, B. J., Ventura, S. J., & Slater, B. K. (1997). Fuzzy and isodata classification of landform elements from digital terrain data in Pleasant Valley, Wisconsin. *Geoderma*, *77*(2-4), 137–154. doi:10.1016/S0016-7061(97)00019-0

Islam, K., Jashimuddin, M., Nath, B., & Nath, T. K. (2018). Land use classification and change detection by using multi-temporal remotely sensed imagery: The case of Chunati wildlife sanctuary, Bangladesh. *The Egyptian Journal of Remote Sensing and Space Sciences*, *21*(1), 37–47. doi:10.1016/j.ejrs.2016.12.005

Ismail, R., Mutanga, O., Kumar, L., & Bob, U. (2008). Determining the optimal spatial resolution of remotely sensed data for the detection of sirex noctilio infestations in pine plantations in kwazulu-natal, south africa. *The South African Geographical Journal*, *90*(1), 22–31. doi:10.1080/03736245.2008.9725308

Jaber, J. O., Jaber, Q. M., Sawalha, S. A., & Mohsen, M. S. (2008). Evaluation of conventional and renewable energy sources for space heating in the household sector. *Renewable & Sustainable Energy Reviews*, *12*, 278–289.

Jackson, R. D., & Huete, A. R. (1991). Interpreting vegetation indices. *Preventive Veterinary Medicine*, *11*(3-4), 185–200. doi:10.1016/S0167-5877(05)80004-2

Jacobs, S. (2002). Land reform: Still a goal worth pursuing for rural women? *Journal of International Development*, *14*(6), 887–898. doi:10.1002/jid.934

Jadkowski, M. A., Howard, R. R., & Brostuen, D. E. (1990). Application of SPOT data for regional growth analysis and local planning. *Photogrammetric Engineering and Remote Sensing*, *56*(2), 175–180.

Jahncke, R., Leblon, B., Bush, P., & Larocque, A. (2018). Mapping wetlands in Nova Scotia with multi-beam RADARSAT-2 Polarimetric SAR, optical satellite imagery, and Lidar data. *Int J Appl Earth Obs Geoinformation*, *68*, 139–156. doi:10.1016/j.jag.2018.01.012

Jain, V. J., Pandey, R. P., Jain, M. K., & Byun, H. R. (2015). Comparison of drought indices for appraisal of drought characteristics in the Ken River Basin. *Weather and Climate Extremes*, *8*, 1–11. doi:10.1016/j.wace.2015.05.002

Jaiswal, R., Mukherjee, S., Krishnamurthy, J., & Saxena, R. (2003). Role of remote sensing and GIS techniques for generation of groundwater prospect zone towards rural development – an approach. *International Journal of Remote Sensing*, *24*(5), 993–108. doi:10.1080/01431160210144543

Jamir, I., Gupta, V., Kumar, V., & Thong, G. T. (2017). Evaluation of potential surface instability using finite element method in Kharsali Village, Yamuna Valley, Northwest Himalaya. *Journal of Mountain Science*, *14*(8), 1666–1676. doi:10.100711629-017-4410-3

Jamir, I., Gupta, V., Thong, G. T., & Kumar, V. (2019). Litho-tectonic and precipitation implications on landslides, Yamuna valley, NW Himalaya. *Physical Geography*, 1–24. Advance online publication. doi:10.1080/02723646.2019.1672024

Jamwal, P. S., Chandan, P., Rattan, R., Anand, A., Kannan, P. M., & Parsons, M. H. (2017). Survey of avifauna of the Gharana wetland reserve: Implications for conservation in a semi-arid agricultural setting on the Indo-Pakistan border. *BMC Zoology*, *2*(1), 1–9. doi:10.118640850-017-0016-z

JAPCC (Jharkhand- Action Plan on Climate Change). (2014). Government of Jharkhand.

Jayakumar, K., Selvam, V., & Prabavathy, V. R. (2019). A WebGIS-Based Study for Managing Mangroves of Godavari Wetland, Andhra Pradesh, India. *Photonirvachak (Dehra Dun)*, *47*(12), 2027–2045. doi:10.100712524-019-01046-x

Jensen, J. R. (2005). *Introductory Digital Image Processing A Remote Sensing Perspective* (3rd ed.). Prentice Hall.

Jha, C. S., Dutt, C. B. S., & Bawa, K. S. (2000). Deforestation and land use changes in Western Ghats, India. *Current Science*, 231–238.

Jhala, Y. V., Gopal, R., & Qureshi, Q. (Eds.). (2008). Status of the tigers, co-predators, and prey in India. National Tiger Conservation Authority, Govt. of India, New Delhi, and Wildlife Institute of India, Dehradun. TR 08/001, 151.

Jia, K., Li, Q., Tian, Y., Wu, B., Zhang, F., & Meng, J. (2012). Crop classification using multi-configuration SAR data in the North China Plain. *International Journal of Remote Sensing*, *33*(1), 170–183. doi:10.1080/01431161.2011.587844

Ji, L., & Peters, A. J. (2003). Assessing vegetation response to drought in the northern Great Plains using vegetation and drought indices. *Remote Sensing of Environment*, *87*(1), 85–98. doi:10.1016/S0034-4257(03)00174-3

Jones, G., Coops, N. C., & Sharma, T. (2010). Assessing the utility of airborne hyperspectral and LiDAR data for species distribution mapping in the coastal Pacific Northwest, Canada. *Remote Sensing of Environment*, *114*(12), 2841–2852. doi:10.1016/j.rse.2010.07.002

Joshi, P. K., Lele, N., & Agarwal, S. P. (2006). Entropy as an indicator of fragmented landscape. *Current Science*, *91*(3), 276–278.

Joshi, P. K., Roy, P. S., Singh, S., Agrawal, S., & Yadav, D. (2006). Vegetation cover mapping in India using multi-temporal IRS Wide Field Sensor (WiFS) data. *Remote Sensing of Environment*, *103*(2), 190–202. doi:10.1016/j.rse.2006.04.010

JSADP (Jharkhand State Agriculture Development Plan). (2012). *Report by Nabard Consultancy Services*. Jharkhand Regional Office Ranchi.

JSDMP (A draft on Jharkhand State Disaster Management Plan). (2011). Disaster Management Department MDI Building, Dhurwa, Ranchi, Jharkhand.

Judd, C., Steinberg, S., Shaughnessy, F., & Crawford, G. (2007). Mapping salt marsh vegetation using aerial hyperspectral imagery and linear unmixing in Humboldt Bay, California mapping salt marsh vegetation using aerial hyperspectral. *BioOne*, *27*(4), 1144–1152.

Kahraman, C., Cebeci, U., & Ruan, D. (2004). Multi-attribute comparison of catering service companies using fuzzy AHP: The case of Turkey. *International Journal of Production Economics, 87*, 171–184.

Kalambe, G. N., & Upasani, V. N. (2017). *Isolation and Identification of Haloalkaliphilic Archaeal Isolates from a Soda Lake in India.* doi:10.15680/IJIRSET.2017.0606048

Kamal, V., Mukherjee, S., Singh, P., Sen, R., Vishwakarma, C. A., Sajadi, P., Asthana, H., & Rena, V. (2017). Flood frequency analysis of Ganga river at Haridwar and Garhmukteshwar. *Applied Water Science, 7*(4), 1979–1986. doi:10.100713201-016-0378-3

Kamilaris, A., Kartakoullis, A., & Prenafeta-Boldú, F. X. (2017). A review on the practice of big data analysis in agriculture. *Computers and Electronics in Agriculture, 143*, 23–37. doi:10.1016/j.compag.2017.09.037

Kamvar, K., Sepandar, S., Klein, K., Dan, D., Manning, M., & Christopher, C. (2003). *Spectral learning.* Paper presented at the International Joint Conference of Artificial Intelligence.

Kangas, J. (1992). Multiple-use planning of forest resources by using the analytic hierarchy process. *Scandinavian Journal of Forest Research, 7*, 259–268.

Karimi, A. R., Mehrdadi, N., Hashemian, S. J., & Nabi-Bidhendi, Gh. R., & Tavakkloi-Moghaddam, R. (2011). Using of the fuzzy TOPSIS and fuzzy AHP methods for wastewater treatment process selection. *International Journal of Academic Research, 3*, 737–745.

Kasim, O. F., & Agbola, S. B. (2018). Strategies and challenges of land reform in Nigeria. *Journal of Public Administration, 53*(2-1), 541-564.

Kaur, H., & Attri, R. (2015). Morphological and molecular characterization of Henneguya bicaudi n. sp. (Myxosporea: Myxobolidae) infecting gills of Cirrhinus mrigala (Ham.) in Harike Wetland, Punjab (India). *Parasitology Research, 114*(11), 4161–4167. doi:10.100700436-015-4647-0 PMID:26239799

Kaur, H., Attri, R., & Joshi, J. (2016). Molecular identification of a new myxozoan, *Myxobolus dermiscalis n. sp.* (Myxosporea) infecting scales of Labeo rohita Hamilton in Harike Wetland, Punjab (India). *International Journal for Parasitology. Parasites and Wildlife, 5*(2), 139–144. doi:10.1016/j.ijppaw.2015.10.003 PMID:27330981

Kaur, J., Walia, H., Mabwoga, S. O., & Arora, S. (2017). Water quality monitoring of an international wetland at Harike, Punjab and its impact on biological systems. *Applied Water Science, 7*(3), 1107–1115. doi:10.100713201-015-0347-2

Kauth, R., & Thomas, G. 1976. The tasselled cap-a graphic description of the spectral-temporal development of agricultural crops as seen by LANDSAT. In *Proceedings of the Symposium on Machine Processing of Remotely Sensed Data.* Purdue University.

Keesstra, S. D., van Huissteden, J., Vandenberghe, J., Van Dam, O., de Gier, J., & Pleizier, I. D. (2005). Evolution of the morphology of the river Dragonja (SW Slovenia) due to land-use changes. *Geomorphology, 69*(1–4), 191–207. doi:10.1016/j.geomorph.2005.01.004

Keller, E. A., & Pinter, N. (2002). *Active Tectonics: Earthquakes, Uplift, and Landscape* (2nd ed.). Prentice Hall.

Keller, M., Palace, M., & Hurtt, G. (2001). Biomass estimation in the Tapajos National Forest, Brazil: Examination of sampling and allometric uncertainties. *Forest Ecology and Management, 154*(3), 371–382. doi:10.1016/S0378-1127(01)00509-6

Kerfoot, W. C., Hobmeier, M. M., Regis, R., Raman, V. K., Brooks, C. N., Shuchman, R., … Reif, M. (2019). Lidar (light detection and ranging) and benthic invertebrate investigations: Migrating tailings threaten Buffalo Reef in Lake Superior. *Journal of Great Lakes Research.* doi:10.1016/j.jglr.2019.07.009

Ketterings, Q. M., Coe, R., van Noordwijk, M., & Palm, C. A. (2001). Reducing uncertainty in the use of allometric biomass equations for predicting above-ground tree biomass in mixed secondary forests. *Forest Ecology and Management, 146*(1-3), 199–209. doi:10.1016/S0378-1127(00)00460-6

Khadim, F. K., Su, H., Xu, L., & Tian, J. (2019). Soil Salinity Mapping in Everglades National Park Using Remote Sensing Techniques and Vegetation Salt Tolerance. *Physics and Chemistry of the Earth, 110*, 31–50. Advance online publication. doi:10.1016/j.pce.2019.01.004

Khalefa, E. I., Smit, P. J., Nickless, A., Archibald, S., Comber, A., & Balzter, H. (2013). Retrieval of Savanna Vegetation Canopy Height from ICESat-GLAS Spaceborne LiDAR With Terrain Correction. *IEEE Geoscience and Remote Sensing Letters, 10*(6), 1439–1443. doi:10.1109/LGRS.2013.2259793

Khamis, Afan, Nur Muhammad, & ElShafie. (2017). *Development of a spatial hydrologic soil map using spectral reflectance band recognition and a multiple – output artificial neural network model.* doi:10.5194/hess-2017-13

Khoshand, A., Rahimi, K., Ehteshami, M., & Gharaei, S. (2019). Fuzzy AHP approach for prioritizing electronic waste management options: A case study of Tehran, Iran. *Environmental Science and Pollution Research International, 26*(10), 9649–9660. PMID:30729431

Kim, C., Jeong, J., Kim, R.-H., Son, Y.-M., Lee, K. H., Kim, J.-S., & Park, I.-H. (2011). Allometric equations and biomass expansion factors of Japanese red pine on the local level. *Landscape and Ecological Engineering, 7*(2), 283–289. doi:10.100711355-010-0131-2

Kingsford, R. T. (2000). Ecological impacts of dams, water diversions and river management on floodplain wetlands in Australia. *Austral Ecology, 25*(2), 109–127. doi:10.1046/j.1442-9993.2000.01036.x

Kirby, E., & Whipple, K. (2001). Quantifying differential rock-uplift rates via stream profile analysis. *Geology, 29*(5), 415–418. doi:10.1130/0091-7613(2001)029<0415:QDRURV>2.0.CO;2

Kogan, F. N. (1997a). *Contribution of Remote Sensing to Drought Early Warning.* https://drought.unl.edu/monitor/EWS/ch7_Kogan.pdf

Kogan, F. N. (1995). Application of vegetation index and brightness temperature for drought detection. *Advances in Space Research, 15*(11), 91–100. doi:10.1016/0273-1177(95)00079-T

Kogan, F. N. (1997b). Global drought watch from space. *Bulletin of the American Meteorological Society, 78*(4), 621–636. doi:10.1175/1520-0477(1997)078<0621:GDWFS>2.0.CO;2

Kogan, F. N. (2000). Satellite-observed sensitivity of world land ecosystems to El Nino/La Nina. *Remote Sensing of Environment, 74*(3), 445–462. doi:10.1016/S0034-4257(00)00137-1

Kolstad, I., & Søreide, T. (2009). Corruption in natural resource management: Implications for policy makers. *Resources Policy, 34*(4), 214–226. doi:10.1016/j.resourpol.2009.05.001

Kong, F., & Liu, H. (2005). Applying fuzzy analytic hierarchy process to evaluate success factors of e-commerce. *International Journal of Information and Systems Sciences, 1*(3–4), 406–412.

Kora, A. J., Rastogi, L., Kumar, S. J., & Jagatap, B. N. (2017). Physico-chemical and bacteriological screening of Hussain Sagar lake: An urban wetland. *Water Science, 31*(1), 24–33. doi:10.1016/j.wsj.2017.03.003

Kordi, M. (2008). *Comparison of fuzzy and crisp analytic hierarchy process (AHP) methods for spatial multicriteria decision analysis in GIS* (MSc dissertation). University of Gavle.

Kothapalli Venkata, R., Poloju, S., Mullapudi Venkata Rama, S. S., Gogineni, A., Prabir Kumar, D., Allakki Venkata, R., & ... & KRP, S. (. (2017). Multi-incidence angle RISAT-1 Hybrid Polarimetric SAR data for large area mapping of maize crop–a case study in Khagaria district, Bihar, India. *International Journal of Remote Sensing, 38*(20), 5487–5501. doi:10.1080/01431161.2017.1338783

Krishnakumar, A., Saranya, P., Prasad, P. G., & Rakhi, C. (2015). Holocene Records of Human Driven Geological Impacts in a Ramsar Wetland of India. *Aquatic Procedia, 4*, 373–380. doi:10.1016/j.aqpro.2015.02.050

Kuhlbusch, T. A., & Crutzen, P. J. (1996). Black carbon, the global carbon cycle, and atmospheric carbon dioxide. *Biomass Burning and Global Change, 1*, 160-169.

Kulak, O., & Kahraman, C. (2005). Fuzzy multi-criterion selection among transportation companies using axiomatic design and analytic hierarchy process. *Information Sciences, 170*, 191–210.

Kumar, A., Kishore, B. S. P. C., Saikia, P., Deka, J., Bharali, S., Singha, L. B., Tripathi, O. P., & Khan, M. L. (2019). Tree diversity assessment and above ground forests biomass estimation using SAR remote sensing: A case study of higher altitude vegetation of North-East Himalayas, India. *Physics and Chemistry of the Earth Parts A/B/C, 111*, 53–64. doi:10.1016/j.pce.2019.03.007

Kumar, D. (2011). Monitoring forest cover changes using remote sensing and GIS: A global prospective. *Research Journal of Environmental Sciences, 5*(2), 105–123. doi:10.3923/rjes.2011.105.123

Kumar, K. C. V. N., Demudu, G., Malini, B. H., Rao, K. N., & Kubo, S. (2016). Geospatial Analysis of the Changing Environment of Kolleru Lake, the Largest Freshwater Wetland in India. *Wetlands*, *36*(4), 745–758. doi:10.100713157-016-0787-y

Kumar, P., & Sharma, A. (2019). Wetland Birds Assemblages in Man-Made Sacred Ponds of Kurukshetra, India. *Proceedings of the Zoological Society*, *72*(1), 61–73. doi:10.100712595-018-0259-x

Kumar, R., Jain, V., Prasad Babu, G., & Sinha, R. (2014). Connectivity structure of the Kosi megafan and role of rail-road transport network. *Geomorphology*, *227*, 73–86. doi:10.1016/j.geomorph.2014.04.031

Kumar, S., & Devadas, V. (2016). Integrated Planning for Sustainable Development of Kuttanad Wetland Region, Kerala State. *Procedia Technology*, *24*, 1660–1667. doi:10.1016/j.protcy.2016.05.185

Kumar, V., Gupta, V., & Jamir, I. (2018). Hazard evaluation of progressive Pawari landslide zone, Satluj valley, Himachal Pradesh, India. *Natural Hazards*, *93*(2), 1029–1047. doi:10.100711069-018-3339-3

Kumar, V., Gupta, V., Jamir, I., & Chattoraj, S. L. (2019a). Evaluation of potential landslide damming: Case study of Urni landslide, Kinnaur, Satluj valley, India. *Geoscience Frontiers*, *10*(2), 753–767. doi:10.1016/j.gsf.2018.05.004

Kumar, V., Gupta, V., & Sundriyal, Y. P. (2019b). Spatial interrelationship of landslides, litho-tectonics, and climate regime, Satluj valley, Northwest Himalaya. *Geological Journal*, *54*(1), 537–551. doi:10.1002/gj.3204

Kundzewicz, Z. W., Pińskwar, I., & Robert Brakenridge, G. (2013). Large Floods in Europe, 1985–2009. *Hydrological Sciences Journal*, *58*(1), 1–7. doi:10.1080/02626667.2012.745082

Kurvonen, L., Pulliainen, J., & Hallikainen, M. (1999). Retrieval of biomass in boreal forests from multitemporal ERS-1 and JERS-1 SAR images. *IEEE Transactions on Geoscience and Remote Sensing*, *37*(1), 198–205. doi:10.1109/36.739154

Kushwaha, S. P. S., Nandy, S., Shah, M. A., Agarwal, R., & Mukhopadhyay, S. (2018). Forest cover monitoring and prediction in a Lesser Himalayan elephant landscape. *Current Science, 115*(3).

Kushwaha, S. P. S. (2005). Geospatial Technology for Forest Resources Management. *International Journal of Ecology and Environmental Sciences*, *31*(1), 9–20.

Kushwaha, S. P. S., Munkhtuya, S., & Roy, P. S. (2000). Mountain goat habitat evaluation in Rajaji National Park using remote sensing and GIS. *Photonirvachak (Dehra Dun)*, *28*, 293–303.

La, H. P., Eo, Y. D., Chang, A., & Kim, C. (2015). Extraction of Individual Tree Crown using Hyperspectral Image and LiDAR Data. *KSCE Journal of Civil Engineering*, *19*(4), 1078–1087. doi:10.100712205-013-1178-z

Lallianthanga, R. K., & Sailo, R. L. (2012). Monitoring of bamboo flowering using satellite remote sensing and GIS techniques in Mizoram, India. *Science Vision, 12*(4), 147.

Lang, A., Moya, J., Corominas, J., Schrott, L., & Dikau, R. (1999). Classic and new dating methods for assessing the temporal occurrence of mass movements. *Geomorphology, 30*(1), 33–52. doi:10.1016/S0169-555X(99)00043-4

Lasko, K., Vadrevu, K. P., Tran, V. T., & Justice, C. (2018). Mapping double and single crop paddy rice with Sentinel-1A at varying spatial scales and polarizations in Hanoi, Vietnam. *IEEE Journal of Selected Topics in Applied Earth Observations and Remote Sensing, 11*(2), 498–512. doi:10.1109/JSTARS.2017.2784784 PMID:30151066

Laurin, G. V., Chen, Q., & Lindsell, J. A. (2014). Above ground biomass estimation in an African tropical forest with lidar and hyperspectral data. *ISPRS Journal of Photogrammetry and Remote Sensing, 89*, 49–58. doi:10.1016/j.isprsjprs.2014.01.001

Lawrence, R. L., Wood, S. D., & Sheley, R. L. (2006). Mapping invasive plants using hyperspectral imagery and breiman cutler classifications (randomforest). *Remote Sensing of Environment, 100*(3), 356–362. doi:10.1016/j.rse.2005.10.014

Le Toan, T., Quegan, S., Davidson, M., Balzter, H., Paillou, P., Papathanassiou, K., ... Shugart, H. (2011). The BIOMASS mission: Mapping global forest biomass to better understand the terrestrial carbon cycle. *Remote Sensing of Environment, 115*(11), 2850–2860. doi:10.1016/j.rse.2011.03.020

Lee, K.-S., Cohen, W. B., Kennedy, R. E., Maiersperger, T. K., & Gower, S. T. (2004). Hyperspectral versus multispectral data for estimating leaf area index in four different biomes. *Remote Sensing of Environment, 91*(3-4), 508–520. doi:10.1016/j.rse.2004.04.010

Lefsky, M. A., Cohen, W. B., Acker, S. A., Spies, T. A., Parker, G. G., & Harding, D. (1998). LiDAR remote sensing of forest canopy structure and related biophysical parameters at the H.J. Andrews Experimental Forest, Oregon, USA. *Geoscience and Remote Sensing Symposium Proceedings, IGARSS*, 1252–1254.

Lefsky, M.A., Harding, D.J., Keller, M., 2005. Estimates of forest canopy height and aboveground biomass using ICESat. *Geophysical Research Letters, 32*(L22S02), 1-4. doi:10.1029/2005GL023971

Lefsky, M. A., Cohen, W. B., Harding, D. J., Parker, G. G., Acker, S. A., & Gower, S. T. (2002). Lidar remote sensing of above-ground biomass in three biomes. *Global Ecology and Biogeography, 11*(5), 393–399. doi:10.1046/j.1466-822x.2002.00303.x

Lehner, B., Liermann, C. R., Revenga, C., Vörösmarty, C., Fekete, B., Crouzet, P., Döll, P., Endejan, M., Frenken, K., & Magome, J. (2011b). Global reservoir and dam (GRanD) database. *Technical Documentation. Version, 1*, 1–14.

Lehner, B., Liermann, C. R., Revenga, C., Vörösmarty, C., Fekete, B., Crouzet, P., Döll, P., Endejan, M., Frenken, K., Magome, J., Nilsson, C., Robertson, J. C., Rödel, R., Sindorf, N., & Wisser, D. (2011a). High-resolution mapping of the world's reservoirs and dams for sustainable river-flow management. *Frontiers in Ecology and the Environment*, *9*(9), 494–502. doi:10.1890/100125

Lehner, B., Reidy Liermann, C., Revenga, C., Vorosmarty, C., Fekete, B., Crouzet, P., Doll, P., Endejan, M., Frenken, K., & Magome, J. (2011c). Global Reservoir and Dam Database, Version 1 (GRanDv1): Dams. *Revision*, *01*. Advance online publication. doi:10.7927/H4N877QK

Lele, N., & Joshi, P. K. (2009). Analyzing deforestation rates, spatial forest cover changes and identifying critical areas of forest cover changes in North-East India during 1972–1999. *Environmental Monitoring and Assessment*, *156*(1-4), 159–170. doi:10.100710661-008-0472-6 PMID:18677545

Lenderink, G., & Van Meijgaard, E. (2008). Increase in hourly precipitation extremes beyond expectations from temperature changes. *Nature Geoscience*, *1*(8), 511–514. doi:10.1038/ngeo262

Leung, L. C., & Chao, D. (2000). On consistency and ranking of alternatives in fuzzy AHP. *European Journal of Operational Research*, *124*, 102–113.

Lewis, A., Oliver, S., Lymburner, L., Evans, B., Wyborn, L., Mueller, N., Raevksi, G., Hooke, J., Woodcock, R., Sixsmith, J., Wu, W., Tan, P., Li, F., Killough, B., Minchin, S., Roberts, D., Ayers, D., Bala, B., Dwyer, J., ... Wang, L.-W. (2017). The Australian geoscience data cube-foundations and lessons learned. *Remote Sensing of Environment*, *202*, 276–292. doi:10.1016/j.rse.2017.03.015

Lian, J.-J., Guo, X.-Y., Ma, C., & Xu, K. (2019). Optimal Reservoir Flood Control Operation Using a Hedging Model and Considering the Near-Field Vibrations Induced by Flood Release. *Water Resources Management*, *33*(8), 2645–2663. doi:10.100711269-019-02231-w

Liao, W., Gao, L., Zhang, B., & Chanusso, J. (2018). Deep Learning for Fusion of APEX Hyperspectral and Full-Waveform LiDAR Remote Sensing Data for Tree Species Mapping. *IEEE Access : Practical Innovations, Open Solutions*, *6*, 68716–68729.

Lillesand, T., Kiefer, R. W., & Chipman, J. (2015). *Remote sensing and image interpretation*. John Wiley & Sons.

Lim, C.H., Kim, S.H., Chun, J.A., Kafatos, M.C., & Lee, W.K. (2019). *Assessment of Agricultural Drought Considering the Hydrological Cycle and Crop Phenology in the Korean Peninsula.* . doi:10.3390/w11051105

Limin, Y., & George, X. (2003, September). Urban Land-Cover Change Detection through Sub-Pixel Imperviousness Mapping Using Remotely Sensed Data. *Photogrammetric Engineering and Remote Sensing*, *69*(9), 1003–1010. doi:10.14358/PERS.69.9.1003

Lim, K., Treitz, P., Wulder, M., St-Onge, B., & Flood, M. (2003). LiDAR remote sensing of forest structure. *Progress in Physical Geography*, *27*(1), 88–106. doi:10.1191/0309133303pp360ra

Lin, Q. L., & Wang, D. (2019). Facility layout planning with SHELL and fuzzy AHP method based on human reliability for operating theatre. *Journal of Healthcare Engineering*, *2019*, 1–12. PMID:30792832

Liu, L., Coops, N. C., Aven, N. W., & Pang, Y. (2017). Mapping urban tree species using integrated airborne hyperspectral and LiDAR remote sensing data. *Remote Sensing of Environment*, *200*, 170–182. doi:10.1016/j.rse.2017.08.010

Liu, M., Tian, H., Chen, G., Ren, W., Zhang, C., & Liu, J. (2008). Effects of Land-Use and Land-Cover Change on Evapotranspiration and Water Yield in China During 1900-2000 1. *JAWRA Journal of the American Water Resources Association*, *44*(5), 1193–1207. doi:10.1111/j.1752-1688.2008.00243.x

Lone, J. M., Sivasankar, T., Sarma, K. K., Qadir, A., & Raju, P. L. N. (2017). Influence of slope aspect on above ground biomass estimation using ALOS-2 data. *International Journal of Scientific Research*, *6*(6), 1422–1428.

Loushambam, R. S., Singh, N. R., Taloh, A., & Mayanglambam, S. (2017). Bamboo in North East India. *Indian Journal of Hill Farming*, *30*(2), 181–185.

Lucas, R. M., Cronin, N., Lee, A., Moghaddam, M., Witte, C., & Tickle, P. (2006). Empirical relationships between AIRSAR backscatter and LiDAR-derived forest biomass, Queensland, Australia. *Remote Sensing of Environment*, *100*(3), 407–425. doi:10.1016/j.rse.2005.10.019

Lucas, R. M., Lee, A. C., & Bunting, P. J. (2008). Retrieving forest biomass through integration of CASI and LIDAR data. *International Journal of Remote Sensing*, *29*(5), 1553–1577. doi:10.1080/01431160701736497

Lucht, W., Schaphoff, S., Erbrecht, T., Heyder, U., & Cramer, W. (2006). Terrestrial vegetation redistribution and carbon balance under climate change. *Carbon Balance and Management*, *1*(1), 6. doi:10.1186/1750-0680-1-6 PMID:16930462

Luckman, A., Baker, J., Kuplich, T. M., Yanasse, C. C. F., & Frery, A. C. (1997). A study of the relationship between radar backscatter and regenerating tropical forest biomass for spaceborne SAR instruments. *Remote Sensing of Environment*, *60*(1), 1–13. doi:10.1016/S0034-4257(96)00121-6

Lu, D., Mausel, P., Brondízio, E., & Moran, E. (2004). Change detection techniques. *International Journal of Remote Sensing*, *25*(12), 2365–2401. doi:10.1080/0143116031000139863

Ludwig, C., Walli, A., Schleicher, C., Weichselbaum, J., & Riffler, M. (2019). A highly automated algorithm for wetland detection using multi-temporal optical satellite data. *Remote Sensing of Environment*, *224*, 333–351. doi:10.1016/j.rse.2019.01.017

Lugo, A. E., & Brown, S. (1992). Tropical forests as sinks of atmospheric carbon. *Forest Ecology and Management*, *54*(1-4), 239–255. doi:10.1016/0378-1127(92)90016-3

Luna, E., Jouany, C., & Castañeda, C. (2018, February). 2019. Soil composition and plant nutrients at the interface between crops and saline wetlands in arid environments in NE Spain. *Catena*, *173*, 384–393. doi:10.1016/j.catena.2018.10.032

Maavara, T., Chen, Q., Van Meter, K., Brown, L. E., Zhang, J., Ni, J., & Zarfl, C. (2020). River dam impacts on biogeochemical cycling. *Nature Reviews Earth & Environment, 1*(2), 103–116. doi:10.103843017-019-0019-0

Mabogunje, A. L. (2009). New land reform in Nigeria. *Science, 325*(5942), 793.

Mabogunje, A. L. (2010, April). Land Reform in Nigeria: progress, problems & prospects. In *Annual Conference on Land Policy and Administration*. World Bank.

Macchione, F., Costabile, P., Costanzo, C., & De Santis, R. (2019). Moving to 3-D flood hazard maps for enhancing risk communication. *Environmental Modelling & Software, 111*, 510–522. doi:10.1016/j.envsoft.2018.11.005

Madhavan Unni, N. V. (1979). *Satellite remote sensing survey of natural resources of Nagaland. Report*. National Remote Sensing Agency.

Madu, I. A. (2012). Spatial Impacts of Rural Population Pressure on Agricultural Land Use in Nigeria. *Applied Spatial Analysis and Policy, 5*(2), 123–135. doi:10.100712061-011-9064-x

Magilligan, F. J., & Goldstein, P. S. (2001). El Niño Floods and Culture Change: A Late Holocene Flood History for the Rio Moquegua, Southern Peru. *Geology, 29*(5), 431–434. doi:10.1130/0091-7613(2001)029<0431:ENOFAC>2.0.CO;2

Maire, G., François, C., Soudani, K., Berveiller, D., Pontailler, J.-Y., Bréda, N., Genet, H., Davi, H., & Dufrêne, E. (2008). Calibration and validation of hyperspectral indices for the estimation of broadleaved forest leaf chlorophyll content, leaf mass per area, leaf area index and leaf canopy biomass. *Remote Sensing of Environment, 112*(10), 3846–3864. doi:10.1016/j.rse.2008.06.005

Majumder, M., Shukla, A. K., & Arunachalam, A. (2010). Agricultural practices in Northeast India and options for sustainable management. In *Biodiversity, Biofuels, Agroforestry and Conservation Agriculture* (pp. 287–315). Springer. doi:10.1007/978-90-481-9513-8_10

Malczewski, J. (1999). *GIS and multicriteria decision analysis*. John Willey and Sons Inc.

Manjusree, P., Prasanna Kumar, L., & Bhatt, C. M. (2012). Optimization of threshold ranges for rapid flood inundation mapping by evaluating backscatter profiles of high incidence angle SAR images. *International Journal of Disaster Risk Science, 3*, 113–122. doi:10.100713753-012-0011-5

Man, Q., Dong, P., Guo, H., Liu, G., & Shi, R. (2014). Light detection and ranging and hyperspectral data for estimation of forest biomass: A review. *Journal of Applied Remote Sensing, 8*(1), 081598. doi:10.1117/1.JRS.8.081598

Martinis, S., Kersten, J., & Twele, A. (2015). A fully automated TerraSAR-X based flood service. *ISPRS Journal of Photogrammetry and Remote Sensing, 104*, 203–212. doi:10.1016/j.isprsjprs.2014.07.014

Martin, K. S., & Wing, J. (2007). The discourse and discipline of GIS. *Cartographica: The International Journal for Geographic Information and Geovisualization, 42*(3), 235–248. doi:10.3138/carto.42.3.235

Martin, M. E., & Aber, J. D. (1997). High spectral resolution remote sensing of forest canopy lignin, nitrogen, and ecosystem processes. *Ecological Applications*, *7*(2), 431–443. doi:10.1890/1051-0761(1997)007[0431:HSRRSO]2.0.CO;2

Martin, M. E., Newman, S. D., Aber, J. D., & Congalton, J. C. (1998). Determining forest species composition using high spectral resolution remote sensing data. *Remote Sensing of Environment*, *65*(3), 249–254. doi:10.1016/S0034-4257(98)00035-2

Mashi, S. A., & Shuaibu, H. S. (2018). People and sustainable land management: Assessment of stakeholders' knowledge of the nature of land use/cover change in Abuja, Nigeria. *GeoJournal*, *83*(3), 545–562. doi:10.100710708-017-9782-y

Mason, D. C., Giustarini, L., Garcia-Pintado, J., & Cloke, H. L. (2014). Detection of flooded urban areas in high resolution Synthetic Aperture Radar images using double scattering. *International Journal of Applied Earth Observation and Geoinformation*, *28*, 150–159. doi:10.1016/j.jag.2013.12.002

Matemilola, S., Adedeji, O. H., & Enoguanbhor, E. C. (2018). Land Use/Land Cover Change in Petroleum-Producing Regions of Nigeria. In The Political Ecology of Oil and Gas Activities in the Nigerian Aquatic Ecosystem (pp. 257-276). Academic Press.

Matgen, P., Hostache, R., Schumann, G., Pfister, L., Hoffmann, L., & Savenije, H. H. G. (2011). Towards an automated SAR-based flood monitoring system: Lessons learned from two case studies. *Physics and Chemistry of the Earth Parts A/B/C*, *36*(7-8), 241–252. doi:10.1016/j.pce.2010.12.009

Matgen, P., Schumann, G., & Henry, J.-B. (2007). Integration of SAR-derived river inundation areas, high-precision topographic data and a river flow model toward near real-time flood management. *International Journal of Applied Earth Observation and Geoinformation*, *9*, 247–263. doi:10.1016/j.jag.2006.03.003

Mather, P., & Tso, B. (2016). *Classification methods for remotely sensed data*. CRC press. doi:10.1201/9781420090741

Matsuki, T., Yokoya, N., & Iwasaki, A. (2015). Hyperspectral tree species classification of Japanese complex mixed forest with the aid of LiDAR data. *IEEE Journal of Selected Topics in Applied Earth Observations and Remote Sensing*, *8*(5), 2177–2187. doi:10.1109/JSTARS.2015.2417859

Matthews, M. W., Bernard, S., & Winter, K. (2010). Remote sensing of cyanobacteria-dominant algal blooms and water quality parameters in Zeekoevlei, a small hypertrophic lake, using MERIS. *Remote Sensing of Environment*, *114*(9), 2070–2087. doi:10.1016/j.rse.2010.04.013

Mbuyi, K. (2019). Kinshasa: problems of land management, infrastructure, and food supply. In African Cities in Crisis (pp. 148-175). Routledge.

McFeeters, S. K. (1996). The use of the Normalized Difference Water Index (NDWI) in the delineation of open water features. *International Journal of Remote Sensing*, *17*(7), 1425–1432. doi:10.1080/01431169608948714

McGrath, H., Bourgon, J.-F., Proulx-Bourque, J.-S., Nastev, M., & Abo El Ezz, A. (2018). A comparison of simplified conceptual models for rapid web-based flood inundation mapping. *Natural Hazards*, *93*(2), 905–920. doi:10.100711069-018-3331-y

Mcinerney, D. O., Suarez-Minguez, J., Valbuena, R., & Nieuwenhuis, M. (2010). Forest canopy height retrieval using LiDAR data, medium-resolution satellite imagery and kNN estimation in Aberfoyle, Scotland. *Forestry*, *83*(2), 1–12. doi:10.1093/forestry/cpq001

McMichael, A. J. (2013). Globalization, climate change, and human health. *The New England Journal of Medicine*, *368*(14), 1335–1343. doi:10.1056/NEJMra1109341 PMID:23550671

McNairn, H., Champagne, C., Shang, J., Holmstrom, D., & Reichert, G. (2009). Integration of optical and Synthetic Aperture Radar (SAR) imagery for delivering operational annual crop inventories. *ISPRS Journal of Photogrammetry and Remote Sensing*, *64*(5), 434–449. doi:10.1016/j.isprsjprs.2008.07.006

McNairn, H., Shang, J., Champagne, C., & Jiao, X. (2009, July). TerraSAR-X and RADARSAT-2 for crop classification and acreage estimation. In *2009 IEEE International Geoscience and Remote Sensing Symposium* (Vol. 2, pp. II-898). IEEE. 10.1109/IGARSS.2009.5418243

Meena, D.K., Lianthuamluaia, L., Mishal, P., Swain, H. S., Naskar, B. K., Saha, S., Das, B. K. (2019). Assemblage patterns and community structure of macro-zoobenthos and temporal dynamics of eco-physiological indices of two wetlands, in lower gangetic plains under varying ecological regimes: A tool for wetland management. *Ecological Engineering, 130*, 1–10. doi:10.1016/j.ecoleng.2019.02.002

Merritt, D. M., & Cooper, D. J. (2000). Riparian vegetation and channel change in response to river regulation: A comparative study of regulated and unregulated streams in the Green River Basin, USA. *Regulated Rivers: Research and Management*, *16*(6), 543–564. doi:10.1002/1099-1646(200011/12)16:6<543::AID-RRR590>3.0.CO;2-N

Middleton, B. A., Van der Valk, A. G., & Davis, C. B. (2015). Responses to water depth and clipping of twenty-three plant species in an Indian monsoonal wetland. *Aquatic Botany*, *126*, 38–47. doi:10.1016/j.aquabot.2015.06.004

Miehle, P., Grote, R., Battaglia, M., Feikema, P. M., & Arndt, S. K. (2010). Evaluation of a process-based ecosystem model for long-term biomass and stand development of Eucalyptus globulus plantations. *European Journal of Forest Research*, *129*(3), 377–391. doi:10.100710342-009-0343-x

Mikhailov, L., & Tsvetinov, P. (2004). Evaluation of services using a fuzzy analytic hierarchy process. *Applied Soft Computing*, *5*, 23–33.

Milja, T. E., Prathish, K. P., & Rao, T. P. (2011). Synthesis of surface imprinted nanospheres for selective removal of uranium from simulants of Sambhar Salt Lake and ground water. *Journal of Hazardous Materials*, *188*(1-3), 384–390. doi:10.1016/j.jhazmat.2011.01.121 PMID:21345587

Mmom, P. C., & Arokoyu, S. B. (2010). Mangrove forest depletion, biodiversity loss and traditional resources management practices in the Niger Delta, Nigeria. *Research Journal of Applied Sciences, Engineering and Technology*, *2*(1), 28–34.

Modrick, T. M., & Georgakakos, K. P. (2015). The character and causes of flash flood occurrence changes in mountainous small basins of Southern California under projected climatic change. *Journal of Hydrology. Regional Studies*, *3*, 312–336. doi:10.1016/j.ejrh.2015.02.003

Mondal, B., Dolui, G., Pramanik, M., Maity, S., & Sarathi, S. (2017). Urban expansion and wetland shrinkage estimation using a GIS-based model in the East Kolkata Wetland, India. *Ecological Indicators*, *83*(July), 62–73. doi:10.1016/j.ecolind.2017.07.037

Montes, N., Gauquelin, T., Badri, W., Bertaudiere, V., & Zaoui, E. H. (2000). A non-destructive method for estimating above-ground forest biomass in threatened woodlands. *Forest Ecology and Management*, *130*(1-3), 37–46. doi:10.1016/S0378-1127(99)00188-7

Moore, J. N. (2016). Recent desiccation of Western Great Basin saline lakes: Lessons from Lake Abert, Oregon, USA. *The Science of the Total Environment*, *554*, 142–154. doi:10.1016/j.scitotenv.2016.02.161 PMID:26950628

Moran, M. S., Vidal, A., Troufleau, D., Qi, J., Clarke, T. R., Pinter, P. J. Jr, ... Neale, C. M. U. (1997). Combining multifrequency microwave and optical data for crop management. *Remote Sensing of Environment*, *61*(1), 96–109. doi:10.1016/S0034-4257(96)00243-X

Murali, K., Ravikumar, G., & Krishnaveni, M. (2008). Remote sensing based agricultural drought assessment in Palar basin of Tamil Nadu State, India. *Photonirvachak (Dehra Dun)*, *37*, 9–20.

Mure-Ravaud, M., Binet, G., Bracq, M., Perarnaud, J.-J., Fradin, A., & Litrico, X. (2016). A web based tool for operational real-time flood forecasting using data assimilation to update hydraulic states. *Environmental Modelling & Software*, *84*, 35–49. doi:10.1016/j.envsoft.2016.06.002

Murray, N. J., Phinn, S. R., Clemens, R. S., Roelfsema, C. M., & Fuller, R. A. (2012). Continental scale mapping of tidal flats across East Asia using the Landsat archive. *Remote Sensing*, *4*(11), 3417–3426. doi:10.3390/rs4113417

Murthy, C. S., Sai, M. V. R. S., Kumari, V. B., & Roy, P. S. (2007). Agricultural drought assessment at disaggregated level using AWiFS/WiFS data of Indian Remote Sensing satellites. *Geocarto International*, *22*(2), 127–140. doi:10.1080/10106040701205039

Murthy, I. K., Sharma, N., & Ravindranath, N. H. (2013). Harnessing REDD+ opportunities for forest conservation and carbon stock enhancement in the Northeastern States of India. *Nature and Science*, *5*(3), 349–358.

Mutanga, O., & Skidmore, A. K. (2004). Narrow band vegetation indices overcome the saturation problem in biomass estimation. *International Journal of Remote Sensing*, *25*(19), 3999–4014. doi:10.1080/01431160310001654923

Muthumanickam, D., Kannan, P., Kumaraperumal, R., Natarajan, S., Sivasamy, R., & Poongodi, C. (2011). Drought assessment and monitoring through remote sensing and GIS in western tracts of Tamil Nadu, India. *International Journal of Remote Sensing*, *32*(18), 5157–5176. doi:10.1080/01431161.2010.494642

Nandeesha & Ramu, D. (2015). Agricultural drought assessment in central dry zone of agro climatic zones of Karnataka using Gis and Remote Sensing. *International Journal of Information Technology and Computer Science, 3*(2), 575-581. http://eprints.uni-mysore.ac.in/id/eprint/4842

Narumalani, S., Mishra, D. R., Wilson, R., Reece, P., & Kohler, A. (2009). Detecting and mapping four invasive species along the floodplain of North Platte River, Nebraska. *Weed Technology*, *23*(1), 99–107. doi:10.1614/WT-08-007.1

Nasirzadehdizaji, R., Balik Sanli, F., Abdikan, S., Cakir, Z., Sekertekin, A., & Ustuner, M. (2019). Sensitivity Analysis of Multi-Temporal Sentinel-1 SAR Parameters to Crop Height and Canopy Coverage. *Applied Sciences (Basel, Switzerland)*, *9*(4), 655. doi:10.3390/app9040655

Nasta, P., Palladino, M., Ursino, N., Saracino, A., Sommella, A., & Romano, N. (2017). Assessing long-term impact of land-use change on hydrological ecosystem functions in a Mediterranean upland agro-forestry catchment. *The Science of the Total Environment*, *605–606*, 1070–1082. doi:10.1016/j.scitotenv.2017.06.008 PMID:28715855

Navale, A., & Haldar, D. (n.d.). Evaluation of machine learning algorithms to Sentinel SAR data. *Spatial Information Research*, 1-11. Doi:10.100741324-019-00296-8

Navalgund, R. R., & Ray, S. S. (2000). Geomatics in natural resources management. Proceedings of Geomatics-2000, NR1-NR14.

Navalgund, R. R., Jayaraman, V., & Roy, P. S. (2007). Remote sensing applications: An overview. *Current Science*, *93*(12), 1747–1766.

Navar, J. (2009). Allometric equations for tree species and carbon stocks for forests of northwestern Mexico. *Forest Ecology and Management*, *257*(2), 427–434. doi:10.1016/j.foreco.2008.09.028

Nayar, M. P. (2002). Hot spots of endemic plants of India, Nepal and Bhutan. Tropical Botanical Garden Research Institute.

Negi, G., Rikhari, H., & Singh, S. (1992). Phenological features in relation to growth forms and biomass accumulation in an alpine meadow of the Central Himalaya. *Vegetatio*, *101*(2), 161–170.

Nelson, J. C. (1999). A review of the efficacy of serotonergic and noradrenergic reuptake inhibitors for treatment of major depression. *Biological Psychiatry*, *46*(9), 1301–1308. doi:10.1016/S0006-3223(99)00173-0 PMID:10560035

Nelson, R., Oderwald, R., & Gregoire, T. G. (1997). Separating the ground and airborne laser sampling phases to estimate tropical forest basal area, volume, and biomass. *Remote Sensing of Environment*, *60*(3), 311–326. doi:10.1016/S0034-4257(96)00213-1

Nelson, R., Ranson, K. J., Sun, G., Kimes, D. S., Kharuk, V., & Montesano, P. (2009). Estimating Siberian timber volume using MODIS and ICESat/GLAS. *Remote Sensing of Environment*, *113*(3), 691–701. doi:10.1016/j.rse.2008.11.010

Neuenschwander, A. L., & Magruder, L. A. (2016). The Potential Impact of Vertical Sampling Uncertainty on ICESat-2/ATLAS Terrain and Canopy Height Retrievals for Multiple Ecosystems. *Remote Sensing*, *8*(12), 1039. doi:10.3390/rs8121039

Nilsson, M. (1996). Estimation of tree heights and stand volume using an airborne LiDAR system. *Remote Sensing of Environment*, *56*(1), 1–7. doi:10.1016/0034-4257(95)00224-3

NITI Aayog. (2018). *Report of Working Group III, Shifting Cultivation: Towards a Transformational Approach*. Author.

Nkwunonwo, U. C. (2015). Digital mapping of soil chemical properties for erosion hazard management in Bayelsa state, Nigeria. *Chemical Science Review and Letters*, *4*(13), 209–217.

Nkwunonwo, U. C., & Okeke, F. I. (2013). GIS-based production of digital soil map for Nigeria. *Ethiopian Journal of Environmental Studies and Management*, *6*(5), 498–506. doi:10.4314/ejesm.v6i5.7

Nkwunonwo, U. C., Okeke, F. I., Chiemelu, E. N., & Ebinne, E. S. (2019). Geospatial Technology Potentials in Reawakening the Consciousness of Soil Distribution in Nigeria's North-Central Region and Mediating the Herdsmen-Farmers Conflicts. *Journal of Geoscience and Environment Protection*, *7*(02), 156–175. doi:10.4236/gep.2019.72011

Nkwunonwo, U. C., Whitworth, M., & Baily, B. (2016). Review article: A review and critical analysis of the efforts towards urban flood risk management in the Lagos region of Nigeria. *Natural Hazards and Earth System Sciences*, *16*(2), 349–369. doi:10.5194/nhess-16-349-2016

Nohro, S., & Jayakumar, S. (2020). Tree species diversity and composition of the Pala Wetland Reserve Forest, Mizoram, Indo-Burma hotspot, India. *Biocatalysis and Agricultural Biotechnology*, *23*, 101474. doi:10.1016/j.bcab.2019.101474

Nongkynrih, J. M., Pohshna, C., & Sarma, K. K. (2018). Dynamics of shifting cultivation in relation to slope and elevation in parts of Nagaland, India. *Current Science*, *114*(5), 1094. doi:10.18520/cs/v114/i05/1094-1099

O'Neil, G. L., Saby, L., Band, L. E., & Goodall, J. L. (2019). Effects of LiDAR DEM Smoothing and Conditioning Techniques on a Topography-Based Wetland Identification Model. *Water Resources Research*, *55*(5), 4343–4363. doi:10.1029/2019WR024784

O'Neill, A. R. (2019). Evaluating high-altitude Ramsar wetlands in the Eastern Himalayas. *Global Ecology and Conservation*, *20*, e00715. doi:10.1016/j.gecco.2019.e00715

Obioha, E. E. (2009). Climate variability, environment change and food security nexus in Nigeria. *Journal of Human Ecology (Delhi, India)*, *26*(2), 107–121. doi:10.1080/09709274.2009.11906172

Ocampo, L. A. (2019). Decision modeling for manufacturing sustainability with fuzzy Analytical Hierarchy Process. *Global Business Review*, *20*(1), 25–41.

Ogunjinmi, A. A., Aiyeloja, A. A., Osunsina, I. O. O., & Ogunjinmi, K. O. (2014). Understanding perceptions of protected area management in developing countries: The case of Nigeria National Parks. *Global Environment*, *7*(2), 610–628. doi:10.3197/ge.2014.070213

Olley, J. M., & Wasson, R. J. (2003). Changes in the flux of sediment in the Upper Murrumbidgee catchment, Southeastern Australia, since European settlement. *Hydrological Processes*, *17*(16), 3307–3320. doi:10.1002/hyp.1388

Olorunfemi, M. A., & Olorunfemi, M. O. (1987). Geophysical investigations for groundwater in Precambrian terrains: A case history from Ikare, southwestern Nigeria. *Journal of African Earth Sciences*, *6*(6), 787–796. doi:10.1016/0899-5362(87)90035-2

Opricovic, S., & Tzeng, G.-H. (2003). Defuzzification within a multicriteria decision model. *International Journal of Uncertainty, Fuzziness and Knowledge-based Systems*, *11*(5), 635–652.

Otubu, A. (2018). The land use act and land administration in 21st century Nigeria: Need for reforms. *Journal of Sustainable Development Law and Policy (The)*, *9*(1), 80–108.

Ozdagoglu, A., & Ozdagoglu, G. (2007). Comparison of AHP and fuzzy AHP for the multicriteria decision-making processes with linguistic evaluation. *Istanbul Ticaret Universitesi Fen Bilimleri Dergisi*, *6*, 65–85.

Padalia, H., Ghosh, S., Reddy, C. S., Nandy, S., Singh, S., & Kumar, S. A. (2019). Assessment of historical forest cover loss and fragmentation in Asian elephant ranges in India. *Environmental Monitoring and Assessment*, *191*(3), 802. doi:10.100710661-019-7696-5 PMID:31989279

Pal, S., & Saha, T. K. (2018). Identifying dam-induced wetland changes using an inundation frequency approach: The case of the Atreyee River basin of Indo-Bangladesh. *Ecohydrology & Hydrobiology*, *18*(1), 66–81. doi:10.1016/j.ecohyd.2017.11.001

Pal, S., & Talukdar, S. (2018). Drivers of vulnerability to wetlands in Punarbhaba river basin of India- Bangladesh. *Ecological Indicators*, *93*(April), 612–626. doi:10.1016/j.ecolind.2018.05.043

Pandey, M. K. (2014). *Confluence Dynamics of Lower Son River with River Ganga: Tectonic and Climatic Perspective* (Unbpublished Ph.D. Thesis). Banaras Hindu University.

Pandey, M., Ray, Y., & Arora, A. (2020). Structural control on the landscape Evolution of Son Alluvial Fan System in Ganga foreland basin. In Advances in Remote Sensing for Natural Resource Monitoring. Wiley and Sons.

Pandey, A. C., Singh, S. K., & Nathawat, M. S. (2010). Waterlogging and flood hazards vulnerability and risk assessment in Indo Gangetic plain. *Natural Hazards*, *55*(2), 273–289. doi:10.100711069-010-9525-6

Pandey, P. C., Anand, A., & Srivastava, P. K. (2019). Spatial distribution of mangrove forest species and biomass assessment using field inventory and earth observation hyperspectral data. *Biodiversity and Conservation*, 1–20.

Pandey, R. K., Crétaux, J.-F., Bergé-Nguyen, M., Tiwari, V. M., Drolon, V., Papa, F., & Calmant, S. (2014). Water level estimation by remote sensing for the 2008 flooding of the Kosi River. *International Journal of Remote Sensing*, *35*(2), 424–440. doi:10.1080/01431161.2013.870678

Panigrahy, S., Murthy, T. V. R., Patel, J. G., & Singh, T. S. (2012). Wetlands of India: Inventory and assessment at 1:50,000 scale using geospatial techniques. *Curr. Sci., 102*, 852–856.

Panigrahy, S., & Ray, S. S. (2006). Remote Sensing. In K. L. Chadha & M. S. Swaminathan (Eds.), *Environment and Agriculture* (pp. 361–375). Malhotra Publishing House.

Paola, J. D., & Schowengerdt, R. A. (1995). A detailed comparison of backpropagation neural network and maximum-likelihood classifiers for urban land use classification. *IEEE Transactions on Geoscience and Remote Sensing*, *33*(4), 981–996. doi:10.1109/36.406684

Papworth, A., Maslin, M., & Randalls, S. (2015). Is climate change the greatest threat to global health? *The Geographical Journal*, *181*(4), 413–422. doi:10.1111/geoj.12127

Parida, B. R., & Oinam, B. (2015). Unprecedented drought in North East India compared to Western India. *Current Science*, *109*(11), 2121–2126. doi:10.18520/cs/v109/i11/2121-2126

Parida, B. R., & Ranjan, A. K. (2019a). Wheat Acreage Mapping and Yield Prediction Using Landsat-8 OLI Satellite Data: A Case Study in Sahibganj Province, Jharkhand (India). *Remote Sensing in Earth Systems Sciences*, *2*(2-3), 96–107. doi:10.100741976-019-00015-9

Parida, B. R., & Ranjan, A. K. (2019b). Up-scaling paddy yield at satellite-footprint scale using satellite data in conjunction with CCE data in Sahibganj district, Jharkhand. *The International Archives of the Photogrammetry, Remote Sensing and Spatial Information Sciences*, *17*(W6), 235–238. doi:10.5194/isprs-archives-XLII-3-W6-235-2019

Parkinson, B. W., & Gilbert, S. W. (1983). NAVSTAR: Global positioning system—Ten years later. *Proceedings of the IEEE*, *71*(10), 1177–1186. doi:10.1109/PROC.1983.12745

Patel, P., & Srivastava, H. S. (2007, May). Polarimetric SAR classification using physical based scattering mechanism: comparative evaluation of L and P bands. In Conference of Joint Experiment Project towards microwave Remote Sensing Data Utilisation, Ahmedabad (pp. 15-16). Academic Press.

Patel, P., & Srivastava, H. S. (2013). RADARSAT-2 announcement of opportunity project on soil moisture, surface roughness and vegetation parameter retrieval using SAR polarimetry. SAC/EPSA/MPSG/CVD/TDP R&D/01/13, SOAR International Closing and Reporting–2013.

Patel, N. R., Parida, B. R., Venus, V., Saha, S. K., & Dadhwal, V. K. (2012). Analysis of Agricultural Drought Using Vegetation Temperature Condition Index (VTCI) from Terra/MODIS Satellite Data. *Environmental Monitoring and Assessment*, *184*(12), 7153–7163. doi:10.100710661-011-2487-7 PMID:22200944

Patel, P., Srivastava, H. S., Panigrahy, S., & Parihar, J. S. (2006). Comparative evaluation of the sensitivity of multi-polarized multi-frequency SAR backscatter to plant density. *International Journal of Remote Sensing*, *27*(2), 293–305. doi:10.1080/01431160500214050

Patil, A. N. (2018). Fuzzy AHP methodology and its sole applications. *International Journal of Management Research & Review*, *8*(5), 24–32.

Patra, A., Tushar, J., & Dubey, B. (2017). Modeling and simulation of a wetland park: An application to Keoladeo National Park, India. *Mathematics and Computers in Simulation*, *134*, 54–78. doi:10.1016/j.matcom.2016.10.001

Pavlovskaya, M. (2006). Theorizing with GIS: A tool for critical geographies? *Environment & Planning A*, *38*(11), 2003–2020. doi:10.1068/a37326

Peerbhay, K., Mutanga, O., & Ismail, R. (2016). The identification and remote detection of alien invasive plants in commercial forests: An Overview. *South African Journal of Geomatics*, *5*(1).

Peerbhay, K. Y., Mutanga, O., & Ismail, R. (2013). Commercial tree species discrimination using airborne AISA Eagle hyperspectral imagery and partial least squares discriminant analysis (PLS-DA) in KwaZulu–Natal, South Africa. *SPRS J Photogramm Remote Sens.*, *79*, 19–28. doi:10.1016/j.isprsjprs.2013.01.013

Peerbhay, K. Y., Mutanga, O., & Ismail, R. (2015). Random forests unsupervised classification: The detection and mapping of solanum mauritianum infestations in plantation forestry using hyperspectral data. *IEEE Journal of Selected Topics in Applied Earth Observations and Remote Sensing*, *8*(6), 3107–3122. doi:10.1109/JSTARS.2015.2396577

Peng, D., Zhang, H., Yu, L., Wu, M., Wang, F., Huang, W., Liu, L., Sun, R., Li, C., Wang, D., & Xu, F. (2018). Assessing spectral indices to estimate the fraction of photosynthetically active radiation absorbed by the vegetation canopy. *International Journal of Remote Sensing*, *39*(22), 1–19. doi:10.1080/01431161.2018.1479795

Pengra, B. W., Johnston, C. A., & Loveland, T. R. (2007). Mapping an invasive plant, *Phragmites australis*, in coastal wetlands using the EO-1 Hyperion hyperspectral sensor. *Remote Sensing of Environment*, *108*(1), 74–81. doi:10.1016/j.rse.2006.11.002

Penman, J., Gytarsky, M., Hiraishi, T., Krug, T., Kruger, D., Pipatti, R., . . . Tanabe, K. (2003). *Good practice guidance for land use, land-use change and forestry*. IPCC National Greenhouse Gas Inventories Programme and Institute for Global Environmental Strategies (IGES), Kanagawa, Japan. Intergovernmental Panel on Climate Change. www. ipcc-nggip. iges. or. j p/public/ gpglulucf/gpglulucf. htm

Peterson, B., Dubayah, R., Hyde, P., Hofton, M., Blair, J. B., & Fites-Kaufman, J. 2005. Use of LIDAR for Forest Inventory and Forest Management Application. *Proceedings of the Seventh Annual Forest Inventory and Analysis Symposium, 193*-202.

Peterson, B., & Nelson, K. J. (2014). Mapping Forest Height in Alaska Using GLAS, Landsat Composites, and Airborne LiDAR. *Remote Sensing*, *6*(12), 12409–12426. doi:10.3390/rs61212409

Pfister, S., Bayer, P., Koehler, A., & Hellweg, S. (2011). Environmental impacts of water use in global crop production: Hotspots and trade-offs with land use. *Environmental Science & Technology*, *45*(13), 5761–5768. doi:10.1021/es1041755 PMID:21644578

Pingali, P., Aiyar, A., Abraham, M., & Rahman, A. (2019). *Transforming food systems for a rising India*. Springer. doi:10.1007/978-3-030-14409-8

Pourrahmati, M. R. (2016). *Investigating the possibility of forest height/volume estimation using lidar, radar and optical images: case study: Nowshahr Forests in Mazindaran*. Electronics. Université Montpellier.

Pourrahmati, M. R., Baghdadi, N., Darvishsefat, A. A., Namiranian, M., Gond, V., Bailly, J.-S., & Zargham, N. (2018). Mapping Lorey's height over Hyrcanian forests of Iran using synergy of ICESat/ GLAS and optical images. *European Journal of Remote Sensing*, *51*(1), 100–115. doi: 10.1080/22797254.2017.1405717

Pozo, K., Sarkar, S. K., Estellano, V. H., Mitra, S., Audi, O., Kukucka, P., & Corsolini, S. (2017). Passive air sampling of persistent organic pollutants (POPs) and emerging compounds in Kolkata megacity and rural mangrove wetland Sundarban in India: An approach to regional monitoring. *Chemosphere*, *168*, 1430–1438. doi:10.1016/j.chemosphere.2016.09.055 PMID:27912912

Prasad, P. R. C., Reddy, C. S., Rajasekhar, G., & Dutt, C. B. S. (2008). Mapping and analyzing vegetation types of North Andaman Islands, India. *GIS Development*.

Prasad, G. M., Narang, J. L., & Singhal, S. (2007). *Design aspects of the dam*. Water and Energy International.

Project Tiger Report–National Tiger Conservation Authority, Government of India. (2009). *Project Tiger. Stripes*, *1*(1), 1–20.

Promentilla, M. A. B., Antonio, M. R. S., Chuaunsu, R. M. M. II, & Serra, A. J. (2016). *A calibrated fuzzy AHP approach to derive priorities in a decision model for low carbon technologies*. In *DLSU Research Congress*.

Puyravaud, J. P., Davidar, P., & Laurance, W. F. (2010). Cryptic destruction of India's native forests. *Conservation Letters*, *3*(6), 390–394. doi:10.1111/j.1755-263X.2010.00141.x

Qi, H., Paz-Kagan, T., Karnieli, A., Jin, X., & Li, S. (2018). Evaluating calibration methods for predicting soil available nutrients using hyperspectral VNIR data. *Soil and Tillage Research*, *175*, 267–275. doi:10.1016/j.still.2017.09.006

Qin, H., Wang, C., Zhao, K., & Xi, X. (2018). Estimation of the fraction of absorbed photosynthetically active radiation (fPAR) in maize canopies using LiDAR data and hyperspectral imagery. *PLoS One*, *13*(5), e0197510. doi:10.1371/journal.pone.0197510

Ragab, R., Bromley, J., Rosier, P., Cooper, J. D., & Gash, J. H. C. (2003). Experimental study of water fluxes in a residential area: 1. Rainfall, roof runoff and evaporation: the effect of slope and aspect. *Hydrological Processes*, *17*(12), 2409–2422. doi:10.1002/hyp.1250

Raghunath, H. M. (2006). *Hydrology: principles, analysis and design*. New Age International.

Raghunath, H. M. (2009). *Hydrology: principle analysis and design* (revised edition). New age international publishers.

Ramachandra, T. V., Sudarshan, P. B., Mahesh, M. K., & Vinay, S. (2018). Spatial patterns of heavy metal accumulation in sediments and macrophytes of Bellandur wetland, Bangalore. *Journal of Environmental Management*, *206*, 1204–1210. doi:10.1016/j.jenvman.2017.10.014 PMID:29157887

Ramakrishnan, P. S. (1998). Sustainable development, climate change and tropical rain forest landscape. *Climatic Change*, *39*(2-3), 583–600. doi:10.1023/A:1005398606125

Raman, R., & Punia, M. (2012). The application of GIS-based bivariate statistical methods for landslide hazards assessment in the upper Tons river valley, Western Himalaya, India. Georisk. *Assessment and Management of Risk for Engineered Systems and Geohazards*, *6*(3), 145–161. doi:10.1080/17499518.2011.637504

Ranasinghe, R., Wu, C. S., Conallin, J., Duong, T. M., & Anthony, E. J. (2019). Disentangling the relative impacts of climate change and human activities on fluvial sediment supply to the coast by the world's large rivers: Pearl River Basin, China. *Scientific Reports*, *9*(1), 9236. doi:10.103841598-019-45442-2 PMID:31239484

Ranjan, A. K., Prakash, S., Anand, A., Verma, S. K., Murmu, L., & Kumar, P. B. (2017). Ground water prospect variability analysis with spatio temporal changes in Ranchi City, Jharkhand India using Geospatial Technology. *International Journal of Earth Sciences and Engineering*, *10*(3), 616–625.

Ranjan, A.K., Anand, A., Vallisree, S., & Singh, R.K. (2016b). LU/LC Change Detection and Forest Degradation Analysis in Dalma Wildlife Sanctuary Using Geospatial Technology: A Case Study in Jamshedpur-India. *AIMS Geosciences, 2*(4), 273-285. doi:10.3934/geosci.2016.4.273

Ranjan, A. K., Anand, A., Kumar, P. B. S., Verma, S. K., & Murmu, L. (2018). Prediction of Land Surface Temperature within Sun City Jodhpur (Rajasthan) in India Using Integration of Artificial Neural Network and Geoinformatics Technology. *Asian Journal of Geoinformatics*, *17*(3).

Ranjan, A. K., & Parida, B. R. (2019). *Paddy Acreage Mapping and Yield Prediction Using Sentinel-Based Optical and SAR Data in Sahibganj District*. Spatial Information Research. doi:10.100741324-019-00246-4

Ranjan, A. K., Vallisree, S., & Singh, R. K. (2016a). Role of Geographic Information System and Remote Sensing in Monitoring and Management of Urban and Watershed Environment: Overview. *J of Remote Sens & GIS*, *7*, 1–14.

Ranson, K. J., Saatchi, S., & Sun, G. (1995). Boreal forest ecosystem characterization with SIR-C/XSAR. *IEEE Transactions on Geoscience and Remote Sensing*, *33*(4), 867–876. doi:10.1109/36.406673

Ranson, S., & Stewart, J. (1994). *Management for the public domain: enabling the learning society*. Macmillan International Higher Education. doi:10.1007/978-1-349-23787-6

Rapinel, S., Hubert-moy, L., & Clément, B. (2015). Combined use of LiDAR data and multispectral earth observation imagery for wetland habitat mapping. *International Journal of Applied Earth Observation and Geoinformation*, *37*, 56–64. doi:10.1016/j.jag.2014.09.002

Rasula, A. (2015). *Heiko Balztera, Claire Smitha, 2015, Spatial variation of the daytime Surface Urban Cool Island during the dry season in Erbil, Iraqi Kurdistan, from Landsat 8*. Urban Climate. doi:10.1016/j.uclim.2015.09.001

Ravindranath, N. H., Rao, S., Sharma, N., Nair, M., Gopalakrishnan, R., Rao, A., ... Krishna, N. (2011). Climate change vulnerability profiles for North East India. *Current Science*, *101*(3).

Ray, T. W. (1994). *A FAQ on Vegetation in Remote Sensing*. Accessed March 23, 2014. http://www. yale.edu/ceo/Documentation/rsvegfaq.html

Reddy, C. S., Dutta, K., & Jha, C. S. (2013). Analysing the gross and net deforestation rates in India. *Current Science*, *105*(11), 1492–1500.

Reddy, C. S., Hari Krishna, P., & Ravikiran, A. (2011). Mapping the vegetation types of Rajasthan, India using remote sensing data. *Journal of Environmental Research and Management*, *2*(1), 1–9.

Reddy, C. S., Jha, C. S., Dadhwal, V. K., Krishna, P. H., Pasha, S. V., Satish, K. V., ... Diwakar, P. G. (2016). Quantification and monitoring of deforestation in India over eight decades (1930–2013). *Biodiversity and Conservation*, *25*(1), 93–116. doi:10.100710531-015-1033-2

Reddy, C. S., Jha, C. S., Diwakar, P. G., & Dadhwal, V. K. (2015). Nationwide classification of forest types of India using remote sensing and GIS. *Environmental Monitoring and Assessment*, *187*(12), 777. doi:10.100710661-015-4990-8 PMID:26615560

Redfern, T. W., Macdonald, N., Kjeldsen, T. R., Miller, J. D., & Reynard, N. (2016). Current understanding of hydrological processes on common urban surfaces. *Progress in Physical Geography: Earth and Environment*, *40*(5), 699–713. doi:10.1177/0309133316652819

Richter, B. D., & Thomas, G. A. (2007). Restoring Environmental Flows by Modifying Dam Operations. *Ecology and Society*, *12*(1), art12. doi:10.5751/ES-02014-120112

Robock, A., & Liu, Y. (1994). The volcanic signal in Goddard Institute for Space Studies and 3d model simulations. *Journal of Climate*, *7*, 44–55. doi:10.1175/1520-0442(1994)007<0044:TVSIGI>2.0.CO;2

Rodríguez-Iruretagoiena, A., Chowdhury, R., Gredilla, A., deVallejuelo, S. F. O., de Diego, A., Sarkar, S. K., & Venkatachalam, P. (2016). Uptake and Distribution of Trace Elements in Dominant Mangrove Plants of the Indian Sundarban Wetland. *Bulletin of Environmental Contamination and Toxicology*, *97*(5), 721–727. doi:10.100700128-016-1919-z PMID:27650375

Rollet, R., Benie, G., Li, W., Wang, S., & Boucher, J. (1998). Image classification algorithm based on the RBF neural network and K-means. *International Journal of Remote Sensing*, *19*(15), 3003–3009. doi:10.1080/014311698214398

Romijn, E., Ainembabazi, J. H., Wijaya, A., Herold, M., Angelsen, A., Verchot, L., & Murdiyarso, D. (2013). Exploring different forest definitions and their impact on developing REDD+ reference emission levels: A case study for Indonesia. *Environmental Science & Policy*, *33*, 246–259. doi:10.1016/j.envsci.2013.06.002

Rosette, J. A. B., North, P. R. J., & Suárez, J. C. (2008). Vegetation height estimates for a mixed temperate forest using satellite laser altimetry. *International Journal of Remote Sensing*, *29*(5), 1475–1493.

Roy, M. B., Roy, P. K., Samal, N. R., & Mazumdar, A. (2016). A Comparative Study of Labour Participation in Three Wetland Regions of West Bengal, India. *Journal of The Institution of Engineers (India): Series A, 97*(2), 81–91. doi:10.100740030-016-0150-6

Roy, P. S., & Joshi, P. K. (2002). Landscape fragmentation and biodiversity conservation. *Map India*, 281-285.

Roy, P. S., Kushwaha, S. P. S., Murthy, M. S. R., Roy, A., Kushwaha, D., Reddy, C. S., . . . Singh, S. (2012). Biodiversity characterisation at landscape level: national assessment. Indian institute of remote sensing, Dehradun, India, 140.

Roy, P. S., Behera, M. D., Murthy, M. S. R., Roy, A., Singh, S., Kushwaha, S. P. S., ... Gupta, S. (2015). New vegetation type map of India prepared using satellite remote sensing: Comparison with global vegetation maps and utilities. *International Journal of Applied Earth Observation and Geoinformation*, *39*, 142–159. doi:10.1016/j.jag.2015.03.003

Roy, P. S., Joshi, P. K., Singh, S., Agarwal, S., Yadav, D., & Jegannathan, C. (2006). Biome mapping in India using vegetation type map derived using temporal satellite data and environmental parameters. *Ecological Modelling*, *197*(1-2), 148–158. doi:10.1016/j.ecolmodel.2006.02.045

Roy, P. S., & Tomar, S. (2000). Biodiversity characterization at landscape level using geospatial modelling technique. *Biological Conservation*, *95*(1), 95–109. doi:10.1016/S0006-3207(99)00151-2

Roy, P., & Ravan, S. A. (1996). Biomass estimation using satellite remote sensing data—An investigation on possible approaches for natural forest. *Journal of Biosciences*, *21*(4), 535–561. doi:10.1007/BF02703218

Roy, P., Roy, A., Joshi, P., Kale, M., Srivastava, V., Srivastava, S., ... Sharma, Y. (2015). Development of decadal (1985–1995–2005) land use and land cover database for India. *Remote Sensing*, *7*(3), 2401–2430. doi:10.3390/rs70302401

Saaty, T. L. (1980). *The Analytic Hierarchy Process*. McGraw- Hill.

Saaty, T. L. (1980). *The analytic hierarchy process*. McGraw-Hill International.

Saaty, T. L. (1988). *Multicriteria decision making: the analytic hierarchy process*. RWS Publications.

Saaty, T. L. (1996). *Decision making with dependence and feedback, The Analytic Network Process*. RWS Publications.

Saaty, T. L. (1999). Fundamentals of the analytic network process. *International Symposium of the Analytic Hierarchy Process (ISAHP)*, Kobe, Japan.

Saaty, T. L. (2004). Fundamentals of the analytic network process – multiple networks with benefits, costs, opportunities and risks. *Journal of Systems Science and Systems Engineering*, *13*(3), 348–379. doi:10.100711518-006-0171-1

Saaty, T. L., & Tran, L. T. (2007). On the invalidity of fuzzifying numerical judgments in the analytic hierarchy process. *Mathematical and Computer Modelling*, *46*, 962–975.

Sabins, F. F. (2007). *Remote sensing: principles and applications*. Waveland Press.

Saha, S., Chakraborty, D., Choudhury, B. U., Singh, S. B., Chinza, N., Lalzarliana, C., ... Singh, A. R. (2015). Spatial variability in temporal trends of precipitation and its impact on the agricultural scenario of Mizoram. *Current Science*, *109*(12), 2278–2282. doi:10.18520/cs/v109/i12/2278-2282

Sahay, H., Mahfooz, S., & Singh, A. K. (2012). *Exploration and characterization of agriculturally and industrially important haloalkaliphilic bacteria from environmental samples of hypersaline Sambhar lake*. doi:10.100711274-012-1131-1

Sahoo, A. K., Pan, M., Troy, T. J., Vinukollu, R. K., Sheffield, J., & Wood, E. F. (2011). Reconciling the global terrestrial water budget using satellite remote sensing. *Remote Sensing of Environment*, *115*(8), 1850–1865. doi:10.1016/j.rse.2011.03.009

Sahu, R., & Patel, N. R. (2016). Agricultural drought assessment using satellite derived vegetation condition index and rainfall anomaly index over Bundelkhand region. Academic Press.

Saini, S., Bowser, J., Enck, R., Luciani, V., Heim, P., & Dagenais, M. (2004). *A semiconductor optical amplifier with high saturation power, low noise figure and low polarization dependent gain over the C-band.* Paper presented at the 17th Annual Meeting of the IEEE Lasers and Electro-Optics Society. 10.1109/LEOS.2004.1363131

Sajinkumar, K. S., Revathy, A., & Rani, V. R. (2017). Hydrogeochemistry and spatio-temporal changes of a tropical coastal wetland system: Veli-Akkulam Lake, Thiruvananthapuram, India. *Applied Water Science*, *7*(3), 1521–1534. doi:10.100713201-015-0333-8

Sakamoto, T., Van Nguyen, N., & Kotera, A. (2007). Detecting temporal changes in the extent of annual flooding within the Cambodia and the Vietnamese Mekong Delta from MODIS time-series imagery. *Remote Sensing of Environment*, *109*, 295–313. doi:10.1016/j.rse.2007.01.011

Sala-i-Martin, X., & Subramanian, A. (2013). Addressing the natural resource curse: An illustration from Nigeria. *Journal of African Economies*, *22*(4), 570–615. doi:10.1093/jae/ejs033

Saluja, R., Prasad, S., & Garg, J. K. (2018). Field spectroradiometry for discrimination of wetland components: A case study of a tropical inland wetland in India. *Wetlands Ecology and Management*, *26*(5), 915–930. doi:10.100711273-018-9620-0

Sam, K., Coulon, F., & Prpich, G. (2016). Working towards an integrated land contamination management framework for Nigeria. *The Science of the Total Environment*, *571*, 916–925. doi:10.1016/j.scitotenv.2016.07.075 PMID:27443458

Sampson, P. H., Zarco-Tejada, P. J., Mohammed, G. H., Miller, J. R., & Noland, T. L. (2003). Hyperspectral remote sensing of forest condition: Estimating chlorophyll content in tolerant hardwoods. *For. Sci.*, *49*, 381–391. doi:10.1093/forestscience/49.3.381

Sanchez, G., Rolland, Y., Corsini, M., Braucher, R., Bourlès, D., Arnold, M., & Aumaître, G. (2010). Relationships between tectonics, slope instability and climate change: Cosmic ray exposure dating of active faults, landslides and glacial surfaces in the SW Alps. *Geomorphology*, *117*(1), 1–13. doi:10.1016/j.geomorph.2009.10.019

Santos, J., Lacruz, M. P., Araujo, L., & Keil, M. (2002). Savanna and tropical rainforest biomass estimation and spatialization using JERS-1 data. *International Journal of Remote Sensing*, *23*(7), 1217–1229. doi:10.1080/01431160110092867

Sarkar, S., Parihar, S. M., & Dutta, A. (2016). Fuzzy risk assessment modelling of East Kolkata Wetland Area: A remote sensing and GIS based approach. *Environmental Modelling & Software*, *75*, 105–118. doi:10.1016/j.envsoft.2015.10.003

Sarma, K., & Barik, S. K. (2011). Coal mining impact on vegetation of the Nokrek Biosphere Reserve, Meghalaya, India. *Biodiversity (Nepean)*, *12*(3), 154–164. doi:10.1080/14888386.2011.629779

Schaphoff, S., Heyder, U., Ostberg, S., Gerten, D., Heinke, J., & Lucht, W. (2013). Contribution of permafrost soils to the global carbon budget. *Environmental Research Letters*, *8*(1), 014026. doi:10.1088/1748-9326/8/1/014026

Schilling, K. E., Chan, K.-S., Liu, H., & Zhang, Y.-K. (2010). Quantifying the effect of land use land cover change on increasing discharge in the Upper Mississippi River. *Journal of Hydrology (Amsterdam)*, *387*(3–4), 343–345. doi:10.1016/j.jhydrol.2010.04.019

Schlaffer, S., Matgen, P., Hollaus, M., & Wagner, W. (2015a). Flood detection from multi-temporal SAR data using harmonic analysis and change detection. *International Journal of Applied Earth Observation and Geoinformation*, *38*, 15–24. doi:10.1016/j.jag.2014.12.001

Schlerf, M., & Atzberger, C. (2006). Inversion of a forest reflectance model to estimate structural canopy variables from hyperspectral remote sensing data. *Remote Sensing of Environment*, *100*(3), 281–294. doi:10.1016/j.rse.2005.10.006

Schlerf, M., Atzberger, C., & Hill, J. (2005). Remote sensing of forest biophysical variables using HyMap imaging spectrometer data. *Remote Sensing of Environment*, *95*(2), 177–194. doi:10.1016/j.rse.2004.12.016

Schmidt, J. C., & Wilcock, P. R. (2008). Metrics for assessing the downstream effects of dams. *Water Resources Research*, *44*(4). Advance online publication. doi:10.1029/2006WR005092

Schmidt-Thomé, P., Greiving, S., Kallio, H., Fleischhauer, M., & Jarva, J. (2006). Economic Risk Maps of Floods and Earthquakes for European Regions. *Quaternary International*, *150*(1), 103–112. doi:10.1016/j.quaint.2006.01.024

Schroeder, P., Brown, S., Mo, J., Birdsey, R., & Cieszewski, C. (1997). Biomass estimation for temperate broadleaf forests of the United States using inventory data. *Forest Science*, *43*(3), 424–434.

Schumann, G. J.-P., & Domeneghetti, A. (2016). Exploiting the proliferation of current and future satellite observations of rivers: Invited Commentary. *Hydrological Processes*, *30*(16), 2891–2896. doi:10.1002/hyp.10825

Schumann, G. J.-P., Frye, S., Wells, G., Adler, R., Brakenridge, R., Bolten, J., Murray, J., Slayback, D., Policelli, F., Kirschbaum, D., Wu, H., Cappelaere, P., Howard, T., Flamig, Z., Clark, R., Stough, T., Chini, M., Matgen, P., Green, D., & Jones, B. (2016). Unlocking the full potential of Earth observation during the 2015 Texas flood disaster: EO imagery assisting the texas flood disaster response. *Water Resources Research*, *52*(5), 3288–3293. doi:10.1002/2015WR018428

Schuurman, N. (2000). Trouble in the heartland: GIS and its critics in the 1990s. *Progress in Human Geography*, *24*(4), 569–590. doi:10.1191/030913200100189111

Science Learning Hub. (2016). *What is clay*. University of Waikato.

Seiler, R. A., Kogan, F., & Sullivan, J. (1998). AVHRR-Based vegetation and temperature condition indices for drought detection in Argentina. *Advances in Space Research*, *21*(3), 481–484. doi:10.1016/S0273-1177(97)00884-3

Serrano, L., Peñuelas, J., & Ustin, S. L. (2002). Remote sensing of nitrogen and lignin in Mediterranean vegetation from AVIRIS data: Decomposing biochemical from structural signals. *Remote Sensing of Environment*, *81*(2-3), 355–364. doi:10.1016/S0034-4257(02)00011-1

Sfeir-Younis, A. (2019). *Land and soil management: Technology, economics, and institutions*. CRC Press. doi:10.1201/9780429041631

Sforna, G. (2019). Climate change and developing countries: From background actors to protagonists of climate negotiations. *International Environmental Agreement: Politics, Law and Economics*, *19*(3), 273–295. doi:10.100710784-019-09435-w

Shakya, A., & Kumar, A. (2019). Noise Clustering-Based Hypertangent Kernel Classifier for Satellite Imaging Analysis. *Photonirvachak (Dehra Dun)*, *47*, 2009–2025. doi:10.100712524-019-01044-z

Sharma, L. K., Kanga, S., Nathawat, M. S., Sinha, S., & Pandey, P. C. (2012). Fuzzy AHP for forest fire risk modeling. *Disaster Prevention and Management*, *21*(2), 160–171.

Sharma, P. K., Kumar, D., Srivastava, H. S., Patel, P., & Sivasankar, T. (2019). Soil moisture retrieval under wheat crop using RISAT-1 hybrid polarimetric SAR data. *Journal of Agrometeorology*, *21*(1), 58–62.

Sharma, S., Kaur, I., & Nagpal, A. K. (2018). Estimation of arsenic, manganese and iron in mustard seeds, maize grains, groundwater and associated human health risks in Ropar wetland, Punjab, India, and its adjoining areas. *Environmental Monitoring and Assessment*, *190*(7), 385. Advance online publication. doi:10.100710661-018-6763-7 PMID:29884925

Sharma, S., Nagpal, A. K., & Kaur, I. (2019). Appraisal of heavy metal contents in groundwater and associated health hazards posed to human population of Ropar wetland, Punjab, India and its environs. *Chemosphere*, *227*, 179–190. doi:10.1016/j.chemosphere.2019.04.009 PMID:30986600

Sharma, S., & Shivakoti, G. (2017). En route to effective management of natural resources for conservation and livelihood advances in central Vietnam. In *Redefining Diversity & Dynamics of Natural Resources Management in Asia* (Vol. 3, pp. 237–248). Elsevier. doi:10.1016/B978-0-12-805452-9.00016-3

Sheet, I. R. N. F. (2003). Tehri dam. *Environment*.

Shen, G., Sakai, K., & Hoshino, Y. (2010). High Spatial Resolution Hyperspectral Mapping for Forest Ecosystem at Tree Species Level. *Agric. Inf. Res.*, *19*(3), 71–78. doi:10.3173/air.19.71

Sholihah, R. I., Trisasongko, B. H., Shiddiq, D., Iman, L. O. S., Kusdaryanto, S., Manijo, & Panuju, D. R. (2016). Identification of agricultural drought extent based on vegetation health indices of Landsat data: Case of Subang and Karawang, Indonesia. *Procedia Environmental Sciences*, *33*, 14–20. doi:10.1016/j.proenv.2016.03.051

Shrestha, P. M., & Dhillion, S. S. (2003). Medicinal plant diversity and use in the highlands of Dolakha district, Nepal. *Journal of Ethnopharmacology*, *86*(1), 81–96. doi:10.1016/S0378-8741(03)00051-5 PMID:12686446

Shukla, P. (2014). *Cumulative Impact Assessment on River Ganga from Hardiwar to Garhmukteshwar with Focus on Point Source Discharges*. GRIN Verlag.

Simard, M., Pinto, N., Fisher, J. B., & Baccini, A. (2011). Mapping forest canopy height globally with spaceborne lidar. *Journal of Geophysical Research*, *116*(G4), G04021. Advance online publication. doi:10.1029/2011JG001708

Sinclair, M., Ghermandi, A., & Sheela, A. M. (2018). A crowdsourced valuation of recreational ecosystem services using social media data: An application to a tropical wetland in India. *The Science of the Total Environment*, *642*, 356–365. doi:10.1016/j.scitotenv.2018.06.056 PMID:29906727

Singh, R. L. (1971). India: A Regional Geography. National Geographical Society of India.

Singh, A. K., Sathya, M., Verma, S., & Jayakumar, S. (2018). Health risk assessment of heavy metals in crop grains grown on open soils of Kanwar wetland, India. *Euro-Mediterranean Journal for Environmental Integration*, *3*(1), 29. Advance online publication. doi:10.100741207-018-0073-x

Singh, D., & Kathpalia, A. (2007). An efficient modeling with GA approach to retrieve soil texture, moisture and roughness from ERS-2 SAR data. *Progress In Electromagnetics Research, 77*, 121–136. doi:10.2528/PIER07071803

Singh, M. P., Singh, P., & Singh, P. (2019). Fuzzy AHP-based multi-criteria decision-making analysis for route alignment planning using geographic information system (GIS). *Journal of Geographical Systems, 21*, 395–432.

Singh, M., Singh, I. B., & Müller, G. (2007). Sediment characteristics and transportation dynamics of the Ganga River. *Geomorphology, 86*(1–2), 144–175. doi:10.1016/j.geomorph.2006.08.011

Singh, M., & Sinha, R. (2019). Evaluating dynamic hydrological connectivity of a floodplain wetland in North Bihar, India using geostatistical methods. *The Science of the Total Environment, 651*, 2473–2488. doi:10.1016/j.scitotenv.2018.10.139 PMID:30336437

Singh, M., & Sinha, R. (2020). Distribution, diversity, and geomorphic evolution of floodplain wetlands and wetland complexes in the Ganga plains of north Bihar, India. *Geomorphology, 351*, 106960. doi:10.1016/j.geomorph.2019.106960

Singh, P. K., & Sarkar, P. (2019). A framework based on fuzzy AHP-TOPSIS for prioritizing solutions to overcome the barriers in the implementation of ecodesign practices in SMEs. *International Journal of Sustainable Development and World Ecology, 26*(6), 506–521.

Singh, P., Javed, S., Shashtri, S., Singh, R. P., Vishwakarma, C. A., & Mukherjee, S. (2017). Influence of changes in watershed landuse pattern on the wetland of Sultanpur National Park, Haryana using remote sensing techniques and hydro-chemical analysis. *Remote Sensing Applications: Society and Environment, 7*, 84–92. doi:10.1016/j.rsase.2017.07.002

Singh, S. K., Pandey, A. C., & Nathawat, M. S. (2011). Rainfall Variability and Spatio Temporal Dynamics of Flood Inundation during the 2008 Kosi Flood in Bihar State, India. *Asian Journal of Earth Sciences, 4*, 9–19. doi:10.3923/ajes.2011.9.19

Sinha, R., Bapalu, G. V., Singh, L. K., & Rath, B. (2008). Flood risk analysis in the Kosi river basin, north Bihar using multi-parametric approach of analytical hierarchy process (AHP). *Photonirvachak (Dehra Dun), 36*(4), 335–349. doi:10.100712524-008-0034-y

Sinha, R., Mohanta, H., Jain, V., & Tandon, S. K. (2017). Geomorphic diversity as a river management tool and its application to the Ganga River, India. *River Research and Applications, 33*(7), 1156–1176. doi:10.1002/rra.3154

Sinha, S. (2014). Integrating fuzziness to wildlife relocation and habitat analysis in Rajasthan, India. *International Journal of Earth Science and Engineering, 7*(1), 280–288.

Sinha, S. (2015). Bland-Altman analysis for evaluating AHP-based wildlife habitat suitability models. *Research & Reviews: Journal of Space Science & Technology, 4*(2), 11–18.

Sinha, S. (2016). Integrated GIS and Multi Criteria approach for the wildlife habitat suitability. *Journal of Advanced Research in Geo Sciences & Remote Sensing, 3*, 1–16.

Sinha, S., Jeganathan, C., Sharma, L., & Nathawat, M. (2015). A review of radar remote sensing for biomass estimation. *International Journal of Environmental Science and Technology*, *12*(5), 1779–1792. doi:10.100713762-015-0750-0

Sinha, S., Mohan, S., Das, A. K., Sharma, L. K., Jeganathan, C., Santra, A., Mitra, S. S., & Nathawat, M. S. (2020). Multi-sensor approach integrating optical and multi-frequency synthetic aperture radar for carbon stock estimation over a tropical deciduous forest in India. *Carbon Management*. 11(1), 39-55. doi:10.1080/17583004.2019.1686931

Sinha, S., & Santra, A. (2019). Estimation of change in forest aboveground carbon in Bhimbandh Wildlife Sanctuary, Bihar, India between 2007 and 2016. *Current Science*, *117*, 1090–1094.

Sinha, S., Sharma, L. K., & Nathawat, M. S. (2011b). Retrieving tiger habitats: Conserving wildlife geospatially. *Applied Remote Sensing Journal*, *2*(1), 1–5.

Sinha, S., Sharma, L. K., & Nathawat, M. S. (2012). Tigers losing grounds: Impact of anthropogenic occupancy on tiger habitat suitability using integrated geospatial-fuzzy techniques. *The Ecoscan*, *1*, 259–263.

Sinha, S., Sharma, L. K., Pandey, P. C., Nathawat, M. S., & Kanga, S. (2011a). Impact of human intrusion on tiger habitat and conservation using integrated geospatial techniques. *International Journal of Earth Sciences and Engineering*, *4*(3), 39–45.

Sivasankar, T., Kumar, D., Srivastava, H. S., & Patel, P. (2018). Advances in radar remote sensing of agricultural crops: A review. *International Journal on Advanced Science. Engineering and Information Technology*, *8*(4), 1126–1137.

Sivasankar, T., Kumar, D., Srivastava, H. S., & Patel, P. (2020). Wheat leaf area index retrieval using RISAT-1 hybrid polarized SAR data. *Geocarto International*, *35*(8), 905–915. doi:10.1080/10106049.2019.1566404

Sivasankar, T., Lone, J. M., Sarma, K. K., Qadir, A., & Raju, P. L. N. (2018). The potential of multi-frequency multi-polarized ALOS-2/PALSAR-2 and Sentinel-1 SAR data for aboveground forest biomass estimation. *International Journal of Engineering and Technology*, *10*(3), 797–802. doi:10.21817/ijet/2018/v10i3/181003095

Sivasankar, T., Lone, J. M., Sarma, K. K., Qadir, A., & Raju, P. L. N. (2019). Estimation of Above Ground Biomass Using Support Vector Machines and ALOS/PALSAR data. *Vietnam Journal of Earth Sciences*, *41*(2), 95–104. doi:10.15625/0866-7187/41/2/13690

Skowronek, S., Asner, G. P., & Feilhauer, H. (2017). Performance of one-class classifiers for invasive species mapping using airborne imaging spectroscopy. *Ecological Informatics*, *37*, 66–76. doi:10.1016/j.ecoinf.2016.11.005

Skriver, H., Mattia, F., Satalino, G., Balenzano, A., Pauwels, V. R., Verhoest, N. E., & Davidson, M. (2011). Crop classification using short-revisit multitemporal SAR data. *IEEE Journal of Selected Topics in Applied Earth Observations and Remote Sensing*, *4*(2), 423–431. doi:10.1109/JSTARS.2011.2106198

Smith, M.-L., Ollinger, S. V., Martinm, M. E., Aber, J. D., Hallett, R. A., & Goodale, C. L. (2002). Direct estimation of aboveground forest productivity through hyperspectral remote sensing of canopy nitrogen. *Ecological Applications*, *12*(5), 1286–1302. doi:10.1890/1051-0761(2002)012[1286:DEOAFP]2.0.CO;2

Snyder, N. P., Whipple, K. X., Tucker, G. E., & Merritts, D. J. (2000). Landscape response to tectonic forcing: Digital elevation model analysis of stream profiles in the Mendocino triple junction region, northern California. *Geological Society of America Bulletin*, *112*(8), 1250–1263. doi:10.1130/0016-7606(2000)112<1250:LRTTFD>2.0.CO;2

Sodhi, M. S., & Tang, C. S. (2014). Buttressing Supply Chains against Floods in Asia for Humanitarian Relief and Economic Recovery. *Production and Operations Management*, *23*(6), 938–950. doi:10.1111/poms.12111

Sonobe, R., Tani, H., Wang, X., Kobayashi, N., & Shimamura, H. (2014a). Random forest classification of crop type using multi-temporal TerraSAR-X dual-polarimetric data. *Remote Sensing Letters*, *5*(2), 157–164.

Sonobe, R., Tani, H., Wang, X., Kobayashi, N., & Shimamura, H. (2014b). Parameter tuning in the support vector machine and random forest and their performances in cross-and same-year crop classification using TerraSAR-X. *International Journal of Remote Sensing*, *35*(23), 7898–7909. doi:10.1080/01431161.2014.978038

Speckhann, G. A., Borges Chaffe, P. L., Fabris Goerl, R., Abreu, J. J., & Altamirano Flores, J. A. (2018). Flood hazard mapping in Southern Brazil: A combination of flow frequency analysis and the HAND model. *Hydrological Sciences Journal*, *63*(1), 87–100. doi:10.1080/02626667.2017.1409896

Sredjevic, B., & Medeiros, Y. D. P. (2008). Fuzzy AHP assessment of water management plans. *Water Resources Management*, *22*, 877–894.

Srivastava, H. S., Patel, P., & Navalgund, R. R. (2006). How far SAR has fulfilled its expectation for soil moisture retrieval. In *Microwave Remote Sensing of the Atmosphere and Environment V* (Vol. 6410, p. 641001). International Society for Optics and Photonics.

Srivastava, H. S., Patel, P., Navalgund, R. R., & Sharma, Y. (2008). Retrieval of surface roughness using multi-polarized Envisat-1 ASAR data. *Geocarto International*, *23*(1), 67–77. doi:10.1080/10106040701538157

Srivastava, S., Singh, T. P., Singh, H., Kushwaha, S. P. S., & Roy, P. S. (2002). Assessment of large-scale deforestation in Sonitpur district of Assam. *Current Science*, 1479–1484.

Srivastava, V., Srivastava, H. B., & Lakhera, R. C. (2010). Fuzzy gamma based geomatic modelling for landslide hazard susceptibility in a part of Tons river valley, northwest Himalaya, India. *Geomatics, Natural Hazards & Risk*, *1*(3), 225–242. doi:10.1080/19475705.2010.490103

Stähly, S., Franca, M. J., Robinson, C. T., & Schleiss, A. J. (2019). Sediment replenishment combined with an artificial flood improves river habitats downstream of a dam. *Scientific Reports*, *9*(1), 5176. doi:10.103841598-019-41575-6 PMID:30914717

Stein, S. E. (1999). An integrated method for spectrum extraction and compound identification from gas chromatography/mass spectrometry data. *Journal of the American Society for Mass Spectrometry*, *10*(8), 770–781. doi:10.1016/S1044-0305(99)00047-1

Stern, N. H. (2006). *The Economics of Climate Change:The Stern Review*. Cambridge University Press.

Stoll-Kleemann, S., de la Vega-Leinert, A. C., & Schultz, L. (2010). The role of community participation in the effectiveness of UNESCO Biosphere Reserve management: Evidence and reflections from two parallel global surveys. *Environmental Conservation*, *37*(3), 227–238. doi:10.1017/S037689291000038X

Stratoulias, D., Balzter, H., Zlinszky, A., & Tóth, V. R. (2015). Assessment of ecophysiology of lake shore reed vegetation based on chlorophyll fluorescence, field spectroscopy and hyperspectral airborne imagery. *Remote Sensing of Environment*, *157*, 72–84. doi:10.1016/j.rse.2014.05.021

Sun, T., Lin, W., Chen, G., Guo, P., & Zeng, Y. (2016). Wetland ecosystem health assessment through integrating remote sensing and inventory data with an assessment model for the Hangzhou Bay, China. *The Science of the Total Environment*, *566–567*, 627–640. doi:10.1016/j.scitotenv.2016.05.028 PMID:27236628

Sun, Y., Luo, J., Wu, T., Zhou, Y. N., Liu, H., Gao, L., ... Wang, L. (2019). Synchronous Response Analysis of Features for Remote Sensing Crop Classification Based on Optical and SAR Time-Series Data. *Sensors (Basel)*, *19*(19), 4227. doi:10.339019194227 PMID:31569430

Surian, N., & Rinaldi, M. (2003). Morphological response to river engineering and management in alluvial channels in Italy. *Geomorphology*, *50*(4), 307–326. doi:10.1016/S0169-555X(02)00219-2

Swer, S., & Singh, O. P. (2004). Water quality, availability and aquatic life affected by coal mining in ecologically sensitive areas of Meghalaya. In *Proceedings of the 3rd National Seminar on Inland Water Resources and Environment*. University of Kerala.

Swer, S., & Singh, O. P. (2003). Coal mining impacting water quality and aquatic biodiversity in Jaintia Hills district of Meghalaya. *Himalayan Ecology*, *11*(2), 29.

Tàbara, J. D., Jäger, J., Mangalagiu, D., & Grasso, M. (2019). Defining transformative climate science to address high-end climate change. *Regional Environmental Change*, *19*(3), 807–818. doi:10.100710113-018-1288-8

Tai, A. P., Martin, M. V., & Heald, C. L. (2014). Threat to future global food security from climate change and ozone air pollution. *Nature Climate Change*, *4*(9), 817–821. doi:10.1038/nclimate2317

Tan, C. P., Ewe, H. T., & Chuah, H. T. (2011). Agricultural crop-type classification of multi-polarization SAR images using a hybrid entropy decomposition and support vector machine technique. *International Journal of Remote Sensing, 32*(22), 7057–7071. doi:10.1080/01431161.2011.613414

Tan, C. P., Koay, J. Y., Lim, K. S., Ewe, H. T., & Chuah, H. T. (2007). Classification of multi-temporal SAR images for rice crops using combined entropy decomposition and support vector machine technique. *Progress In Electromagnetics Research, 71*, 19–39.

Tan, C., Wang, D., Zhou, J., Du, Y., Luo, M., Zhang, Y., & Guo, W. (2018). Remotely Assessing Fraction of Photosynthetically Active Radiation (FPAR) for Wheat Canopies Based on Hyperspectral Vegetation Indexes. *Front. Plant Sci., 9*(776), 1–9. doi:10.3389/fpls.2018.00776

Tang, H., Armston, J., Hancock, S., Marselis, S., Goetz, S., & Dubayah, R. (2019). Characterizing global forest canopy cover distribution using spaceborne lidar. *Remote Sensing of Environment, 231*, 111262. doi:10.1016/j.rse.2019.111262

Taylor, B. W. (2004). *Introduction to management science*. Pearson Education Inc.

Taylor, M., & Yin, A. (2009). Active structures of the Himalayan-Tibetan orogen and their relationships to earthquake distribution, contemporary strain field, and Cenozoic volcanism. *Geosphere, 5*(3), 199–214. doi:10.1130/GES00217.1

Teng, J., Jakeman, A. J., Vaze, J., Croke, B. F. W., Dutta, D., & Kim, S. (2017). Flood inundation modelling: A review of methods, recent advances and uncertainty analysis. *Environmental Modelling & Software, 90*, 201–216. doi:10.1016/j.envsoft.2017.01.006

Thapa, G. B., & Yila, O. M. (2012). Farmers' land management practices and status of agricultural land in the Jos Plateau, Nigeria. *Land Degradation & Development, 23*(3), 263–277. doi:10.1002/ldr.1079

Thenkabail, P. S., & Lyon, J. G. (2011). *Hyperspectral Remote Sensing of Vegetation*. CRC Press.

Thenkabail, P. S., Schull, M., & Turral, H. (2005). Ganges and Indus river basin land use/land cover (LULC) and irrigated area mapping using continuous streams of MODIS data. *Remote Sensing of Environment, 95*(3), 317–341.

Thenkabail, P. S., Smith, R. B., & de Pauw, E. (2000). Hyperspectral vegetation indices and their relationships with agricultural crop characteristics. *Remote Sensing of Environment, 71*(2), 158–182. doi:10.1016/S0034-4257(99)00067-X

Thiede, R. C., Ehlers, T. A., Bookhagen, B., & Strecker, M. R. (2009). Erosional variability along the northwest Himalaya. *Journal of Geophysical Research. Earth Surface, 114*, F101015.

Thiemann, S., & Kaufmann, H. (2002). Lake water quality monitoring using hyperspectral airborne data – a semi empirical multisensory and multi-temporal approach for the Ecklenburg Lake District, Germany. *Remote Sensing of Environment, 81*(2-3), 228–237. doi:10.1016/S0034-4257(01)00345-5

Thi, N. B. D., Kumar, G., & Lin, C. Y. (2015). An overview of food waste management in developing countries: Current status and future perspective. *Journal of Environmental Management, 157*, 220–229. doi:10.1016/j.jenvman.2015.04.022 PMID:25910976

Thompson, W. D., & Walter, S. D. (1988). A reappraisal of the kappa coefficient. *Journal of Clinical Epidemiology, 41*(10), 949–958. doi:10.1016/0895-4356(88)90031-5 PMID:3057117

Timm, B. C., & McGarigal, K. (2012). Fine-scale remotely-sensed cover mapping of coastal dune and salt marsh ecosystems at Cape Cod National Seashore using Random Forests. *Remote Sensing of Environment, 127*, 106–117. doi:10.1016/j.rse.2012.08.033

Tong, Q., Xue, Y., & Zhang, L. (2014). Progress in Hyperspectral Remote Sensing Science and Technology in China Over the Past Three Decades. *IEEE Journal of Selected Topics in Applied Earth Observations and Remote Sensing, 1*(1), 70–90. doi:10.1109/JSTARS.2013.2267204

Tonolli, S., Dalponte, M., Neteler, M., Rodeghiero, M., Vescovo, L., & Gianelle, D. (2011). Fusion of airborne LiDAR and satellite multispectral data for the estimation of timber volume in the Southern Alps. *Remote Sensing of Environment, 115*(10), 2486–2498. doi:10.1016/j.rse.2011.05.009

Tripathi, A., Tripathi, D. K., Chauhan, D. K., Kumar, N., & Singh, G. S. (2016). Paradigms of climate change impacts on some major food sources of the world: A review on current knowledge and future prospects. *Agriculture, Ecosystems & Environment, 216*, 356–373. doi:10.1016/j.agee.2015.09.034

Tripathi, P., Parida, B. R., & Pandey, A. C. (2019). Spatio-Temporal Rainfall Variability and Flood Prognosis Analysis using Satellite Data over North Bihar during the August 2017 Flood Event. *Hydrology, 6*(2), 38. doi:10.3390/hydrology6020038

Tsai, J. S., Venne, L. S., McMurry, S. T., & Smith, L. M. (2010). Vegetation and land use impact on water loss rate in playas of the Southern High Plains, USA. *Wetlands, 30*(6), 1107–1116. doi:10.100713157-010-0117-8

Tsyganskaya, V., Martinis, S., Marzahn, P., & Ludwig, R. (2018). SAR-based detection of flooded vegetation – a review of characteristics and approaches. *International Journal of Remote Sensing, 39*, 2255–2293. doi:10.1080/01431161.2017.1420938

Turpie, K. R., Klemas, V. V., Byrd, K., Kelly, M., & Jo, Y. H. (2015). Prospective HyspIRI global observations of tidal wetlands. *Remote Sensing of Environment, 167*, 206–217. doi:10.1016/j.rse.2015.05.008

Tusa, E., Laybros, A., Monnet, J.-M., & Mura, M. D. (2020). *Fusion of hyperspectral imaging and LiDAR for forest monitoring* (Vol. 32). Data Handling in Science and Technology. doi:10.1016/B978-0-444-63977-6.00013-4

Twele, A., Cao, W., Plank, S., & Martinis, S. (2016). Sentinel-1-based flood mapping: A fully automated processing chain. *International Journal of Remote Sensing, 37*(13), 2990–3004. doi:10.1080/01431161.2016.1192304

Tyo, J. S., Goldstein, D. L., Chenault, D. B., & Shaw, J. A. (2006). Review of passive imaging polarimetry for remote sensing applications. *Applied Optics*, *45*(22), 5453–5469. doi:10.1364/AO.45.005453 PMID:16855644

Uddin, M. (1581). Meyer (2019) Operational Flood Mapping Using Multi-Temporal Sentinel-1 SAR Images: A Case Study from Bangladesh. *Remote Sensing*, *11*. Advance online publication. doi:10.3390/rs11131581

Umar, G. K., Yusuf, D. A., & Mustapha, A. (2019). Urban land use, planning and historical theories: An overview of Kano Metropolis. *World Scientific News*, *118*, 257–264.

Underwood, E., Ustin, S., & Dipietro, D. (2003). Mapping nonnative plants using hyperspectral imagery. *Remote Sensing of Environment*, *86*(2), 150–161. doi:10.1016/S0034-4257(03)00096-8

UNDP Emergency Analyst. (2008). *Situation Report: Bihar Flood 2008*. Author.

Unganai, L. S., & Kogan, F. N. (1998). Drought monitoring and corn yield estimation in Southern Africa from AVHRR data. *Remote Sensing of Environment*, *63*(3), 219–232. doi:10.1016/S0034-4257(97)00132-6

United Nations, Department of Economics and Social Affairs, Population Division. (2019). *World Population Prospects 2019: Highlights (ST/ESA/SER.A/423)*. Author.

Uniyal, A., & Prasad, C. (2006). Disaster management strategy for mass wasting hazard prone Naitwar Bazar and surrounding areas in Upper Tons valley in Uttarkashi district, Uttaranchal (India). *Disaster Prevention and Management: An International Journal*, *15*(5), 821–837. doi:10.1108/09653560610712766

Urbazaev, M., Thiel, C., Cremer, F., Dubayah, R., Migliavacca, M., Reichstein, M., & Schmullius, C. (2018). Estimation of forest aboveground biomass and uncertainties by integration of field measurements, airborne LiDAR, and SAR and optical satellite data in Mexico. *Carbon Balance and Management*, *13*(5), 1–20. doi:10.118613021-018-0093-5

Ustin, S. L., DiPietro, D., Olmstead, K., Underwood, E., & Scheer, G. J. (2002). Hyperspectral remote sensing for invasive species detection and mapping. *IEEE Geoscience and Remote Sensing Symposium*, *3*, 1658 – 1660 10.1109/IGARSS.2002.1026212

Vahidnia, M. H., Alesheikh, A., Alimohammadi, A., & Bassiri, A. (2008). Fuzzy analytical hierarchy process in GIS application. *The International Archives of the Photogrammetry, Remote Sensing and Spatial Information Sciences*, *37*, 593–596.

Vaidya, O. S., & Kumar, S. (2006). Analytic hierarchy process: An overview of applications. *European Journal of Operational Research*, *169*, 1–29.

Van Aardt, J., & Norris-Rogers, M. (2008). Spectral-age interactions in managed, even-aged eucalyptus plantations: Application of discriminant analysis and classification and regression trees approaches to hyperspectral data. *International Journal of Remote Sensing*, *29*(6), 1841–1845. doi:10.1080/01431160701874546

Van Binh, D., Kantoush, S., & Sumi, T. (2020). Changes to long-term discharge and sediment loads in the Vietnamese Mekong Delta caused by upstream dams. *Geomorphology, 353*, 107011. doi:10.1016/j.geomorph.2019.107011

Van Dijk, A., Mueller, N., & Lymburner, L. (2011). *MODIS-based standing water detection for flood and large reservoir mapping: Algorithm development and applications for the Australian continent*. CSIRO. doi:10.4225/08/58518bd176131

Vanlauwe, B., Diels, J., Sanginga, N., & Merckx, R. (2005). Long-term integrated soil fertility management in South-western Nigeria: Crop performance and impact on the soil fertility status. *Plant and Soil, 273*(1-2), 337–354. doi:10.100711104-005-0194-2

Vanni, N., Porchelvan, P. (2018). *Monitoring of agricultural drought using fortnightly variation of vegetation condition index (VCI) for the state of Tamil Nadu, India*. doi:10.5194/isprs-archives-XLII-4-W9-159-2018

Veloso, A., Mermoz, S., Bouvet, A., Le Toan, T., Planells, M., Dejoux, J. F., & Ceschia, E. (2017). Understanding the temporal behavior of crops using Sentinel-1 and Sentinel-2-like data for agricultural applications. *Remote Sensing of Environment, 199*, 415–426. doi:10.1016/j.rse.2017.07.015

Venne, L. S., Tsai, J. S., Cox, S. B., Smith, L. M., & McMurry, S. T. (2012). Amphibian community richness in cropland and grassland playas in the Southern High Plains, USA. *Wetlands, 32*(4), 619–629. doi:10.100713157-012-0305-9

Vibhute, A. D., Kale, K. V., Dhumal, R. K., & Mehrotra, S. (2015). Soil Type Classification and Mapping using Hyperspectral Remote Sensing Data. *International Conference on Man and Machine Interfacing (MAMI), 1*. 10.1109/MAMI.2015.7456607

Vieilledent, G., Vaudry, R., Andriamanohisoa, S. F., Rakotonarivo, O. S., Randrianasolo, H. Z., Razafindrabe, H. N., Rakotoarivony, C. B., Ebeling, J., & Rasamoelina, M. (2012). A universal approach to estimate biomass and carbon stock in tropical forests using generic allometric models. *Ecological Applications, 22*(2), 572–583. doi:10.1890/11-0039.1 PMID:22611855

Vijay, R., Pinto, S.M., Kushwaha, V.K., Pal, S., & Nandy, N. (2016). A multi-temporal analysis for change assessment and estimation of algal bloom in Sambhar Lake, Rajasthan. *Environmental Monitoring and Assessment, 188*(510), 1–10. doi:10.100710661-016-5509-7

Villmann, T., Merényi, E., & Hammer, B. (2003). Neural maps in remote sensing image analysis. *Neural Networks, 16*(3-4), 389–403. doi:10.1016/S0893-6080(03)00021-2 PMID:12672434

Vine, H. (1949). Nigerian soils in relation to parent materials. *Commonwealth Bureau of Soil Science and Technical Communication, 46*, 22–29.

Virkkala, R., Heikkinen, R. K., Kuusela, S., Leikola, N., & Pöyry, J. (2019). Significance of Protected Area Network in Preserving Biodiversity in a Changing Northern European Climate. In *Handbook of Climate Change and Biodiversity* (pp. 377–390). Springer. doi:10.1007/978-3-319-98681-4_23

Vittal, H., Ghosh, S., Karmakar, S., Pathak, A., & Murtugudde, R. (2016). Lack of Dependence of Indian Summer Monsoon Rainfall Extremes on Temperature: An Observational Evidence. *Scientific Reports*, *6*(July), 1–12. doi:10.1038rep31039 PMID:27485661

Vyas, D., Christian, B., & Krishnayya, N. S. R. (2013). Canopy level estimations of chlorophyll and LAI for two tropical species (teak and bamboo) from Hyperion (EO1) data. *International Journal of Remote Sensing*, *34*(5), 1676–1690. doi:10.1080/01431161.2012.725484

Vyas, D., Krishnayya, N. S. R., Manjunath, K. R., Ray, S. S., & Panigrahy, S. (2011). Evaluation of classifiers for processing Hyperion (EO-1) data of tropical vegetation. *International Journal of Applied Earth Observation and Geoinformation*, *13*(2), 228–235. doi:10.1016/j.jag.2010.11.007

Wahab, A. M., & Muhamad Ludin, A. N. (2018) Flood vulnerability assessment using artificial neural networks in Muar Region, Johor Malaysia. *IOP Conference Series: Earth and Environmental Science, 169*. 10.1088/1755-1315/169/1/012056

Wang, D., Lin, H., Chen, J., Zhang, Y., & Zeng, Q. (2010). Application of multi-temporal ENVISAT ASAR data to agricultural area mapping in the Pearl River Delta. *International Journal of Remote Sensing*, *31*(6), 1555–1572. doi:10.1080/01431160903475258

Wang, L., Zhu, J., Xu, Y., & Wang, Z. (2018). Urban Built-Up Area Boundary Extraction and Spatial-Temporal Characteristics Based on Land Surface Temperature Retrieval. *Remote Sensing*, *10*(3), 473. doi:10.3390/rs10030473

Wang, Z., Wang, T., Darvishzadeh, R., Skidmore, A., Jones, S., Suarez, L., Woodgate, W., Heiden, U., Heurich, M., & Hearne, J. (2016). Vegetation Indices for Mapping Canopy Foliar Nitrogen in a Mixed Temperate Forest. *Remote Sensing*, *8*(6), 491. doi:10.3390/rs8060491

Wan, X., Yang, Q., Jiang, P., & Zhong, P. (2019). A Hybrid Model for Real-Time Probabilistic Flood Forecasting Using Elman Neural Network with Heterogeneity of Error Distributions. *Water Resources Management*, *33*(11), 4027–4050. doi:10.100711269-019-02351-3

Ward, R. D., Burnside, N. G., Joyce, C. B., & Sepp, K. (2013). The use of medium point density LiDAR elevation data to determine plant community types in Baltic coastal wetlands. *Ecological Indicators*, *33*, 96–104. doi:10.1016/j.ecolind.2012.08.016

Warran, A., & Patwardhan, A. (2008). Carbon sequestration potential of trees in and around pune city. Case study Department of Environmental Sciences, University of Pune.

Warrick, J. A., Stevens, A. W., Miller, I. M., Harrison, S. R., Ritchie, A. C., & Gelfenbaum, G. (2019). World's largest dam removal reverses coastal erosion. *Scientific Reports*, *9*(1), 13968. doi:10.103841598-019-50387-7 PMID:31562373

Weiss, S. M., Kapouleas, I., & Shavlik, J. (1990). An empirical comparison of pattern recognition, neural nets and machine learning classification methods. *Readings in Machine Learning*, 177-183.

Weng, Q. (2001). A remote sensing? GIS evaluation of urban expansion and its impact on surface temperature in the Zhujiang Delta, China. *International Journal of Remote Sensing*, *22*(10), 1999–2014.

Westmoreland, S., & Stow, D. A. (1992). Category identification of changed land-use polygons in an integrated image processing/geographic information system. *Photogrammetric Engineering and Remote Sensing*, *58*, 1593–1599.

Whelen, T., & Siqueira, P. (2018). Time-series classification of Sentinel-1 agricultural data over North Dakota. *Remote Sensing Letters*, *9*(5), 411–420. doi:10.1080/2150704X.2018.1430393

Wigmore, O., Mark, B., Mckenzie, J., Baraer, M., Lautz, L., & Blanca, C. (2019). Sub-metre mapping of surface soil moisture in proglacial valleys of the tropical Andes using a multispectral unmanned aerial vehicle. *Remote Sensing of Environment, 222*, 104–118. doi:10.1016/j.rse.2018.12.024

Wijitkosum, S., & Sriburi, T. (2019). Fuzzy AHP Integrated with GIS Analyses for Drought Risk Assessment: A Case Study from Upper Phetchaburi River Basin, Thailand. *Water (Basel)*, *11*(5), 939.

Wilkinson, B. H., & McElroy, B. J. (2007). The impact of humans on continental erosion and sedimentation. *Geological Society of America Bulletin*, *119*(1–2), 140–156. doi:10.1130/B25899.1

Wing, M. G., Eklund, A., & Kellogg, L. D. (2005). Consumer-grade global positioning system (GPS) accuracy and reliability. *Journal of Forestry*, *103*(4), 169–173. doi:10.1093/jof/103.4.169

Winjum, J. K., Dixon, R. K., & Schroeder, P. E. (1992). Estimating the global potential of forest and agroforest management practices to sequester carbon. *Water, Air, and Soil Pollution*, *64*(1-2), 213–227. doi:10.1007/BF00477103

Wobus, C., Whipple, K. X., Kirby, E., Snyder, N., Johnson, J., Spyropolou, K., Crosby, B., & Sheehan, D. (2006). *Tectonics from topography: Procedures, promise, and pitfalls* (Vol. 398). Geological Society of America Special Papers.

Wohl, E. (2011). What should these rivers look like? Historical range of variability and human impacts in the Colorado Front Range, USA. *Earth Surface Processes and Landforms*, *36*(10), 1378–1390. doi:10.1002/esp.2180

World Bank – World Development Indicators. (2019). *Internally displaced persons, new displacement associated with disasters (number of cases), 1960-2017*. The Internal Displacement Monitoring Centre. https://Www.Internal-Displacement.Org/

Wright, D. J., Goodchild, M. F., & Proctor, J. D. (1997). Demystifying the persistent ambiguity of GIS as 'tool' versus 'science'. *Annals of the Association of American Geographers*, *87*(2), 346–362. doi:10.1111/0004-5608.872057

Wu, B., Gommes, R., Zhang, M., Zeng, H., Yan, N., Zou, W., Zheng, Y., Zhang, N., Chang, S., Xing, Q., & van Heijden, A. (2015). Global crop monitoring: A satellite-based hierarchical approach. *Remote Sensing*, *7*(4), 3907–3933. doi:10.3390/rs70403907

Wu, H., Qian, H., Chen, J., & Huo, C. (2017). Assessment of Agricultural Drought Vulnerability in the Guanzhong Plain, China. *Water Resources Management*, *31*(5), 1557–1574. doi:10.100711269-017-1594-9

Wu, J., van Aardt, J. A. N., Asner, G. P., Mathieu, R., Kennedy-Bowdoin, T., Knapp, D., Wessels, K., Erasmus, B. F. N., & Smit, I. (2009). Connecting the dots between laser waveform and herbaceous biomass for assessment of land degradation using small-footprint waveform LIDAR data. *Proc. IEEE Int. Geosci. Remote Sens. Symp.*, 2, 334–337. 10.1109/IGARSS.2009.5418078

Wulder, M. (1998). Optical remote-sensing techniques for the assessment of forest inventory and biophysical parameters. *Progress in Physical Geography*, *22*(4), 449–476. doi:10.1177/030913339802200402

Wulder, M. A., Masek, J. G., Cohen, W. B., Loveland, T. R., & Woodcock, C. E. (2012). Opening the archive: How free data has enabled the science and monitoring promise of Landsat. *Remote Sensing of Environment*, *122*, 2–10. doi:10.1016/j.rse.2012.01.010

Wurtsbaugh, W. A., Miller, C., Null, S. E., Derose, R. J., Wilcock, P., Hahnenberger, M., & Moore, J. (2017). Decline of the world' s saline lakes. *Nature Geoscience*, (October), 1–8. doi:10.1038/NGEO3052

Xiang, W., & Whitley, D. L. (1994). Weighting land suitability factors by the PLUS method. *Environment and Planning. B, Planning & Design*, *21*, 273–304.

Xiao, C.-W., & Ceulemans, R. (2004). Allometric relationships for below-and aboveground biomass of young Scots pines. *Forest Ecology and Management*, *203*(1-3), 177–186. doi:10.1016/j.foreco.2004.07.062

Xu, J., Li, Z., Tian, B., Huang, L., Chen, Q., & Fu, S. (2014). Polarimetric analysis of multi-temporal RADARSAT-2 SAR images for wheat monitoring and mapping. *International Journal of Remote Sensing*, *35*(10), 3840–3858. doi:10.1080/01431161.2014.919679

Xu, L., Zhang, H., Wang, C., Zhang, B., & Liu, M. (2019). Crop Classification Based on Temporal Information Using Sentinel-1 SAR Time-Series Data. *Remote Sensing*, *11*(1), 53. doi:10.3390/rs11010053

Yang, Ch.-Ch., & Chen, B.-Sh. (2004). Key quality performance evaluation using fuzzy AHP. *Journal of the Chinese Institute of Industrial Engineers*, *21*(6), 543–550.

Yang, X. (2011). Parameterizing support vector machines for land cover classification. *Photogrammetric Engineering and Remote Sensing*, *77*(1), 27–37. doi:10.14358/PERS.77.1.27

Yanmin, Y., Youqi, C., Yingbin, H. P., & Na, W. (2010). Soil Moisture Monitoring Using Hyper-Spectral Remote Sensing Technology. *International Conference on Geoscience and Remote Sensing*, 373–376. 10.1109/IITA-GRS.2010.5604219

Yeh, A. G. O., & Li, X. (1996). Urban growth management in the Pearl River delta – an integrated remote sensing and GIS approach. *ITC Journal*, *1*, 77–85.

Yu, W., Elleby, C., & Zobbe, H. (2015). Food security policies in India and China: Implications for national and global food security. *Food Security*, *7*(2), 405–414. doi:10.100712571-015-0432-2

Zadeh, L. A. (1973). The concept of linguistic variable and application to approximate reasoning. Berkeley, CA: ERL-M411.

Zadeh, L. A. (1965). Fuzzy sets. *Information and Control*, *8*(3), 338–353.

Zhang, C. (2014). Combining Hyperspectral and Lidar Data for Vegetation Mapping in the Florida Everglades. *Photogrammetric Engineering and Remote Sensing*, *80*(8), 9–19. doi:10.14358/PERS.80.8.733

Zhang, C., & Qiu, F. (2012). Mapping Individual Tree Species in an Urban Forest Using Airborne Lidar Data and Hyperspectral Imagery. *Photogrammetric Engineering and Remote Sensing*, *78*(10), 1079–1087.

Zhang, L., Shao, Z., Liu, J., & Cheng, Q. (2019). Deep Learning Based Retrieval of Forest Aboveground Biomass from Combined LiDAR and Landsat 8 Data. *Remote Sensing*, *11*(12), 1459. doi:10.3390/rs11121459

Zhang, Y. Q., Chen, J. M., Miller, J. R., & Noland, T. L. (2008). Leaf chlorophyll content retrieval from airborne hyperspectral remote sensing imagery. *Remote Sensing of Environment*, *112*(7), 3234–3247. doi:10.1016/j.rse.2008.04.005

Zha, Y., Gao, J., & Ni, S. (2003). Use of normalized difference built-up index in automatically mapping urban areas from TM imagery. *International Journal of Remote Sensing*, *24*(3), 583–594. doi:10.1080/01431160304987

Ziadat, F., Bruggeman, A., Oweis, T., Haddad, N., Mazahreh, S., Sartawi, W., & Syuof, M. (2012). A participatory GIS approach for assessing land suitability for rainwater harvesting in an arid rangeland environment. *Arid Land Research and Management*, *26*(4), 297–311. doi:10.1080/15324982.2012.709214

Zuo, D., Cai, S., Xu, Z., Peng, D., Kan, G., Sun, B. P., & Yang, H. (2019). Assessment of meteorological and agricultural droughts using in-situ observations and remote sensing data. *Agricultural Water Management*, *222*, 125–138. doi:10.1016/j.agwat.2019.05.046

About the Contributors

Suraj Kumar Singh is currently working as an Associate Professor at Centre for Sustainable Development, Suresh Gyan Vihar University, Jaipur. Previously, he worked as an Assistant Professor in the Centre for Land Resource Management, Central University of Jharkhand (CUJ), Ranchi. He has previously served as Research Associate in National Remote Sensing Centre (NRSC), Indian Space Research Organisation (ISRO), GoI. sponsored projects i.e. National Wasteland Mapping and National Urban Information System for Northern parts of India. He did his Ph.D. degree in Technology (Geoinformatics) from Department of Remote Sensing, Birla Institute of Technology, Ranchi in 2012. He has worked in the area of waterlogging and flood hazards, geospatial applications in water resources, disaster management, hydrogeomorphology, urban planning & wasteland mapping. He is currently supervising 06 research scholars, and about 20 M.Tech/ M.Sc. thesis completed under his guidance. He has 50 publications in peer refereed international journal, 05 reference book and 05 book chapters. He has attended and organised around 40 national and international conferences. He was also course coordinator of different modules of EDUSAT based distance learning programme organized by IIRS (Indian Institute of Remote Sensing), ISRO, Govt. of India. He has 12 years of teaching and research experience.

Shruti Kanga is currently working as an Associate Professor at Centre for Climate Change and Water Research, Suresh Gyan Vihar University, Jaipur. Previously, she worked as an Assistant Professor in in the Centre for Land Resource Management, Central University of Jharkhand (CUJ), Ranchi. She had previously served as Research Associate in Ministry of Environment, Forest and Climate Change (MoEF), GoI funded project on Forest Fires and has teaching experience in different universities i.e. Central University of HNB Garhwal and University of Jammu. She completed her Ph.D. degree in Technology (Geoinformatics) from Department of Remote Sensing, Birla Institute of Technology, Ranchi in 2013. She has worked in the area of forest fires risk modelling and management, tourism, resource management etc. she has also 05 M.Sc. thesis completed under her guidance. She has 35 publications in peer refereed national and international journal, 01 book and 02 book chapter. She has attended and organised around 15 national and international conferences. She was also course coordinator of different modules of EDUSAT based distance learning programme organized by IIRS (Indian Institute of Remote Sensing), ISRO (Indian Space Research Organization), Govt. of India. I have 09 years of teaching and research experience.

Varun Narayan Mishra did Ph.D. from Department of Physics, Indian Institute of Technology (BHU), Varanasi. He received M.Tech. degree in Remote Sensing from Birla Institute of Technology, Mesra, Ranchi, after completing M.Sc. degree in Physics. He is currently working as an Assistant Professor at Centre for Climate Change and Water Research (C3WR), Suresh Gyan Vihar University, Jaipur, India. Previously, he worked as a Junior Research Fellow at Centre for the Study of Regional Development (CSRD), School of Social Sciences, JNU, New Delhi under DST Funded Research Project. His main research interests are SAR image processing & analysis, machine learning techniques, land use/land cover mapping, land change analysis & modelling, urban climate, glacier mapping & monitoring. He has published several research papers in various reputed international peer reviewed journals. He is also reviewer of international journals of repute such as Geocarto International, Journal of Earth System Science, and Journal of Urban Management, etc.

* * *

Asraful Alam is Post Doctoral Fellow (PDF), Department of Geography, University of Calcutta, Kolkata, India. He obtained masters in geography from Aligarh Muslim University, India and PhD from Aliah University, Kolkata, India. He has previously worked at Women's College, Calcutta, India as an Asst. coordinator in department of Geography and RA in a major project in the department of Geography, Aliah University, Kolkata, India. Dr. Alam research interests are population geography, agricultural geography, remote sensing and GIS, and development studies. He has thirteen research papers/articles and four book chapter and two book to his credit with national and international publishers and also attend more than thirty nation and international seminars/ conferences/ workshops. He has been the first Vice-President of the Aliah University Research Scholars Association (AURSA).

Akash Anand is post-graduated as a Master of Technology in Geoinformatics from the Department of Geoinformatics, Central University of Jharkhand. His research work involves the use of remote sensing data for various applications in the field of forestry (species, biodiversity, and biomass), precipitation retrieval and climate change. He has published several research articles and conference papers in reputed journals and attended several national/international conferences.

Gajendra Chawla is a Superintendent Engineer, Watershed Development and Soil Conservation Department, Jodhpur, Rajasthan.

Pranaya Diwate is an Assistant Professor in Centre for Climate Change and Water Research, Suresh Gyan Vihar University, Jaipur. His present research interests are reconstruction of paleoclimate variability specially from Holocene period using multi-proxy approach includes Diatoms, Clay mineralogy, Grain size, etc. He is also working on the response of Diatoms to global warming in the NW Himalayan Lake.

Arijit Ghosh is a Research Scholar, Department of Geography, Sidho-Kanho-Birsha University, Purulia. He has previously worked at Sidho-Kanho_Birsha University, India as an Research Associate in a major project in the department of Geography. Mr. Ghosh research interests are hydrology, fluvial geomorphology, remote sensing and GIS. He has five research papers/articles and one book chapter to his credit with national and international publishers.

Kiran Jalem is a Disaster Management Professional having 20 years of experience, which includes Industrial, Scientific, Research, Engineering, Teaching & Training etc. As an Assistant Professor with Shri Krishna Institute of Public Administration (SKIPA), Ranchi, Govt. of Jharkhand under the Non-Plan Scheme of Ministry of Home Affairs, Govt. of India, he had conducted several Training Programs on Disaster Management at SKIPA, Ranchi in collaboration with NIDM, NDMA, MHA, Govt. of India and Govt. of Jharkhand. Dr. (Er.) Kiran Jalem is a Trainer on Disaster Management, Trainer on Gender Issues & Gender Budgeting and a Certified Trainer on Direct Trainer Skills (DTS) and Design of Training (DoT) by DoPT, Govt. of India & Thames University, London. Dr. (Er.) Kiran Jalem Successfully Completed many Short Course Trainings, Advanced Certificate Courses & DST Sponsored Training Courses etc. during his career. Presently, Dr. Kiran Jalem is Head, Centre for Research, Training Coordination & Networking (CRTCN), National Institute of Rural Development & Panchayati Raj (NIRDPR), Hyderabad, Govt. of India.

Akshay Kumar received B.Sc. degree in Geology (with honours) and M.Sc. degree in Geology also from Ranchi University, Ranchi, India. After this he received M.Tech. degree in Remote Sensing from Birla Institute of Technology, Mesra, Ranchi, India. He also completed his Ph.D. degree (Remote Sensing) from the same Institute in 2018. His research interest lies in aerosol monitoring, natural hazards, water resource management, climate change and natural geosciences studies using various soft computing and geoinformatics techniques.

Vipin Kumar has 7 years of experience in teaching and research, primarily in the areas of natural hazards especially landslide studies in the Himalaya. He has published research papers in international and national journals and presented several papers in International and National conferences.

Dipti Ladli is currently pursuing his Integrated M.Tech in Geoinformatics from Department of Geoinformatics, Central University of Jharkhand (India). His research interest surrounds the application of Remote sensing and GIS technique in Agriculture (e.g. crop classification, acreage mapping, yield prediction, drought assessment, etc.).

Kanhaiya Lal, currently working in the capacity of Assistant Professor at the Department of Geoinformatics, Central University of Jharkhand, Ranchi. He completed Ph.D. degree from Jawaharlal Nehru University, New Delhi on a broad theme of land degradation due to coal mining. He has keen interest in linking climate with vegetation and measuring these changes using geoinformatics.

Ugonna Nkwunonwo holds a PhD in Earth and Environmental sciences. He lectures at the faculty of Environmental studies (Geoinformatics and Surveying) of the University of Nigeria, Enugu Campus. His key research developed a new flood simulation schema, GFSP-1, to facilitate modelling of urban flood risk in data sparse locations. His other research interests are in multi-hazard assessment and modelling, vulnerability assessment, geospatial data science, urban heat, renewable energy, tourism and sustainable urban drainage system (SUDS), cinema films and social media data for natural hazard assessment.

Manish Pandey currently works at the University Center for Research & Development (UCRD), Chandigarh University, Mohali. His research interests are in the fields of Geography, Fluvial & Glacial Geomorphology, Glaciology and Remote Sensing & Geoinformatics (GIS). His simple interest is in understanding the process-form relationship in diverse environmental settings. He has developed new interests in the fields of natural hazard susceptibility modeling and landscape evolution modeling.

Avinash Kumar Ranjan has completed his integrated M.Tech in Geoinformatics from the Central University of Jharkhand. He has been broadly involved in Earth and Environmental studies, specifically bone up the environmental issues using the Geoinformatics technologies. His main research interest surrounds the monitoring and assessment of the Agriculture and Forestry Environment using RS and GIS techniques. He has been credited to have several research articles in the reputed journal.

Raghib Raza is a GIS Expert, Watershed Development and Soil Conservation Department, Jodhpur, Rajasthan.

Lakshminarayan Satpati obtained Ph. D. degree in Geography from the University of Calcutta for his thesis entitled 'An Approach Towards Rational Water Resources Management of the Rupnarayan River Basin, West Bengal'. Prof. Satpati has an experience of teaching of geography for about 26 years in schools, colleges, universities and institutes of higher education in West Bengal. He has presented more than 65 research papers and chaired sessions in national and international platforms, of which the 32 nd IGU Conference in Cologne, Germany (2012), 8 th IAG Conference in Paris, France (2013) and PRSCO-2019 of RSAI, Bangkok are notable. Besides, he has been invited to deliver lectures in various faculty development programmes of UGC, ICSSR, DST etc. He has to his credit more than forty publications in the form of research papers and articles published in reputed national and international journals and books. His research and teaching interests include climatology, geomorphology, hydrology, environment, population & development, quantitative geography and curriculum development in geography. He has edited a book on 'Climate and Society- a Contemporary Perspective' published by the University of Calcutta in 2015. Fourteen scholars have already obtained their Ph. D. degrees under his supervision. He has successfully completed two UGC sponsored research projects and organized three national workshops on Climate Science. Prof. Satpati is associated with a large number of academic and professional organizations of teachers, geographers and climate scientists of India. He has been the Hony. Assistant Secretary (Administration) of the Geographical Society of India, Kolkata since 2009. He has been elected as the Hony. Secretary, Indian Meteorological Society (IMS), Kolkata Chapter for the period 2018-20. Prof. Satpati has performed the responsibility of the Chairman, High Powered Committee to restructure NATMO. Currently he is the honorary Editor of UGC recognized peer reviewed biannual 'Indian Journal of Landscape Systems and Ecological Studies' of Indian Institute of Landscape, Ecology and Ekistics, since 2017.

Achala Shakya is pursuing Ph.D. at NIT, Kurukshetra in Department of Computer Engineering.

Masood Siddiqui has an experience of about 30 years of teaching and research . He specializes in Application of Remote sensing and GIS in Watershed Management. Besides, he has interest in Solid waste management and Wetland management. He has more than 40 research papers in reputed journals, two books and four projects to his credit.

Suman Sinha is currently an Assistant Professor in the Department of Geography at Amity University Kolkata. He received M.Sc. in Environmental Science in 2006 from University of Calcutta, India. He received M.Tech. and Ph.D. degrees in Remote Sensing in 2010 and 2017 respectively from Birla Institute of Technology, Mesra, India. He was an INSPIRE Fellow (DST, GoI) during 2011-2016 and a Post-Doctoral researcher under SERB NPDF (DST, GoI) in Haldia Institute of Technology, India during 2016-2018. He is serving as editorial board member and reviewer of reputed international and national journals. His main research interests include geospatial modeling, biomass and carbon modeling, and application of multi-frequency SAR and optical remote sensing in forestry, urban and environmental applications.

Thota Sivasankar is currently working as an Assistant Professor in Geographic Information Systems (GIS) area of NIIT University, India. He received M.Tech. degree in Remote Sensing from Andhra University, and Ph.D. degree on Microwave Remote Sensing from I.I.T. (I.S.M.), Dhanbad while he was working at the Indian Institute of Remote Sensing, ISRO. His primary research focuses on the use of microwave remote sensing for various applications including crop growth monitoring, soil moisture retrieval, forest above ground biomass estimation and, flood mapping and monitoring.

Pithani Venkatesh is working as Sr, Engineer Remote Sensing in Skymet Weather Services Pvt Ltd which was the first private sector entity to provide weather forecasts and solutions to the India. He has 4 years of experience Agricultural crop Health Monitoring, Crop Damage/Loss Assessment, Flood Loss Analysis, Yield calculation, soil moisture analysis using microwave data. Currently, Venkatesh is focusing on Python, scripting, Google Earth-engine and Machine Learning techniques for Microwave Remote Sensing applications. Venkatesh has Bachelor's degree in Electrical and Electronics Engineering and Master's in Remote Sensing from Jawaharlal Nehru Technological University, Kakinada.

Index

A

Above Ground Biomass 180-184, 189-190, 192-197
active floodplain 281
Agricultural drought 122-125, 128-129, 133, 138-139, 142-143
Analytic hierarchy process 1-2, 24
AVIRIS 97, 100, 109

B

Biodiversity 1, 4, 15, 69-70, 72, 207, 213, 215, 253
braiding index 272, 278, 281
Brightness 88-89, 91-92

C

Carbon Stock 180-181, 183-184, 188-192, 195-197, 212
C-band 47, 154, 180, 184, 194, 196, 218
Climate 25, 27, 47, 71-72, 90, 123, 125-126, 169-170, 172-174, 182, 212, 217-218, 226, 245, 270
Climate change 71-72, 123, 212, 217, 226, 245
CRED EM-DAT 228
crop classification 45-46, 54

D

Dartmouth 225

E

Ecosystem 69, 104, 181, 183, 207, 215, 218, 243-244, 246, 248-249, 257, 271

F

Flood 149-164, 225-238, 271, 278-279
Forest 6-7, 15, 33, 46, 62, 69, 97-100, 103, 105-107, 109, 126, 132, 180-184, 189-192, 194-197, 206-211, 213-219, 243
Forest management 15, 69, 103, 105-107, 219
fuzzy 1-5, 8, 10-11, 13, 15

G

Ganga River Basin 225
Geographical information system 89, 98, 124, 209
geo-spatial technologies 62, 64-65, 67, 69-72, 76-79, 163, 206, 209
Greenness 88-89, 91-93
Ground water potential 24, 29, 41-42

H

Himalaya 169-172, 174, 214, 272
Hydrogeomorphic response 270
Hyperspectral 97-101, 103-106, 108-109, 218, 249

I

India 1, 4, 6, 15, 24-26, 44-45, 47, 95, 122-127, 144, 153-154, 162, 176, 181, 206-218, 225-228, 242, 246-247, 250, 252-253, 255, 258, 269-270, 272, 274, 276

L

Land degradation 73
Land resources 62-74, 76, 78-79, 213
Landsat 33, 45, 89-91, 93-94, 143, 151, 174, 184-186, 197, 212, 249, 255-256, 272, 276, 278
Landslides 150, 169-170, 173, 175-176
LIDAR 69, 97-99, 104-109, 217-218, 249
Linear Regression 129, 132, 140-141, 195

M

Maximum Likelihood 7, 182, 185-186
MCDA 1-5, 8, 11, 13, 15
Middle Ganga Plain 225, 227
MODIS 124, 126, 128-129, 132, 134-141, 143-144, 150, 152-154, 157-158, 161-164, 216
MODIS NRT Flood 150, 154, 158, 161, 164
multi-temporal 44-47, 49, 55-56, 149-150, 153, 186, 215

N

Natural resources management 62-64, 69, 71-73, 78
NDVI 88-89, 93, 104, 122, 124-126, 128, 130, 132-135, 137-140, 142-143, 170, 173-175
Nigeria 62-67, 69-72, 74, 76-79
North Bihar 149, 152-154, 158, 160, 229
Northeast India 206-207, 209-211, 213-214, 216-218

P

peak discharge 271-272, 276, 279, 281

R

Ramsar sites 242, 244, 246-248, 252-253, 258
Remote sensing 7, 24-26, 42, 45-46, 65, 67-69, 76, 79, 89, 91, 97-100, 103-106, 108, 124, 143, 150-151, 163, 180-183, 207, 209, 212-213, 216, 218, 249

S

Saline wetlandscapes 242, 258
SAR 44, 46-49, 51, 55-56, 149-154, 156-157, 160-164, 180, 184, 187, 190, 217-218
Sentinal-1A 154, 157, 162, 180
Sentinel-1 44, 46-47, 49, 55-56, 150, 156, 161, 163, 217
Sentinel-2A 154, 161
Sigma Naught 157, 184, 190, 194-195, 197
Spatial data 64-65, 152, 154
Strategic plans 242, 246, 249, 258
Sustainable development 63-64, 71, 78, 248, 258

T

Tasseled Cap 88-89, 91, 93-94
Tectonics 169-170, 175-176, 270
Tehri dam 269-272, 274, 276, 278-280
Tons valley 169

U

Uttarakhand 246, 269, 271, 274

V

Vegetation Indices 122, 124, 142-143

W

Wetland 242-244, 246-248, 252-258

Y

Yield Prediction 122-123, 129, 140-141

www.ingramcontent.com/pod-product-compliance
Lightning Source LLC
LaVergne TN
LVHW080601300926
849162LV00005B/1016